ATLAS

A 25-Year Insider Story of the LHC Experiment

ADVANCED SERIES ON DIRECTIONS IN HIGH ENERGY PHYSICS

ISSN: 1793-1339

This is the best review series in high energy physics today. It comprehensively reviews the most important developments in each sector of high energy physics and is of lasting use to all researchers. All volumes are edited by eminent physicists — researchers who have themselves made substantial contributions to their respective fields of research.

Published

Vol. 30 – *ATLAS: A 25-Year Insider Story of the LHC Experiment* (by The ATLAS Collaboration)
Vol. 29 – *Multiple Parton Interactions at the LHC* (eds. Paolo Bartalini and Jonathan Richard Gaunt)
Vol. 28 – *The State of the Art of Neutrino Physics: A Tutorial for Graduate Students and Young Researchers* (ed. A. Ereditato)
Vol. 27 – *Technology Meets Research 60 Years of CERN Technology: Selected Highlights* (eds. C. Fabjan, T. Taylor, D. Treille and H. Wenninger)
Vol. 26 – *The Standard Theory of Particle Physics: Essays to Celebrate CERN's 60th Anniversary* (eds. L. Maiani and L. Rolandi)
Vol. 25 – *Massive Neutrinos: Flavor Mixing of Leptons and Neutrino Oscillations* (ed. H. Fritzsch)
Vol. 24 – *The High Luminosity Large Hadron Collider: The New Machine for Illuminating the Mysteries of Universe* (eds. O. Brüning and L. Rossi)
Vol. 23 – *60 Years of CERN Experiments and Discoveries* (eds. H. Schopper and L. Di Lella)
Vol. 22 – *Perspectives on String Phenomenology* (eds. B. Acharya, G. L. Kane and P. Kumar)
Vol. 21 – *Perspectives on Supersymmetry II* (ed. G. L. Kane)
Vol. 20 – *Lepton Dipole Moments* (eds. B. Lee Roberts and William J. Marciano)
Vol. 19 – *Linear Collider Physics in the New Millennium* (eds. K. Fujii, D. J. Miller and A. Soni)
Vol. 18 – *Perspectives on Supersymmetry* (ed. G. L. Kane)
Vol. 17 – *Perspectives on Higgs Physics II* (ed. G. L. Kane)
Vol. 16 – *Electroweak Symmetry Breaking and New Physics at the TeV Scale* (eds. T. L. Barklow, S. Dawson, H. E. Haber and J. L. Siegrist)
Vol. 15 – *Heavy Flavours II* (eds. A. J. Buras and M. Lindner)
Vol. 14 – *Precision Tests of the Standard Electroweak Model* (ed. P. Langacker)
Vol. 13 – *Perspectives on Higgs Physics* (ed. G. L. Kane)
Vol. 12 – *Advances of Accelerator Physics and Technologies* (ed. H. Schopper)
Vol. 11 – *Quantum Fields on the Computer* (ed. M. Creutz)
Vol. 10 – *Heavy Flavours* (eds. A. J. Buras and M. Lindner)

The complete list of titles in the series can be found at
http://www.worldscientific.com/series/asdhep

Advanced Series on
Directions in High Energy Physics — Vol. 30

ATLAS

A 25-Year Insider Story of the LHC Experiment

by **The ATLAS Collaboration**

World Scientific

NEW JERSEY • LONDON • SINGAPORE • BEIJING • SHANGHAI • HONG KONG • TAIPEI • CHENNAI • TOKYO

Published by

World Scientific Publishing Co. Pte. Ltd.
5 Toh Tuck Link, Singapore 596224
USA office: 27 Warren Street, Suite 401-402, Hackensack, NJ 07601
UK office: 57 Shelton Street, Covent Garden, London WC2H 9HE

Library of Congress Cataloging-in-Publication Data
Names: ATLAS Collaboration, author.
Title: ATLAS : a 25-year insider story of the LHC experiment / by the ATLAS Collaboration.
Other titles: Advanced series on directions in high energy physics ; v. 30.
Description: Singapore ; Hackensack, NJ : World Scientific Publishing Co. Pte. Ltd., [2019] |
Series: Advanced series on directions in high energy physics, ISSN 1793-1339 ; vol. 30
Identifiers: LCCN 2018049800| ISBN 9789813271791 (hardcover ; alk. paper) |
ISBN 9813271795 (hardcover ; alk. paper)
Subjects: LCSH: ATLAS Collaboration--History. | Large Hadron Collider
(France and Switzerland)--History.
Classification: LCC QC787.P73 A85 2019 | DDC 539.7/36--dc23
LC record available at https://lccn.loc.gov/2018049800

British Library Cataloguing-in-Publication Data
A catalogue record for this book is available from the British Library.

For any available supplementary material, please visit
https://www.worldscientific.com/worldscibooks/10.1142/11030#t=suppl

Desk Editor: Ng Kah Fee

Typeset by Stallion Press
Email: enquiries@stallionpress.com

Printed in Singapore by Mainland Press Pte Ltd.

Foreword

In 2012, the ATLAS and CMS Collaborations working at CERN's Large Hadron Collider (LHC) announced the discovery of the Higgs boson. This landmark achievement was hailed as the "Breakthrough of the Year" in several science magazines. Clarifying the existence of the Higgs boson has always been a major motivation for the construction of the LHC and its experiments, and strongly influenced their design. First planning studies for the ATLAS experiment began in the late 1980s and led to the formation of the Collaboration with the submission of the "Letter of Intent to build a General Purpose pp Experiment" on 1 October 1992.

The importance of the Higgs boson discovery was widely recognised in the scientific world and by the public. It was also realised that this scientific triumph relied on decades of work by thousands of physicists, engineers and technicians. Today, the ATLAS Collaboration counts about 5500 members, with about 2900 scientific authors from 38 countries.

The ATLAS Collaboration has decided to document its more than 25 years of history by writing this book. It is intended to provide an insider story covering all aspects of this global science project, from early discussions on the LHC project to the extraction of the physics results. The book also aims to show the human aspects of the Collaboration and the quest for common goals at the frontiers of physics and technology, with so many people in the stimulating international environment of ATLAS and CERN.

The various chapters and sections in this book have been written by many authors with their individual background and style of representing part of our history. Differences in style also reflect the large diversity in our worldwide collaboration.

The fantastic scientific endeavour of the ATLAS Collaboration is ongoing with the present data taking and analysis, and will continue throughout the High-Luminosity phase of the LHC. The Collaboration will continue to explore fascinating science questions for twenty more years to come.

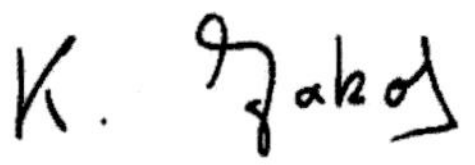

Karl Jakobs
Spokesperson of the ATLAS Collaboration
September 2018

Contents

Chapter 1

Towards the ATLAS Letter of Intent

1.1. The early historical context

The confidence in the Standard Model (SM) of particle physics received a giant boost in the early 1980s with the discovery of the W and Z intermediate vector bosons at the CERN proton-antiproton collider by the UA1 and UA2 experiments. Since then, its predictive power has been experimentally demonstrated with great accuracy in many generations of low- and high-energies experiments. And yet, during the 1980s, there were still two elements missing that would had completed the SM and established it fully as the "Standard Theory" of particle physics: the heaviest of the six quarks — the top quark — and a scalar boson of unknown mass, the Higgs boson. The latter would confirm the proposed spontaneous electroweak symmetry breaking mechanism, as a manifestation of the scalar field that permeates the entire universe. This mechanism, introduced by R. Brout, F. Englert and P. Higgs, gives the W and Z bosons their heavy masses and leaves the photon massless. The interactions of quarks and leptons with the scalar field impart masses to them proportionally to the strength of their couplings to the Higgs boson.

The top quark was expected at that time to show up in existing, or soon-to-operate colliders, and its mass became constrained by electron-positron collider (LEP and SLC) measurements and predicted by the electroweak theory to be around its by now well-known value of 172.5 GeV, where it was eventually discovered in 1995 by the CDF and D0 experiments at the Tevatron. In the 1980s, however, the scalar Higgs boson had no significant mass prediction. It could lie in the range from a few GeV up to about one TeV, where it would be so broad that it would experimentally cease to be visible as a signal peak. The mass could therefore be well beyond the reach of any existing experiment at the time. The hunt for the Higgs boson became a central theme in discussing the future of particle physics, and clearly a primary motivation for the Large Hadron Collider (LHC) and its experiments.

In the late 1970s, the idea was already in the air that the tunnel for the future Large Electron-Positron (LEP) collider should be able to house (at some point) a hadron collider, thanks to the foresight of an ECFA-LEP[1] Working Group. Thankfully, those leading CERN at the time of the LEP approval in 1981, with H. Schopper as CERN's Director-General, had the vision to plan for a tunnel with a large enough circumference to reach very high energies, and with a large enough diameter to eventually accommodate the installation of high-field superconducting magnets for such a hadron collider ring.

In the broader community, enthusiasm for a Large Hadron Collider emerged strongly for the first time in 1984, at a workshop in Lausanne on the "Large Hadron Collider in the LEP Tunnel", which brought together working groups that comprised machine experts, theorists and experimentalists. This workshop is considered in the community, in particular among the experimentalists, as the real starting point for the LHC activities, Quite a number of ATLAS colleagues were already involved at that stage, as individuals concerned about the future of our research. At that time, physics studies and conceptual designs for a very high-energy (40 TeV) proton-proton collider — called the Superconducting Super Collider (SSC) — were already under way in the United States.

With the growing understanding of the great physics potential of a LHC, a series of motivating workshops and conferences followed. They initially included the 1987 workshop in La Thuile organised by the so-called "Rubbia Long-Range Planning Committee", where the physics potential and the feasibility of a Large Hadron Collider and of a high-energy linear electron-positron collider were debated — this workshop became the basis for the CERN Council's decision in 1991 that the LHC was the right project for its future after LEP, thanks to the enthusiastic support of C. Rubbia, CERN's Director-General at that time. The following year, 1988, a dedicated study group about "The feasibility of experiments at high luminosity at the Large Hadron Collider" issued a report to the Long-Range Planning Committee, strengthening the case for the LHC. Detector issues for such a machine were then addressed explicitly during the 1989 study week in Barcelona on the "Instrumentation Technology for High-Luminosity Hadron Colliders". It was

[1]ECFA: European Committee for Future Accelerators.

at that meeting where several colleagues, ending up later in ATLAS, decided to start a concerted and coherent effort, in the form of a proto-collaboration, to prepare for a LHC experiment.

CERN management became aware of such emerging groupings and wanted to make sure that its full community had an opportunity to properly discuss the physics and detector issues in a large, open forum. It initiated, together with the ECFA, the 1990 Aachen "LHC Workshop". This workshop drew a very large attendance, demonstrating the broad support in the community for the LHC. During this time, the design and construction of the 40 TeV SSC was still underway in the United States. The LHC, with its initially planned energy of 16 TeV, could only be competitive when aiming at an order of magnitude higher luminosity, i.e., $10^{34}\,\mathrm{cm}^{-2}\,\mathrm{s}^{-1}$. However, it was at the Aachen workshop that the formidable experimental challenges for detectors at a high-luminosity hadron collider began to appear solvable, provided enough R&D work on instrumentation was carried out. The SSC was ultimately cancelled in October 1993.

As mentioned, it was during the late 1980s and very early 1990s that the seeds for the future ATLAS and CMS Collaborations were planted. There were initially four proto-collaborations emerging in the early 1990s that aimed at exploratory, general-purpose experiments. They presented detector designs in "Expressions of Interest" at the historically well-known conference "Towards the LHC Experimental Programme" that took place in March 1992 in Evian-les-Bains. It is also during these years that R&D and prototyping work for LHC started. The remainder of this chapter will describe these events in greater detail, which led to the ATLAS Letter of Intent of 1 October 1992. Table 1.1 sketches the timeline of the LHC project.

1.2. The physics case for the LHC: A brief history

As mentioned, the starting point for joint theoretical and experimental activities related to the LHC was the workshop in Lausanne in 1984, during the aftermath of the discovery of the W and Z bosons in UA1 and UA2. At that time, the construction of the Fermilab Tevatron was already in full swing, together with the construction of the CDF detector (joined by the D0 detector soon after), which somewhat mirrored those that had been observed at CERN with UA1 and UA2. At the same time, the LEP machine and its four detectors were also being constructed and the theory community was beginning to deal with increasingly precise and complex predictions for multiple physics processes of interest at high-energy hadron colliders.

Table 1.1. Some key dates of the LHC project.

1984	Workshop on a Large Hadron Collider in the LEP tunnel, Lausanne, Switzerland.
1987	Workshop on the Physics at Future Accelerators, La Thuile, Italy. The Rubbia "Long-Range Planning Committee" recommends the Large Hadron Collider as the right choice for CERN's future.
1990	LHC Workshop, Aachen, Germany (discussion of physics, technologies and detector design concepts).
1992	General Meeting on LHC Physics and Detectors, Evian-les-Bains, France (with four general-purpose experiment Expressions of Interest presented).
1993	Three Letters of Intent evaluated by the CERN peer review committee LHCC. ATLAS and CMS selected to proceed to a detailed technical proposal.
1994	The LHC accelerator approved for construction, initially in two stages.
1995	LHC Conceptual Design Report.
1996	ATLAS and CMS Technical Proposals approved.
1996	Approval for the construction of the 14 TeV LHC, to be completed in 2005.
1997	Formal approval for ATLAS and CMS to move to construction (materials cost ceiling of 475 MCHF).
1997	Construction of the experiments commences (after approval of detailed Technical Design Reports of detector subsystems).
2000	Assembly of experiments commences at CERN. The LEP accelerator is closed down to make way for the LHC.
2008	LHC experiments ready for pp collisions. LHC starts operation. An incident stops the LHC operation.
2009	LHC restarts operation, first pp collisions at 900 GeV recorded by the LHC detectors.
2010	LHC collides protons at high energy (centre of mass energy of 7 TeV).
2012	LHC operates at 8 TeV: announcement of the discovery of a Higgs-like boson.
2015	After a shutdown in 2013–2014 the LHC operates at 13 TeV for Run-2.

In this context, it appeared for the first time that it was possible to quantitatively compare the potentials of vastly different accelerators, in terms of answering some of the fundamental questions at the time. These included questions on the existence (or not) of the following: (i) a Higgs boson responsible for the mechanism of the electroweak symmetry breaking, (ii) heavier quarks, including the missing third-generation top quark, and (iii) supersymmetry (see Chap. 9).

So it is not a surprise that the physics potential of ATLAS (and of any LHC experiment) was scrutinised by the community with respect to that of other machines — LEP2 in the short term, Tevatron in the medium term, and SSC and CLIC/NLC (an electron-positron linear collider at 2–3 TeV centre-of-mass energy) in the longer term. The main focus of the comparisons was on the search for the Higgs boson and for supersymmetry. This was the most critical phase for the approval of the LHC project and of its experiments, when the real competition was not between the LHC proto-collaborations but more between soon-to-be-operated and future machines. This is illustrated in the latter case in the historical Figs. 1.1 and 1.2, for comparisons of the basic physics discovery potential of the LHC versus CLIC and SSC, respectively.

The Higgs boson, with its wide variety of possible production mechanisms and decay channels, was an ideal benchmark for physics comparisons, and also for the optimisation of the design of the overall detector in terms of its performance. For the first time in history, theoretical, software and computing tools were available to perform realistic and extensive simulations of the signal and background processes of interest. The development of these tools began in earnest in 1989, and they were all used throughout the various phases of the design, approval and construction of ATLAS.

In the early years of the new century, the painful process of migrating to more modern software tools took place. Today, most of the basic principles of the simulation and many of the algorithms developed then survive in one

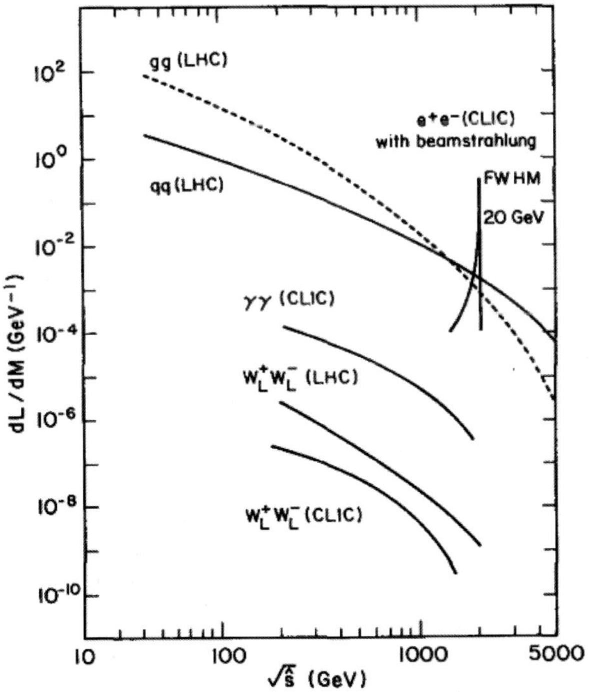

Fig. 1.1. Parton luminosities as a function of the centre-of-mass energy E_{cm} of the constituents (g, q, e, W_L) for LHC and CLIC.

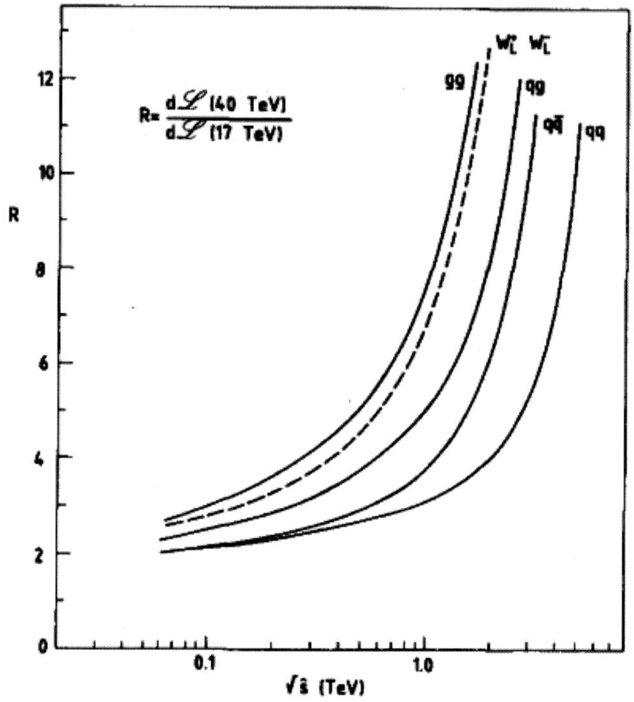

Fig. 1.2. Ratios of parton luminosities as a function of the centre-of-mass energy of the constituents (g, q, W_L) for LHC and SSC.

form or another. This is largely thanks to simulation efforts, which were pursued over a period of twenty years before data taking began, that data and simulation agreed very well in terms of basic detector performance to operate and trigger the experiment with a high efficiency and to publish rapidly important physics results.

In this short survey, only a few examples can be given of the physics and performance simulation work, which led to the ATLAS experiment. Figure 1.3 shows an example of the expected coverage of parameter space in the search for additional Higgs bosons if supersymmetry (MSSM) would be realised in nature. This was shown at the CERN LEP experimental committee in 1996 to compare the LHC potential to that of LEP2, in particular for Higgs boson masses close to that of the Z boson. Figures 1.4 to 1.6 show a few historical examples, without going into detailed explanations, of results from simulations performed during the early design of the experiment in terms of the very critical choices. Figure 1.4 displays the effect of the calorimeter coverage close to the beam (high pseudorapidity

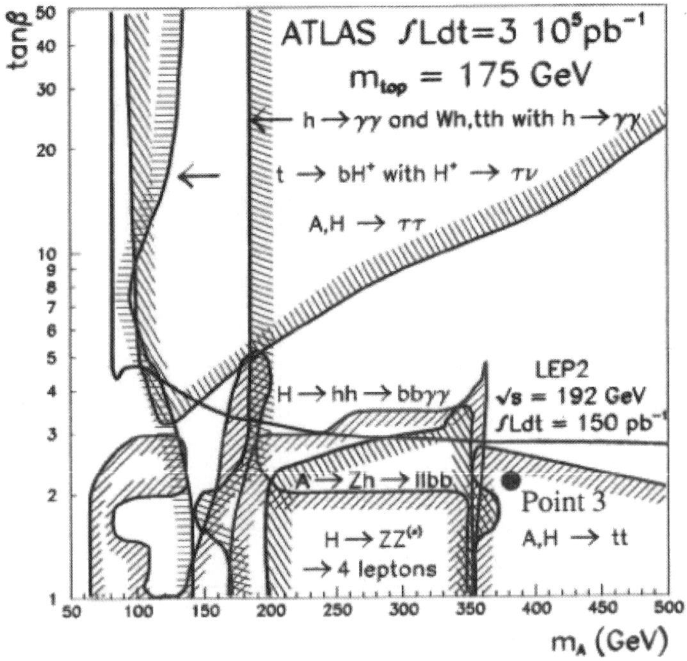

Fig. 1.3. Parameter space for supersymmetric Higgs bosons: mass of the pseudoscalar m_A and the ratio of the vacuum expectation values $\tan\beta$. Expected discovery contours are shown for various decay modes in the case of ATLAS after ten years of data taking, compared to the expected results from LEP2.

η where jets are required to be measured or "tagged" for heavy Higgs boson searches). Figure 1.5 shows one of the impacts of the choice of magnetic field (an option à la UA2/D0 without any magnetic field in the inner detector region was considered initially). Figure 1.6 shows the choice (or not) of a high performance but expensive air-core toroid magnet in the muon spectrometer.

Concluding this short historical overview of how the physics case for ATLAS was developed over the years and how heavily it influenced the detector design, Fig. 1.7 illustrates how the discovery potential for a Standard Model Higgs boson was envisaged more than ten years before data taking began. As will be shown in Chap. 9, the reality is not so different from these early expectations; the main difference arises from the so-called K-factors (higher-order corrections to the dominant Higgs boson production cross section), which were assumed conservatively to be unity during the design phase but turned out later to be much larger. The thrill of a real discovery can only be experienced once in a lifetime, and with real data and the various chapters in this book, is a testimony to that effect.

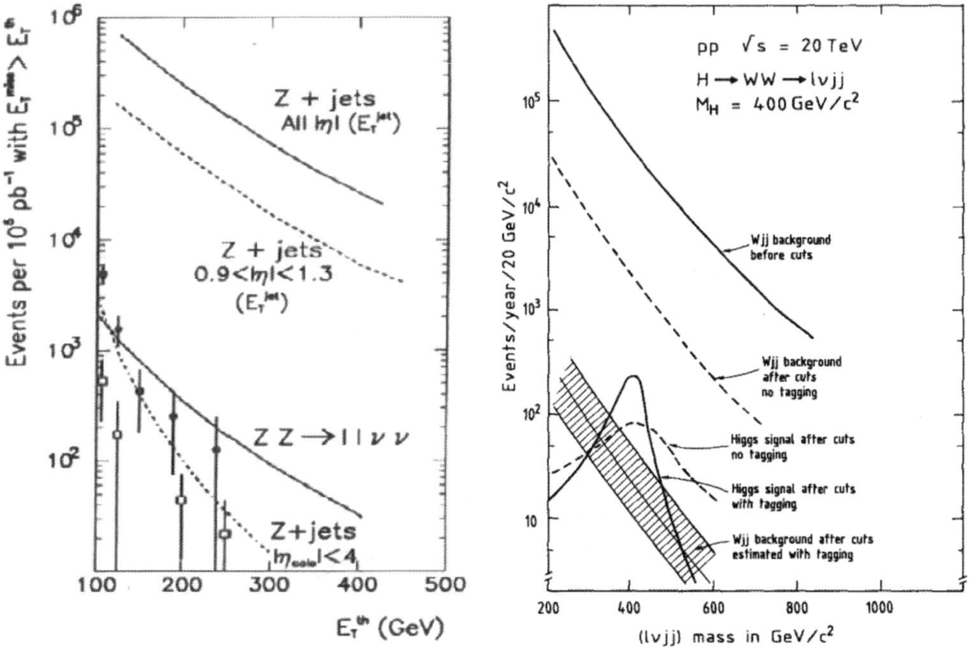

Fig. 1.4. (Left) Expected rates of events with missing transverse energy E_T above a threshold for different calorimeter coverages for the case of events with mismeasured jets in the crack between barrel and endcap cryostats. (Right) Discovery potential for a heavy Higgs boson (400 GeV), expressed as expected rate as a function of the mass of the decay products for the signal and backgrounds with and without jet tagging close to the beam.

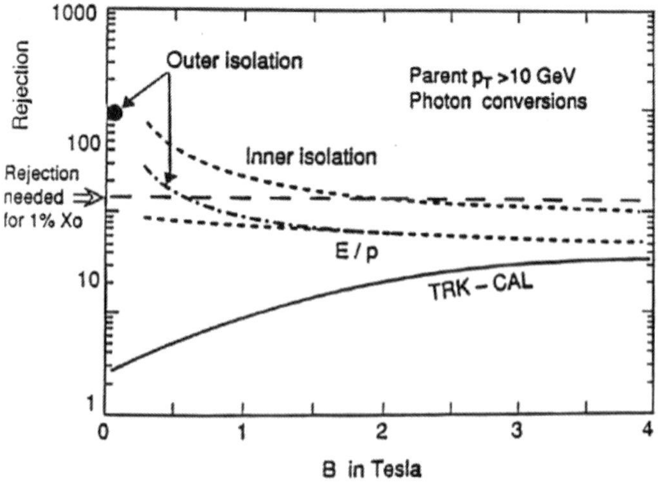

Fig. 1.5. Rejection of the electron identification algorithm (isolation, track-calorimeter matching and energy-momentum measurement matching, E/p) against prompt photon conversions as a function of choice of magnetic field B.

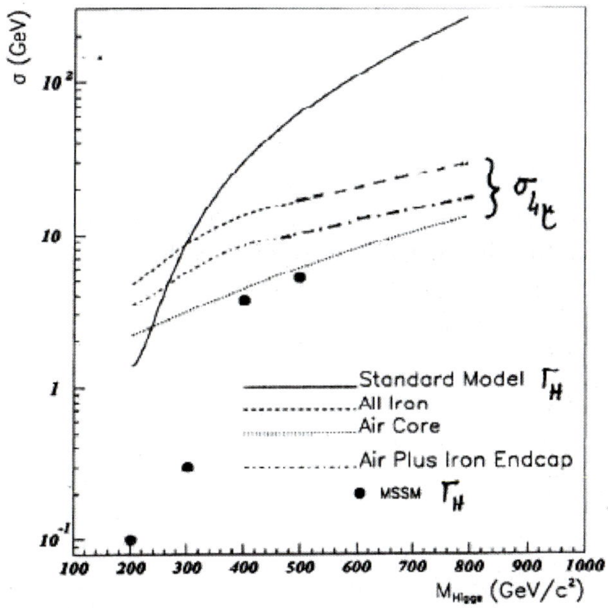

Fig. 1.6. Expected experimental mass resolution as a function of the Higgs boson mass for various muon magnet options. These are compared to the natural width Γ_H of the Higgs boson in the Standard Model and in a supersymmetric scenario (MSSM).

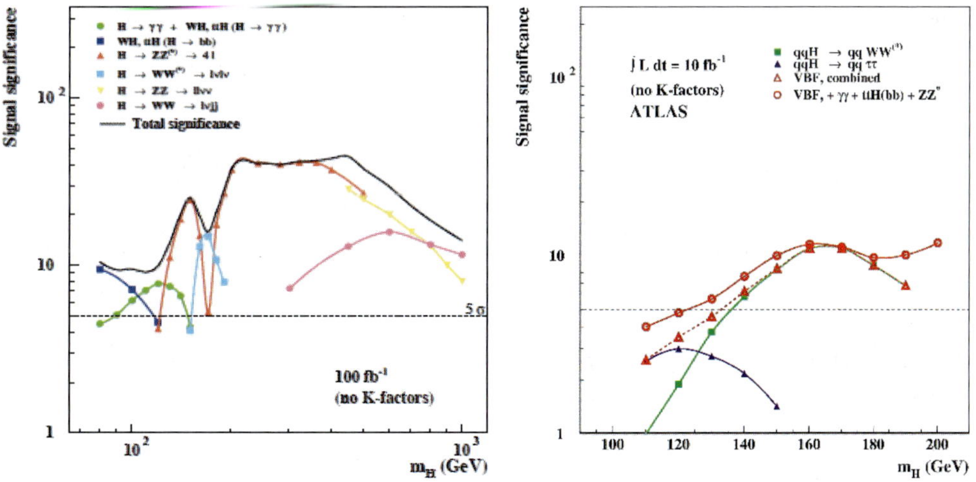

Fig. 1.7. The expected discovery potential as a function of the Higgs boson mass for various decay channels and for their combination. (Left) The entire mass range after ten years of data taking. (Right) The most probable mass range for the vector-boson fusion production modes and for a much smaller integrated luminosity, as expected after 1–2 years of initial data taking.

1.3. The LHC design and its early evolution

The two most important parameters of a proton-proton collider are the energy of the collisions and the luminosity, a parameter proportional to the number of collision events per second. The energy is critical to the discovery potential for new particles of high mass while the luminosity is critical to the production of a relevant number of the desired events over a reasonable running time. The luminosity is a measure of the quality of the beams and their collisions. Achieving a high value for this luminosity is the result of careful design and mastering of several machine physics phenomena, which can limit the collider performance.

The beam energy is proportional to the product of B (the magnetic field of the main dipole magnets) and ρ (the bending radius in the arcs of the collider ring). Since ρ was fixed by the size (27 km circumference) of the LEP tunnel, the LHC magnet design was aimed at the highest practical magnetic field value.

At the 1984 Lausanne workshop, machine experts discussed the first LHC layout, assuming B = 10 T main dipoles in a novel twin-bore configuration in order to fit into the LEP tunnel diameter and to be affordable. The machine was projected to deliver pp collisions at 18 TeV with a luminosity of $1.6 \times 10^{33}\,\text{cm}^{-2}\,\text{s}^{-1}$. By the La Thuile workshop in 1987, much progress had been achieved by the machine experts, under the leadership of G. Brianti, in understanding in greater detail the complexity of the machine with all its components in addition to the main dipole magnets, still assumed to operate at a field of 10 T. The performance was then characterised by collisions at 16 TeV with a luminosity of $1.4 \times 10^{33}\,\text{cm}^{-2}\,\text{s}^{-1}$. However, it was already stated that one could "reasonably assume" that a value of up to $10^{34}\,\text{cm}^{-2}\,\text{s}^{-1}$ would be reached for "special experiments", in response to the growing awareness in the community that this would be necessary to be competitive for the physics reach compared to the SSC.

The LHC conceptual studies evolved further to result in a comprehensive document in 1991 entitled "Design Study of the Large Hadron Collider", broadening its scope by also including chapters on heavy ion (up to PbPb) and ep collisions. The dipole magnets were still intended to have a field of 10 T, resulting for the advanced layout of the full collider lattice to a collision energy of 15.4 TeV and a luminosity of $1.65 \times 10^{34}\,\text{cm}^{-2}\,\text{s}^{-1}$, albeit with a bunch spacing of only 15 ns, as compared to the 25 ns in previous studies.

It took two more iterations to arrive at the collider design we know today, a true marvel of technology. After the so-called LHC White Book in 1993, the

Conceptual Design Report of October 1995 documents the machine as it was then built with dipole fields of 8.4 T, achieved with NbTi superconducting cable cooled at 1.9 K provided by superfluid helium. This left enough margin against quenches, while providing 14 TeV collisions. A bunch spacing of 25 ns was adopted as a compromise given the limitations in detector technologies and read-out electronics, resulting in a design luminosity of $10^{34}\,\mathrm{cm}^{-2}\,\mathrm{s}^{-1}$. The ep option was dropped, enabling the removal of LEP from the tunnel. The same basic structure can be used as a PbPb ion collider with a centre-of-mass energy of 1148 TeV (2.76 TeV per nucleon) and a luminosity of $10^{27}\,\mathrm{cm}^{-2}\,\mathrm{s}^{-1}$. Collisions between protons and Pb ions can also be provided.

In December 1993, a plan was presented to the CERN Council to build the LHC over a period of ten years. Even though the Council was generally positive, a lengthy debate followed because of severe cost constraints facing CERN. It was only finally in December 1994 that the LHC project was approved in a staged "missing-magnet" version (due to costing reasons), which would only have initially 2/3 of the dipoles, limiting the energy to 10 TeV. However, the CERN management, with C. Llewellyn Smith as Director-General and L. Evans as the LHC Project Leader, were able to secure enough contributions from Non-Member States to finally convince the Council in December 1996 to approve the construction of the full 14 TeV LHC, with a projected completion date of 2005. Figure 1.8 shows the layout of the LHC with its major experiments.

1.4. Early ideas about instrumentation and detector requirements

Ideas for detector systems at the LHC were shaped by those at previous colliders, notably the proton-antiproton colliders at CERN and Fermilab, but it was recognised from the outset that LHC conditions would be much more challenging, even when compared to the SSC, as illustrated in Table 1.2.

Conceptually, the detector system for a particle collider can be viewed as a series of devices with complementary functions. Working outwards from the centre of the detector where the colliding beams meet, the trajectories of charged particles are measured, and the charge and momentum can be determined from their deflection in a magnetic field; the energies of electrons, photons and hadrons are measured in calorimeters. The momenta of muons, the only charged particles that traverse the calorimeters, are also determined from their deflection in a magnetic field. Photons and electrons can be distinguished by the absence or presence of a charged particle associated

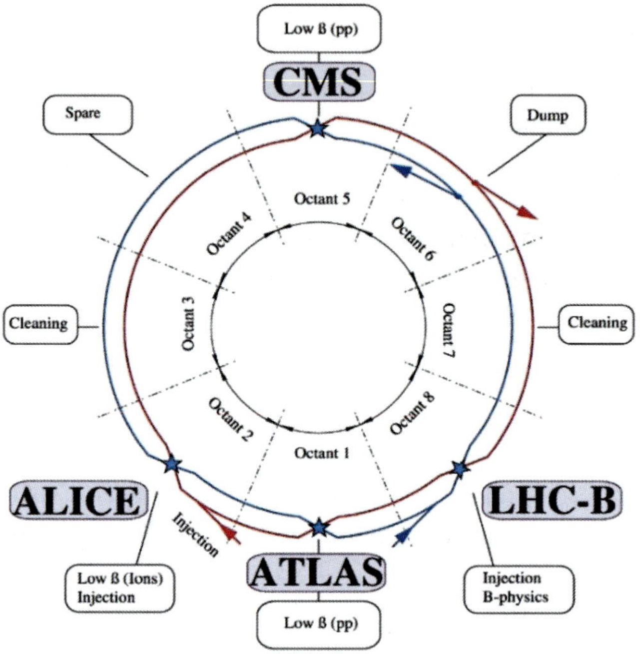

Fig. 1.8. Layout of the LHC Collider and its four major experiments.

Table 1.2. Table comparing approximate parameters for various colliders around 1990. For LHC, the numbers foreseen in 1990 are compared with the actual parameters in 2018.

	SppS	LEP	Tevatron	SSC	LHC 1990	LHC 2018
Particles	proton-antiproton	electron-positron	proton-antiproton	proton-proton	proton-proton	proton-proton
$\sqrt{s}$ [TeV]	0.54–0.63	0.09–0.2	1.8	40	14	13
L[10^{34}/cm^{2}s]	0.0003	0.024–0.1	0.0004	0.1	1.65	2.2
Δt between collisions [ns]	4000	22000	396	16	15	25
Pile-up	0	0	0	0	40	60
Recording rate [events/s]	5	1	2	—	100	1000
P_T max[TeV]	0.1	0.1	0.5	5	3	3
Calo. Cell size $\Delta\eta/\Delta\phi$ [rad]	0.1/3	0.01/0.01	0.1/0.1	—	0.025×0.025	0.003/0.025
Number channels [M]	1.5	0.5	1	—	—	100

with energy deposited in the calorimeter. Hadrons give a different pattern of energy deposition in the calorimeters compared to electrons and photons.

It was understood that the LHC interaction rates would be enormous, with bunches of counter-rotating protons passing through each other 40 million times per second (in 1990 it was 67 million, but this was later reduced), and with each bunch crossing (BC) producing up to about 40 pp collisions at the same time, also referred to as pile-ups, resulting in hundreds of outgoing particles. The implication was that the detectors and any electronics mounted on them would have to be highly resistant to radiation. The decays of new particles, like the Higgs boson, would be accompanied by the products of many other interactions occurring at the same time, thus making it harder to reconstruct the interesting part. The detectors would need to be fast so as not to compound the complexity by combining information from several BCs. They would need fine granularity to avoid more than one particle traversing the same detector element. Using an analogy of a digital camera, the detectors would need unprecedented resolution, an exposure time of less than 25 billionths of a second, about 100 million pixels and special radiation-hard technology.

Although most of the particles produced would have transverse momenta below 1 GeV, the detectors would need to precisely determine the momenta of particles up to ~1000 GeV that might result from "new physics". The calorimeters would need to cover this huge range in energy and also have excellent resolution. The tracking detectors would need fine spatial resolution to measure the tiny deflection of very high-momentum charged particles in the magnetic field.

It was also recognised that only a tiny fraction of the data produced by the detectors could be retained for analysis, even assuming massive improvements in the offline computing infrastructure. A selection needed to be made online, retaining only ~100 "events" per second (Hz) from the 40 MHz BC rate (with ~1 GHz interaction rate) — only around one interaction in ten million could "trigger" the recording of the event data.

Various detector concepts were under consideration in the late 1980s and early 1990s, with the choice of magnet configuration for the muon-detection system as a key feature. Should the muon trajectories be measured in the return yoke of the solenoid magnet (as was adopted for the CMS detector)? Or should there be a separate, dedicated toroid-magnet system. If so, should it have an iron core or an air core? Behind the thinking at the time was the wish to be able to discover the Higgs boson decaying into four muons in a "stand-alone" measurement with the muon system, i.e., without using the inner tracking detector and the calorimeters. This illustrates the enormity of the challenge as perceived at the time — the community was not fully confident that a full detector system could be made to work under the extreme LHC conditions. There was even discussion of a so-called "iron-ball" concept where one would set out only to detect the muons, with the iron ball absorbing all the other particles (with the exception of neutrinos that escape without detection) between the collision point and the muon detectors. This idea was quickly discarded because of its limited physics potential.

Numerous ideas were studied for the inner tracking region, including the possibility of electron identification as well as the determination of electric charge and momentum. The option of having no magnetic field for the inner detector was considered, implying no electron charge-sign determination, as had been done in UA2. In general, the innermost part of the calorimeters would be optimised for identification and energy determination of electrons and photons, followed by an outer part to complete measurements for hadrons.

It was understood that the detector systems would need to be very large; measuring the momenta of very high momentum particles with precision requires a long trajectory to achieve sufficient deflection, even with magnetic field strengths of 2–4 Tesla. The calorimeter would need to be massive to fully contain the energy from high-energy hadrons and prevent leakage into the muon detectors. As a consequence, the muon detectors located far from the interaction point would need to cover a huge surface area, while still achieving fine spatial precision.

At the time of the 1990 Aachen workshop, where the first ideas for LHC detectors were intensively discussed, a consensus emerged that new developments were necessary in order to build a performant LHC detector to discover the Higgs boson in the mass range of 80–1000 GeV.

The physics requirements gave strong motivation for the inclusion of a vertex detector and tracking system with a large angular coverage. The high particle density called for a finely segmented detector, while precise momentum measurements called for high-resolution position measurements within a large detector in a strong magnetic field. Furthermore, in order to resolve particles from different BCs, the time resolution would need to be excellent. All of this could only be achieved with detectors and electronics capable of withstanding many years of operation in an environment with very high radiation. Three basic technologies were considered in which the passage of charged particles would be detected by the ionisation produced in a gas (in small volumes of various geometries) or in semiconductors (Si or GaAs wafers, with detector elements arranged as small strips or pixels), or by light produced in scintillating fibres.

Similar considerations applied to the calorimeters. Fine segmentation would be needed to resolve the characteristic shape of energy deposits from high-momentum electrons and photons, and to minimise associated deposits from other particles that would degrade the resolution and particle-identification performance. The precise absolute energy scale of the measurements would need to be maintained, coupled with excellent resolution, over many years despite exposure to high doses of radiation. Several technologies were considered, including calorimeters where the shower of particles, produced by a primary particle through interactions in a dense material, would be sampled in an active medium. Measurements of the deposited energy would be based on the ionisation produced in noble liquids at cryogenic temperatures (argon, krypton) or warm liquids, or in scintillator plates or fibres. In each case, novel detector geometries were suggested, including those finally adopted for the ATLAS Accordion electromagnetic and Tile hadronic calorimeters, aiming to place electronics, cables, etc., at the outer radius of the detector, thereby minimising inactive material that would degrade the energy measurement or create holes in the calorimeter coverage, and to facilitate access.

The challenges for the muon system were somewhat different. Since almost all particles would be absorbed in the calorimeters, the radiation levels would be comparatively smaller, yet still significant, and there would not be many other particles accompanying the muons nearby in space or in time. On the other hand, with the muon system having an inner radius of several metres, each detector layer would have a huge surface area. Even with a very strong magnetic field, an excellent spatial resolution would be required to obtain a precise momentum measurement for muons with momenta up to $\sim$1 TeV. Achieving a spatial resolution of $\sim$100 μm over a detector area of many hundreds of square metres was identified as a major challenge. Good timing would also be required to uniquely identify the BC of interest. While it was generally understood that the requirements could be met by detecting the ionisation produced by muons in gas volumes, novel ideas would be needed to achieve the required precision in such a large system at an affordable cost.

In addition to the challenges for the detectors themselves, it was clear that new ideas would be needed for the associated electronics, and also for the trigger (online event selection) and data-acquisition systems.

The requirement to minimise confusion due to neighbouring BCs implied measuring signals from each detector element with a time resolution of 25 ns or less. Information derived from the detectors would need to be combined in a central place to make a decision to retain an event for further analysis, or to discard it. Given the detector size, this would unavoidably involve propagating signals over many tens of metres, corresponding to delays far in excess of 25 ns (for reference, signals typically take 50 ns to travel 10 m on an electrical or optical cable). It would therefore be necessary to locally retain all of the information from the detector, pending the arrival of the first stage trigger decision within the chain of a multi-level trigger system. For example, this could be done using a "pipelined" architecture, where digital data or analogue information would, at least logically, move along a pipe (or first-in-first-out memory). Information would enter the pipe with each BC, arriving at the end of it at the time of arrival of the trigger decision. There would need to be millions of such pipelines, one per detector element. Furthermore, similar ideas — so-called pipelined processing — would have to be applied to the aggregation and processing of information to form the trigger decision.

It was recognised that reducing the event rate from the ~1 GHz interaction rate to ~100 Hz would require a combination of technologies. The first step would rely on custom electronics making a trigger decision used to select events at the output of the detector read-out pipelines. The last step could make use of commercial processors, benefitting from advances in commodity computing. The possible need for an intermediate step was under consideration, using dedicated electronics and/or commercial processors accessing a small subset of the data identified by the first step.

1.5. Research and development (R&D) to meet the LHC challenges

Research and developments for the LHC detectors started in 1986 with a special Italian-funded R&D programme called LAA. Directed research towards the emerging LHC proto-collaborations started in July 1990, with the creation of the CERN Detector Research and Development Committee (DRDC) prior to the Aachen workshop. R&D related to sensor technologies, electronic read-out and event selection was needed to meet the new experimental requirements, especially for the inner tracking detector but also for the electromagnetic and hadronic calorimeters, the muon spectrometer, the on-line event selection and the data acquisition.

The Barcelona and Aachen meetings identified the challenges for the inner tracker R&D: extreme granularity of the individual sensors, extreme radiation tolerance of the sensors and their read-out electronics, the ability to associate detector signals with a given LHC bunch crossing, and to store the detector signals until read-out following the signal from a selected event. Optical fibre read-out was essential to keep the detector material low, and the expected read-out rate implied the need for on-detector data reduction. To ensure the dual requirements of pattern recognition and reconstructed track precision, given the large particle multiplicity in a typical pp collision and the collimation expected in hadronic jets, it was thought necessary to ensure a large number of sensor hits along the track, including some hits of very high precision. Physics simulations indicated the need for electron identification in addition to that from the calorimeters. Furthermore, the inner detector material, including all services, had to be minimised to avoid deteriorating the electromagnetic calorimeter performance, a challenge given the requirements for powering and cooling of these detectors.

Fortunately, advances in the late 1980s indicated a clear path forward. The combination of FLUKA (state of the art for radiation) and GEANT (realistic detector geometry) computing codes provided reliable estimates of the radiation dose throughout the detector from both pp collisions and low-energy albedo neutrons (crucial for semiconductor sensors). Initial investigations of the radiation hardness for semiconductor sensors showed promise. The first multi-channel ASICs suitable for microstrip semiconductor read-out had been fabricated, spawning ASIC developments for (slow) first-generation silicon microstrip vertex detectors at the LEP and Fermilab collider experiments. A large-area silicon pad array ($\sim 1\,\mathrm{m}^2$) constructed earlier for the UA2 experiment demonstrated the cost-effectiveness of semi-conductor systems in a hadron collider environment.

At a large radius, and in order to ensure a large number of sensor hits along a charged track with adequate signal speed and low pile-up rates, the R&D concentrated on the development of layers of concentric small-diameter drift tubes detecting ionisation. Filled with a Xe-CO_2-O_2 gas mixture they also detect transition radiation photons from polyethylene foils between the layers, hence its name Transition Radiation Tracker (TRT). Other R&D programmes included the development of large-area assemblies of radiation tolerant micro-strip gas chambers, silicon and gallium arsenide micro-strip sensors, and their associated electronics and mechanical structure. These R&D efforts extended well beyond the ATLAS Letter of Intent (LoI) of October 1992 and the Technical Proposal of December 1994, but provided critical input to choices for the final ATLAS detector. In particular, silicon sensors were shown to operate reliably when cooled to (typically) $-10°C$, after the integrated ionising and non-ionising radiation dose following 10 years of nominal LHC operation. The DMILL radiation-hard silicon-on-oxide technology (eventually replaced by commercial 250 nm CMOS) was used to fabricate ASICs with the required signal speed, radiation hardness (including single event upsets), and read-out buffers. Figure 1.9 shows an early design of a purely silicon central tracker and an early measurement of the effective doping concentration (and hence depletion voltage) for 300 μm thick silicon sensors following non-ionising radiation damage from 1 MeV equivalent neutrons.

Figure 1.10 shows an early prototype of the TRT detector used in test beam studies, as well as the expected efficiency for electron identification as measured in a test beam.

A notable absence from the baseline tracker design at the time of the LoI was the use of pixel semi-conductor position sensors for the innermost tracking layers where the radiation levels are most hostile and the track density at the highest. Nevertheless, pixel detector R&D had been initiated

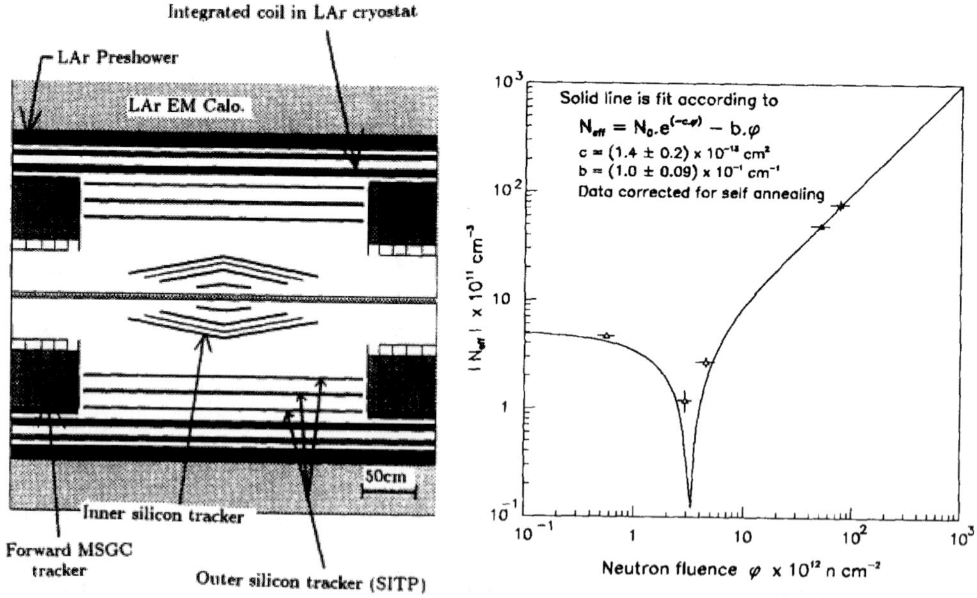

Fig. 1.9. (Left) An early design of a purely silicon central tracking region for the ATLAS experiment (1992). (Right) An early measurement of non-ionising radiation damage (with type inversion) for a 300 μm thick silicon sensor: the effective doping concentration as a function of the equivalent 1 MeV neutron fluence.

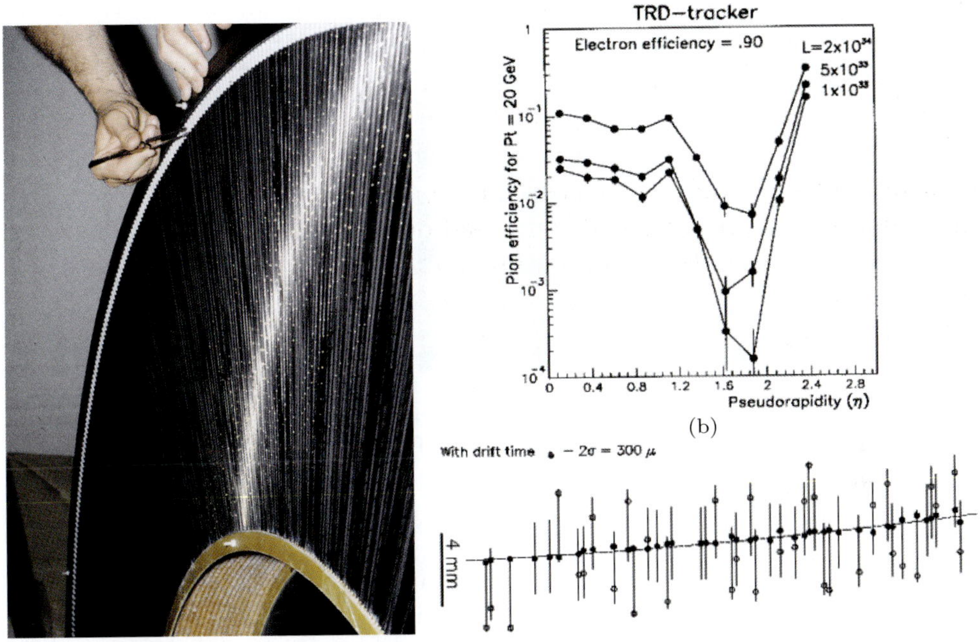

(a) (c)

Fig. 1.10. (a) Photograph of an early TRT prototype showing radial straws assembled in a disk (1993). (b) Early simulation of $\pi^{\pm}$ efficiency using transition radiation as a function of pseudorapidity η (1992). (c) Track display from a 20 GeV $\pi^{\pm}$ test beam recorded in a TRT prototype using drift time measurements (1994).

early with promising results. It was deemed too speculative at the time of the LoI, but was included as an option in the ATLAS Technical Proposal and later on as the baseline technology for the tracking layers closest to the beams. Much later, pixel diamond sensors were studied in a dedicated DRDC project. Their cost (as well as the suitability of silicon sensors) precluded their large-scale use, but they were ideally suited for beam monitors and are used in the ATLAS experiment today.

The calorimeter systems required excellent energy and position resolution over an extended energy range, powerful particle identification (electrons, photons and hadronic jets) and excellent hermeticity over an angular range extending as close as possible to the beam axis. In addition, the calorimeter is an essential detector for event (trigger) selection, implying the need for low electronic noise.

Given such stringent requirements in a harsh radiation environment, the R&D for the electromagnetic (EM) calorimeter was daunting. Noble liquids

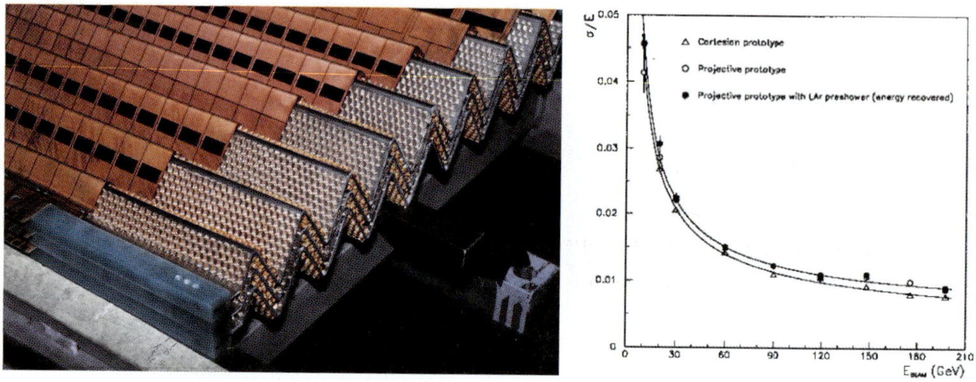

Fig. 1.11. (Left) Photograph of the accordion calorimeter structure, showing copper electrodes separated from the lead plates (1992). (Right) Energy resolution measured for electrons in a multi-cell accordion prototype (1992).

and scintillators were both considered, but a strategic choice was made to prioritise the development of an intrinsically radiation-hard lead/liquid argon (finally chosen) or lead/liquid krypton sampling calorimeter for the ATLAS LoI. A novel longitudinal accordion structure was invented (Fig. 1.11) to allow a longitudinal subdivision of the read-out for improved electron and photon identification, while minimising inactive regions and preserving very fine-grained lateral granularity. In this structure, the front-end read-out amplifiers could be placed at the back of the calorimeter to reduce the impact of radiation while retaining a reasonable electronic noise given the lower inductive load. Given this, the limited occupancy of any cell at even the highest luminosities allowed a signal collection over several bunch crossings, with subsequent bunch identification and signal shaping to optimise the signal-to-noise and determine the energy.

Another key issue was to identify and determine the direction of unconverted photons. This was addressed by developing a highly segmented first calorimeter section after the solenoid coil. The RD3 project and its test beam activities, leading to the ATLAS LoI and Technical Proposal, validated these concepts. Figure 1.11 also shows the electron energy resolution measured in a test beam for a prototype accordion calorimeter of the future ATLAS design.

The R&D challenge for the hadronic calorimeter was different. Initially, the idea of an accordion hadronic calorimeter using steel plates was also considered, but the cost was prohibitive. Lower predicted radiation levels in this region made a deep Fe-plastic scintillator sampling calorimeter feasible, but the requirements of hermeticity as well as lateral and longitudinal granularity remained daunting. An early R&D project had successfully demonstrated the use of scintillating fibres ("spaghetti") implanted in lead for a sampling electromagnetic calorimeter. The RD1 project extended this idea to use scintillating fibres inserted in thousands of precision holes, drilled in a series of large steel plates that were sandwiched together. Tests demonstrated the required energy resolution and spatial granularity, but the logistics and cost for such a large calorimeter remained unsolved. In January 1993, a novel approach was proposed to the DRDC: to use iron and scintillator plates arranged longitudinally in a sandwich, rather than the usual lateral arrangement. In this case, the requirements of granularity and hermeticity could be maintained, while simplifying the construction. Read-out was provided by wavelength-shifting optical fibres running to electronics at the back of the system. After large-scale test beam studies, the concept was proven and subsequently adopted. Furthermore, as with the electromagnetic calorimeter, this solution was well adapted for a trigger signal that could identify interesting energy depositions associated with a single bunch crossing.

At the time of the ATLAS LoI, the choice of a muon magnet technology had not been made. The technology requirements for the detectors of each system differed, but in each case the chambers had to be large (therefore with unprecedented temperature and stability monitoring), with a spatial resolution in the bending plane at the level of less than a hundred microns. This was required over a surface of many hundreds of square metres with several measurement points needed per layer. In addition, a fast trigger signal was required for muons above a given transverse momentum. Existing technologies could fulfil the specifications over a small area, but the size of any detector was limited by the attainable industrial precision. The RD5 project initiated a comprehensive study of many aspects related to the future muon spectrometer. A key challenge was to ensure that the track reconstruction and trigger selection were not adversely affected by secondary particles originating from the collisions inside the detector ("punch-through") or in the material in the cavern. The experimental studies included measurements related to the magnetic field, and the absorber (calorimeter) material.

Detector technologies such as high-pressure drift tubes (an evolution of which was finally chosen), honeycomb strip chambers and multi-wire drift chambers, were all considered at the prototype stage. In parallel, R&D towards an extensive laser-based alignment system was started.

Ignoring the logistic challenge of constructing a large muon detector system and aligning it with micron precision at this stage, the precise measurement of muon trajectories appeared feasible over large areas. What could not be achieved, however, was to simultaneously provide a fast trigger signal (implying an intrinsic time resolution of $\sim$5 ns) with an almost hermetic coverage. The community, therefore, chose to concentrate specific R&D efforts on so-called resistive plate chambers (RPC) for the barrel, consisting of a thin gaseous gap between two parallel resistive plates with a large voltage difference, that provides a fast signal with a limited spatial resolution of several mm; a viable large-scale implementation was achieved many years later. In the forward directions, R&D towards different technologies adapted to the region with higher rates was initiated: cathode strip chambers (CSC) for the precision measurement and thin gap chambers (TGC) for the trigger.

1.6. Proto-collaborations for the Evian meeting

Around and not long after the ECFA Large Hadron Collider Workshop in Aachen in October 1990, it was clearly "in the air" that groups started meeting, more or less publically, in view of discussing possible future LHC experiments. There were two more indications that the time had come for the community to get seriously active on LHC activities if one did not want "to miss the train" for being among the drivers of a future LHC experiment. One was the setting up of the Detector Research Development Committee (DRDC) by CERN in July 1990, to stimulate R&D on detector technologies satisfying the LHC requirements. The other clear signal was the famous statement adopted by CERN Council in December 1991 that "LHC is the right machine for the future of CERN".

Specific physics interests, expertise in and biases towards certain technologies, and quite importantly also, personal experiences and loyalties from previous and current collaborations all played an important role in choosing future partnerships. Everybody realised that there would be strong competition in order to be approved as an experiment at the end of a long process, and that it would be hard to find the necessary resources for all LHC experiments. People were also well aware that committing to developing an LHC experiment, and eventually operating it, would mean almost a lifetime engagement, or at least several decades to come.

Within only a little more than a year after Aachen, the time was mature to discuss these proto-collaboration activities in public, and CERN and the ECFA jointly organised a general meeting ("Towards the LHC Experimental Programme" — it was on LHC physics and detectors) in March 1992 at Evian-les-Bains. This large conference attracted about 650 participants across Europe and other regions, demonstrating once more the broad interest in the LHC project. Proto-collaborations presented their status of studies as well as their ambitions, in a lively atmosphere with lots of discussions, during the sessions and around them. These presentations are referred to as the "Expressions of Interest" (EoIs), and they are documented in proceedings with copies of the original transparencies (Fig. 1.12, left). The Evian conference also marked the birth of the LHC Experiments Committee (LHCC), set up by CERN Management after the meeting in order to review all future experimental proposals for LHC.

There were a total of 12 EoIs presented at Evian, four concerning general-purpose pp experiments, three dedicated to studying B-physics, another three dedicated to the heavy ion running of the LHC, and even two neutrino experiments, having in particular the τ neutrino discovery in mind. The heavy-ion community converged towards one experiment (ALICE) with its LoI submitted in March 1993. The B-physics efforts resulted in a

Fig. 1.12. (Left) The cover of the Evian proceedings (1992). (Right) Poster of the symposium "25 years of the LHC experimental programme" at CERN (2017).

new collaboration, called LHCb, which submitted a LoI in August 1995. However, the LHC neutrino experiment ideas were not carried forward any further.

The four general-purpose detector concepts could be considered as two distinct groups according to their magnetic field configurations. The ASCOT and EAGLE proto-collaborations were proposing detectors incorporating toroidal muon spectrometer magnets. As described in the next section, they joined their efforts to "give birth" to the ATLAS experiment. The other two EoIs, the Compact Muon Solenoid (CMS) and L3+1, were both based on a single large superconducting solenoid magnet structure. There was no convergence in joining efforts in the follow-up discussions of both teams after Evian. They submitted separate LoIs, together with ATLAS on 1 October 1992. The LHCC decided to approve only ATLAS and CMS in June 1993.

A recent symposium at CERN on 15 December 2017 marked the 25th anniversary of launching the LHC experimental programme at Evian (Fig. 1.12, right) and highlighted the successes achieved over these years.

1.7. Merging of toroid concepts, the ATLAS Letter of Intent

As stated above, two of the EoIs for general-purpose detectors presented at the Evian conference in March 1992 were based on a toroid magnet configuration for the muon spectrometer system. It was well understood and made clear by CERN from the onset (of the discussion about the experimental programme for LHC) that there would be resources (money and people) available for at most two large experiments. It was therefore natural for the two teams with toroid magnets to seek for and converge on a joint approach, since it was highly probable that CERN would favour two detectors with complementary magnet configurations in the end.

The merger of the two proto-collaborations, called ASCOT and EAGLE at the time, and of their two toroid-based detector concepts was not straightforward, and it required a lot of goodwill and a carefully planned and transparent strategy on how to proceed towards certain difficult choices. The process succeeded efficiently, thanks to the mutual motivation for LHC physics. Admittedly, it was also helped by the early deadline of 1 October 1992 for the submission of the LoI, which CERN's newly formed LHCC set only half a year after the EoIs.

ASCOT (with some 20 interested institutes) and EAGLE (with some 70 interested institutes) joined forces over summer 1992 and submitted an

LoI under the new name ATLAS, signed by 88 institutes with about 850 scientists on 1 October 1992. The name **ATLAS**, an acronym constructed from "**A** **T**oroidal **LHC** **A**pparatu**S**", was selected by vote in the first official common Collaboration Board meeting of the two merged proto-collaborations, which voted on a shortlist that had been established from a large number of proposals solicited from all members of the future collaboration. In the second round of voting, "ATLAS" obtained the required 2/3 majority over its contender "ALICE". Other proposed names of "ACE", "ALEX" and "LHD" were already eliminated during the first round.

A toroid magnet configuration was, in spite of its complexity, from the onset the common theme, chosen by ATLAS because its field configuration is optimally adapted to a hadron collider environment. However, the two EoI detector concepts initially differed considerably, insofar as there were different priorities on how to approach the LHC physics challenges. One (ASCOT) featured a novel superconducting air-core toroid magnet in the barrel region with an excellent potential for muon measurements (even in stand-alone operation at highest luminosities), complemented by superconducting iron-core toroids in the endcap regions. The other approach (EAGLE), based on a cheaper, clearly less performant, magnetised iron toroid over the full acceptance region, left more resources for a balanced detector concept with excellent electromagnetic and hadronic calorimetry, and a robust and precise tracker.

In both cases, a superconducting solenoid was foreseen around the inner tracking detectors in front of fine-grained electromagnetic and hadronic calorimeters in the field-free region featured by the muon toroid configuration. The original EoI detector drawings are shown in Fig. 1.13.

Combining the best aspects of both approaches was of course attractive and desirable, as it would make a truly high-performance general-purpose LHC detector optimally covering all the ambitious physics goals. Furthermore, pooling the resources within the newly formed ATLAS Collaboration was expected to make such a complete instrument more realistically affordable within the available funding, with the added potential of attracting further collaborators and contributions. Indeed, seen from today's distance, this strategy was successful. However, as will be described later on, it was not always a simple and smooth journey to reach these goals and to converge to the final design of the ATLAS detector, which is so efficiently collecting high-quality data at the LHC now.

During summer 1992, expert groups from the different major detector communities met to discuss possible choices for a common detector concept.

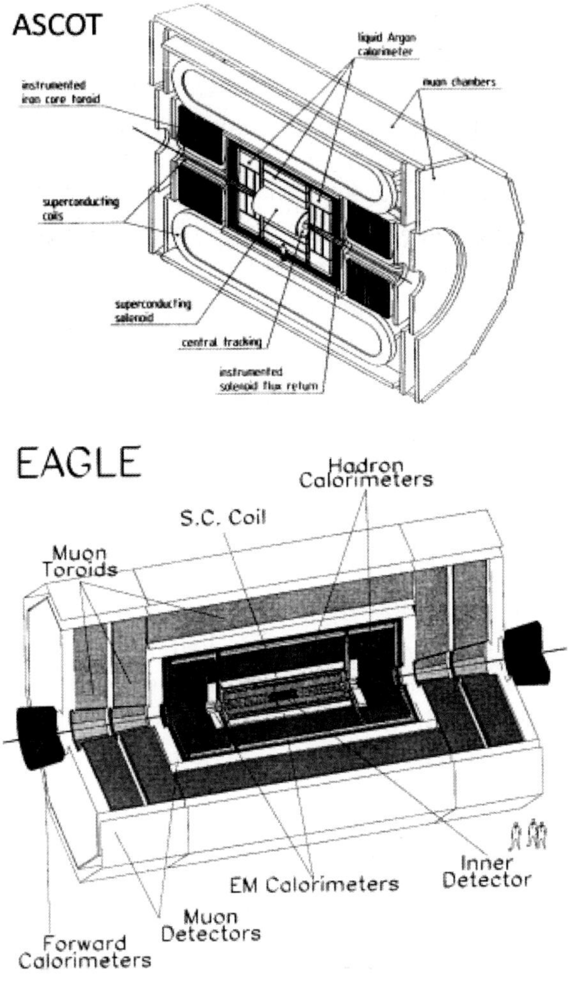

Fig. 1.13. The ASCOT (top) and EAGLE (bottom) detector concepts as presented with their Expressions of Interest at the Evian meeting.

This was an intense period for everyone involved, because in addition to the technical evaluations and discussions, there was also some way to go in building up a collaborative spirit among previously competing physicists. In several cases, people who had already invested a few years of considerable design and development work on early prototypes, had to face a new reality: some of their work would have to be terminated in view of a new common concept. That this process succeeded (in the case of ATLAS) can certainly be taken as a sign of scientific and human strength at the root of the nascent Collaboration.

Fig. 1.14. Cover page of the ATLAS Letter of Intent, submitted on 1 October 1992, which is the first official document carrying the name of the Collaboration.

It was clear that, during the few months available to produce the LoI, not all technological choices could be made; more studies and R&D work was required. Therefore, options were presented for many cases, but more importantly this was done together with a first indication of how these necessary selections could be resolved. It took several rounds of intense encounters between the Collaboration and the LHCC referees before the committee decided at its 7th meeting on 8–9 June 1993 to recommend ATLAS (and CMS in parallel) to proceed to a Technical Proposal (TP). This will be described later together with the technical aspects and the evolution of the ATLAS detector design over the many years from the LoI (the cover is shown in Fig. 1.14) to the start of its operation.

ATLAS first data taking

After many years of construction, the final phase of putting it all together started in spring 2008. Up to 20 detector components, each of them a complex system in itself, had to be integrated and shown to work together coherently before the first beams would pass through the detector. Systems that could work well by themselves might not when combining with others — and often did not! However, towards the end of the summer most problems had been ironed out, and data recording with all detector components could be achieved on a regular basis.

But one big problem remained for the much-anticipated first day of beam from the LHC. Should something go wrong, requiring a restart of the system, more than an hour might pass before we were back online. Today, many automatic procedures allow us to recover from problems in individual components within minutes, without interrupting the operation of the full detector. Not so then! Anything getting stuck would most likely mean shutting down everything and then bringing it back up, and it did not always work the first time around. On the other hand, the projected time for the LHC operators to bring the beam step by step around the accelerator was about one hour — with the media and the world watching. There was not much margin for error.

ATLAS control room on the LHC start up day.

On the evening before the big day, we assembled in a meeting room off the side of the main ATLAS building (our usual room had already been taken over by the news teams) and settled on our strategy for the next 24 hours. We were to perform a complete clean restart of the system very early in the morning, and then not touch anything until we had recorded the beam-induced data we were all waiting for on that day. We knew we could run stably for many hours once we had started up, but any intervention, harmless as it might seem, might have stopped the data taking and lost us the first beam events with all the world watching. So, the motto of the day was — start-up, then hands off, no more changes or tweaks allowed!

Morning came and the operators started to bring the beam around the accelerator, stopping at absorbers at some intervals before opening them and continuing, getting ever closer to ATLAS. The control room was packed full of people, ATLAS colleagues and media. Finally, the absorber just before ATLAS was closed, and the beam directed onto it, sending showers of secondary particles into the ATLAS detector, that should show on our control displays within seconds. Everyone had their eyes glued to the schematic display of the detector projected on the wall, where the detected particles should appear.

After anxious seconds ticking by, finally the first event suddenly appeared on the screen, showing not only that the beam had arrived in ATLAS, but also that our data taking was up and running and working well. The control room erupted with cheers! After two decades of planning, construction and commissioning, the ATLAS detector had recorded its first events induced by the LHC beams — the first of billions more to come!

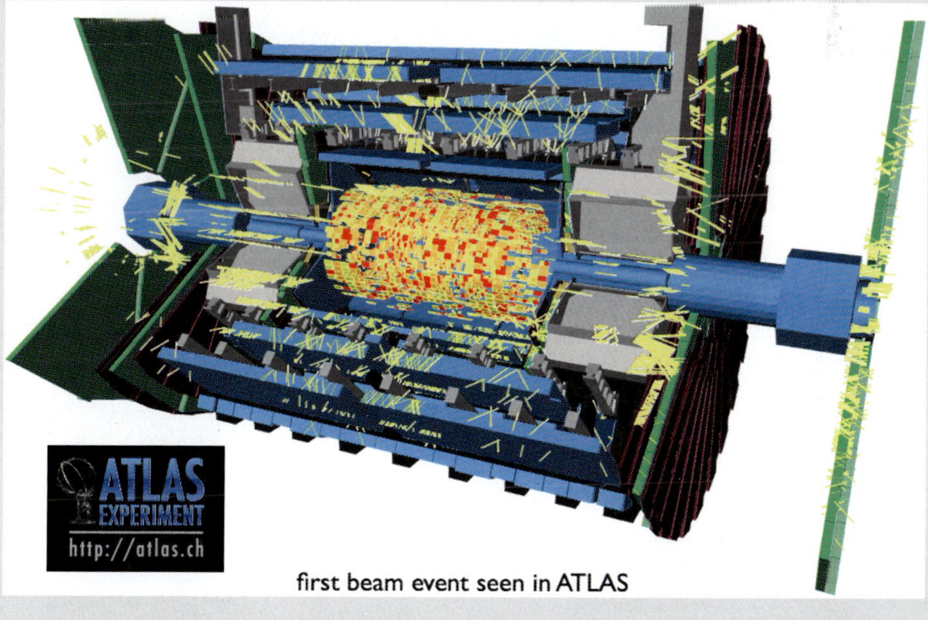

ATLAS on the first day of beams.

Chapter 2

The ATLAS Detector Today

2.1. The ATLAS detector layout

The Large Hadron Collider at CERN provides energetic proton-proton and heavy-ion collisions, normally every 25 ns. The decay products resulting from these collisions need to be detected by the ATLAS particle detection devices. These detectors are designed according to the specific properties exhibited by the particles that they target: their interactions with matter, whether they are electrically charged or neutral, hadronic (containing quarks) or leptonic (e.g., electrons) in origin, etc. Furthermore, these particles are produced in all directions from the central collision point, with energies ranging from the few Tera-electron-volt (TeV) scale down to the Mega-electron-volt (MeV) scale, i.e., an energy range of six orders of magnitude. For this reason, the particle detectors must effectively surround the collision point in order to collect all of the particles produced. The functions of these detectors may be broken down into two broad categories: those that determine the momentum and origin of the charged particles and those that determine the energy of particles.

The dimensions of ATLAS, shown in Fig. 2.1, are 25 m in height and 44 m in length. The overall weight of the detector is approximately 7000 tonnes, similar to that of the Eiffel Tower. It is designed to be nominally forward-backward symmetric with respect to the interaction point, located at the heart of the detector. This nominal interaction point defines the origin of the coordinate system, while the beam direction defines the z-axis and the x–y plane is transverse to the beam direction. The azimuthal angle φ is measured around the beam axis, and the polar angle θ is the angle from the beam axis. The pseudorapidity, η, normally used to describe the coverage of subdetectors, is defined as $\eta = -\ln\tan(\theta/2)$: $\eta = 0$ represents $\theta = 90°$, $\eta = 1$ is $\theta = 40°$, $\eta = 2.5$ is $\theta = 10°$ and $\eta = 5$ is $\theta = 1°$ above the beamline.

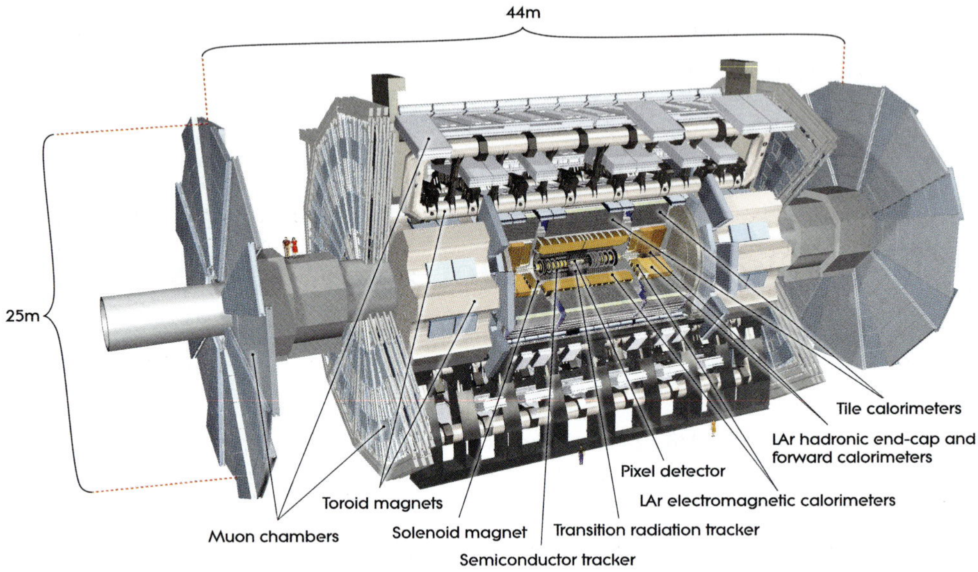

Fig. 2.1. Cut-away view of the ATLAS detector.

The choice of the magnet configuration has driven the design of the rest of the detector; these magnets provide the bending power necessary for the measurement of the momentum of charged particles. A thin superconducting solenoid, which extends over a length of 5.3 m with a diameter of 2.4 m, surrounds the Inner-Detector cavity, immersing it in a magnetic field. Three large superconducting air-core toroid magnets (one barrel and two endcaps) are arranged with their coils in an eight-fold azimuthal symmetry around the calorimeters; the barrel ($0 < |\eta| < 1.4$) and endcap ($1.6 < |\eta| < 2.7$) toroids bend the muon trajectories enabling the measurement of their momentum. The strong bending power in a large volume coupled to a light and open structure minimises the detrimental effects of multiple scattering, as muons emerging from the calorimeters traverse the Muon Spectrometer.

The Muon Spectrometer surrounds the rest of the detector system. This spectrometer is based on the magnetic deflection of muon tracks in the large superconducting air-core toroid magnets, instrumented with separate trigger and tracking chambers. Excellent muon momentum resolution is achieved with three layers of high-precision tracking chambers. The muon instrumentation includes, as a key component, trigger chambers with timing resolution of the order of 1.5–4 ns.

Over most of the η-range, a precision measurement of the track coordinates in the principal bending direction of the magnetic field is provided by Monitored Drift Tubes (MDTs). The mechanical isolation in the drift tubes of each sense wire from its neighbours guarantees a robust and reliable operation. At large pseudorapidities, multiwire proportional chambers with cathodes segmented into strips (Cathode Strip Chambers, CSCs) with higher granularity are used in the innermost plane over $2 < |\eta| < 2.7$, to withstand the demanding rate and background conditions. The trigger chambers for the Muon Spectrometer, which cover the range $|\eta| < 2.4$, serve a threefold purpose: provide bunch-crossing identification and well-defined transverse-momentum thresholds to select events containing muons, and measure the muon coordinate in the direction orthogonal to that determined by the precision-tracking chambers. Resistive Plate Chambers (RPCs) are used in the barrel ($|\eta| < 1.05$) and Thin Gap Chambers (TGCs) in the endcap regions ($1.05 < |\eta| < 2.7$, but only up to 2.4 for triggering).

The electromagnetic and hadronic calorimeters cover the range $|\eta| < 4.9$, using techniques designed to accommodate the diverse requirements of the physics processes of interest, while at the same time mitigating the harsh radiation environment, particularly at high η. The calorimeters provide good containment and measurement of electromagnetic and hadronic showers, and limit punch-through into the muon system.

The high-granularity electromagnetic sampling calorimeter is a lead/liquid-argon (LAr) detector with accordion-shaped kapton electrodes and lead absorber plates. It is divided into a barrel part ($|\eta| < 1.475$) and two endcap components ($1.75 < |\eta| < 3.2$), each housed in their own cryostats, which in the barrel is shared with the central solenoid. Each endcap calorimeter is divided into two coaxial wheels: an outer wheel covering $1.375 < |\eta| < 2.5$ and an inner wheel in $2.5 < |\eta| < 3.2$.

The hadronic calorimetry in the range $|\eta| < 1.7$ is provided by a scintillator-tile calorimeter placed directly outside the electromagnetic calorimeter envelope. Its barrel covers the region $|\eta| < 1.0$, and its two extended barrels the range $0.8 < |\eta| < 1.7$. It is a sampling calorimeter with steel as the absorber and scintillating tiles as the active material, and uses wavelength-shifting fibres for read-out by photomultiplier tubes.

In the endcaps, LAr technology is also used for the hadronic calorimeters. The Hadronic Endcap Calorimeter (HEC) consists of two independent wheels per endcap, located directly behind the endcap electromagnetic calorimeter and sharing the same LAr cryostats. Each wheel has two depth segments. Copper plates interleaved with LAr gaps form the basis of this sampling calorimeter. The HEC η range slightly overlaps that of the Tile Calorimeter by extending to $|\eta| = 1.5$ and extends down to $|\eta| = 3.2$, thereby overlapping with the Forward Calorimeter. The Forward Calorimeter (FCal) provides both electromagnetic and hadronic energy measurements and extends the pseudorapidity coverage to $|\eta| = 4.9$. The FCal is integrated into the endcap cryostats and consists of three modules in each endcap: the first, made of copper, is optimised for electromagnetic measurements, while the other two, made of tungsten, predominantly measure the energy of hadronic interactions. Finally, a set of Minimum-Bias Trigger Scintillators (MBTS) are installed on the front-face of each endcap calorimeter (at $z = \pm 3.6$ m) and cover the region $2.07 < |\eta| < 3.86$.

The Inner-Detector system enables charged-particle tracking, including momentum and vertex measurements as well as some electron identification, in an η range matched to the precision measurements of the electromagnetic calorimeter. It does so with a combination of discrete, high-resolution semiconductor pixel and strip detectors in the inner part of the tracking volume, and straw-tube tracking detectors (TRT) in its outer part, with the capability to generate and detect transition radiation. The first measurement point closest to the interaction is provided by the insertable B-layer (IBL) Pixel Detector which was installed during the long shutdown period between Run-1 and Run-2 (2013–2014).

The precision tracking detectors (IBL, Pixel and SCT) cover the region $0 < |\eta| < 2.5$; their relative locations in radius R are shown in Fig. 2.2. In the barrel region, they are arranged on eight (1 IBL, 3 Pixel and 4 SCT) concentric cylinders around the beam axis while in the endcap regions they are located on 12 (3 Pixel and 9 SCT) disks perpendicular to the beam axis. The highest granularity is achieved around the vertex region with the IBL. The straw-tubes of the TRT, which typically provide 36 hits per track, enable track-following up to $|\eta| = 2.0$. In the barrel region, the straws are parallel to the beam axis and in the endcap region, they are arranged radially in wheels.

Since abnormal beam conditions have the potential to cause serious damage to the Inner-Detector system, a Beam Condition Monitor (BCM) located close to the interaction point is implemented. It consists of two sets of four modules (radiation-hard pCVD diamond sensors) around the beam pipe located symmetrically on either side of the interaction point ($|z| = 184$ cm and radius $= 55$ mm or $|\eta| = 4.2$).

Forward detectors shown in Fig. 2.3 complete the ATLAS coverage in the very high η region. The LUCID (Luminosity Measurement Using Cherenkov Detectors) detectors sit on either side of the interaction point and provide luminosity measurements in the range $5.6 < |\eta| < 5.9$. The ZDC (Zero Degree Calorimeter) detectors are located where the LHC splits into two

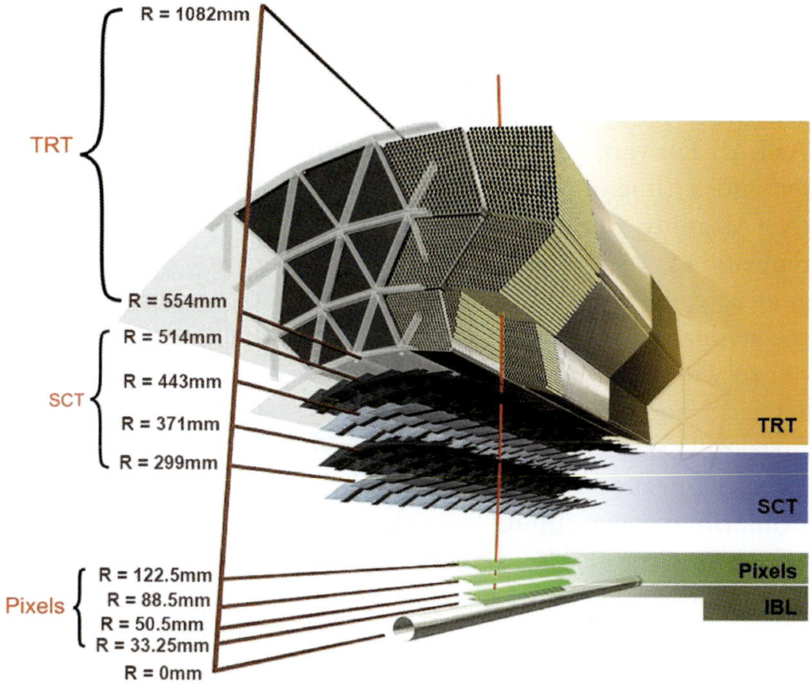

Fig. 2.2. Layout of the ATLAS Inner Detector.

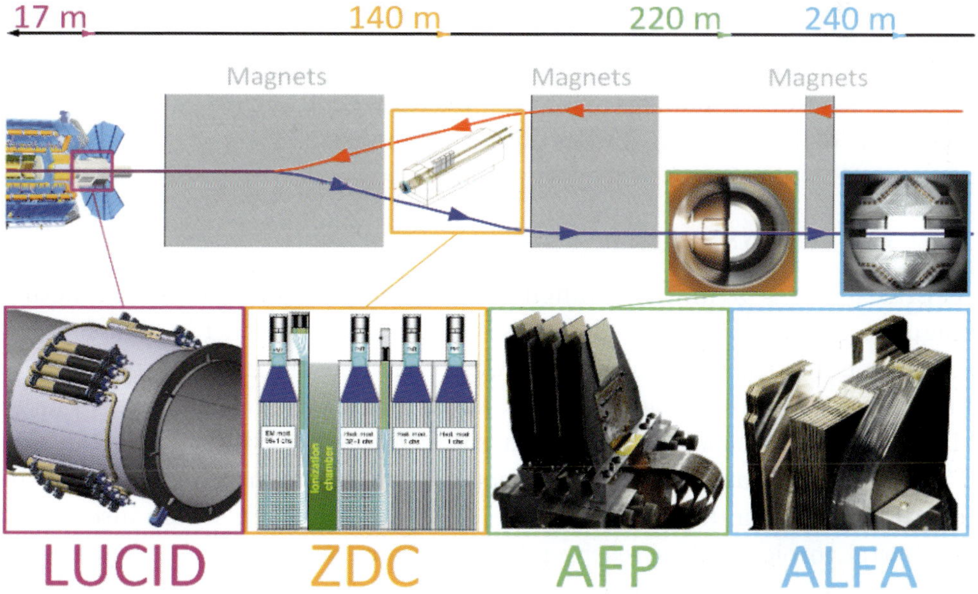

Fig. 2.3. Layout of the forward detectors.

separate beam pipes. Each side of the detector consists of four modules: one electromagnetic calorimeter and three hadronic calorimeters. The ZDC is mainly used during heavy-ion collisions, providing information from neutrons about the centrality of the interaction. Forward protons are detected by the AFP (ATLAS Forward Proton) detector, which consists of two stations — a tracker and a time-of-flight system. Furthest from the interaction point is ALFA (Absolute Luminosity For ATLAS), providing elastic scattering measurements at very small angles. ALFA consists of scintillating-fibre trackers located inside two Roman Pots, four metres apart, and is normally only used under very specific beam conditions which allow the detectors to be moved to within 1 mm of the LHC beam.

The trigger and data acquisition (TDAQ) system consists of a hardware-based first-level trigger (L1) and a software-based high-level trigger (HLT). The L1 trigger decision is formed by the central trigger processor (CTP), which primarily receives inputs from the calorimeter and the muon systems. A new Fast TracKer (FTK) system, which provides global Inner-Detector track reconstruction at the L1 trigger rate, is currently being commissioned. After the L1 trigger acceptance, the events are buffered in the ReadOut System (ROS) and processed by the HLT. The HLT receives Region-of-Interest (RoI) information from L1, which can be used for regional reconstruction in the trigger algorithms. After the events are accepted by the HLT, they are transferred to local storage and subsequently exported to the Tier-0 facility at CERN's computing centre for offline reconstruction. For Run-2 trigger operations, the maximum L1 rate of 100 kHz is set by detector read-out capability while the average HLT physics output rate of 1000 Hz is defined by the offline computing model.

2.2. The ATLAS magnet system

ATLAS features a unique configuration with four large superconducting magnets. A drawing of the magnet system can be seen in Fig. 2.4. It is an arrangement of a central solenoid (CS, not visible in the figure) providing the Inner Detector with a magnetic field, surrounded by a system of three large air-core toroids generating the magnetic field for the Muon Spectrometer. The overall dimensions of the magnet system are approximately 26 m in length and 22 m in diameter. The two endcap toroids (ECT) are inserted in the barrel toroid (BT) at each end and line up with the CS. They have a length of 5 m, an outer diameter of 10.7 m and an inner bore of 1.65 m. The CS coil extends over a length of 5.3 m and has a bore of 2.4 m. The unusual

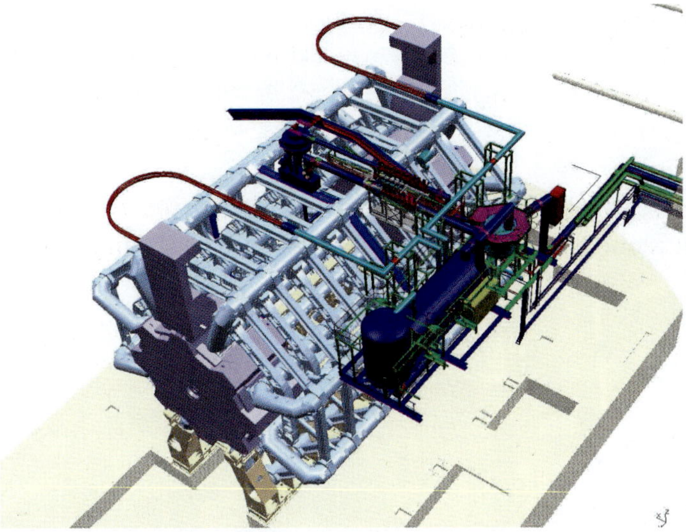

Fig. 2.4. Schematic view of the toroid magnet system including services connections and proximity cryogenics (dark blue). The eight BT coils (light blue) are visible with the two castellated ECTs (violet) inserted on both ends. The central solenoid is not visible.

configuration and large size made the magnet system a considerable challenge requiring careful engineering. The first conceptual design of the magnet system was sketched in the early 1990s and the technical design reports were published in 1997. The toroid cold-mass and cryostat integration work began at CERN in 2001. The first BT coil was lowered into the cavern in Fall 2004, immediately followed by the CS (embedded inside the LAr barrel calorimeter cryostat). The remaining seven BT coils were installed in 2004 and 2005, and the ECTs in 2007.

The CS provides a central field of 2 T with a peak magnetic field of 2.6 T at the superconductor itself. The peak magnetic fields on the superconductors in the BT and ECT are 3.9 and 4.1 T, respectively. The performance in terms of bending power is characterised by the field integral $\int \mathrm{B}dl$, where B is the azimuthal field component and the integral is taken on a straight-line trajectory between the inner and outer radius of the toroids. The BT provides 2 to 6 Tm and the ECTs 1 to 8 Tm in the 0.0–1.4 and 1.6–2.7 absolute pseudorapidity ranges, respectively.

The central solenoid

The position of the CS in front of the EM calorimeter demands a careful minimisation of the material in order to achieve the desired calorimeter

Fig. 2.5. Insertion of the central solenoid into the barrel cryostat with the LAr Calorimeter.

performance. As a consequence, the CS and the LAr Calorimeter share one common vacuum vessel, thereby eliminating two vacuum walls. The solenoid contributes a total of 0.66 radiation lengths at normal incidence. The single-layer coil is wound with high-strength Al-stabilised Nb/Ti conductor (1173 turns, 9.1 km total), specially developed to achieve a high field while optimising thickness, inside a 12 mm thick Al 5083 support cylinder. The coil mass is 5.4 tonnes and the stored energy is 40 MJ. The flux is returned by the steel of the hadronic calorimeter and its girder structure. The CS is energised by an 8 kA power supply. The magnetic field of the CS has been mapped in situ with a relative precision greater than 10^{-4} over the full tracking volume. Figure 2.5 shows the insertion of the solenoid into the barrel cryostat.

The toroid system

Each of the three toroids consists of eight coils assembled radially and symmetrically around the beam axis. The ECT coil system is rotated by 22.5° with respect to the BT coil system in order to provide radial overlap and to optimise the bending power in the interface regions of both coil systems. The conductor and coil-winding technology is essentially the same

Fig. 2.6. Barrel toroid installed in the underground cavern. In the background is the barrel calorimeters including the central solenoid in their common cryostat before insertion into the centre of the detector.

in the BT and ECTs; it is based on winding a pure Al-stabilised Ni/Ti/Cu conductor (81.6 km overall) into pancake-shaped coils, followed by vacuum impregnation.

The eight BT coils are of a flat racetrack type with two double-pancake windings. The windings are housed in an aluminium alloy casing, with a total cold mass of 360 tonnes. The coils are housed in individual cryostats. The magnetic forces are transferred to the warm structure, counteracting the net Lorentz forces of about 1400 tonnes per coil directed inwards. The toroidal structure consists of eight cryostats and the linking elements between them, called voussoirs and struts that provide mechanical stability. Services are brought to the coils through a cryogenic ring linking the eight cryostats to a separate service cryostat. It provides connections to the power supply, the helium refrigerator, the vacuum systems and the control system. Figure 2.6 shows the BT after its installation in the underground cavern.

Each ECT also consists of eight coils in an aluminium alloy housing. They are cold-linked and assembled as a single cold mass of 140 tonnes, housed in one large cryostat. Therefore, the internal forces are taken by the cold supporting structure between the coils, a different design solution than in the BT. Due to the magnetic forces, the ECT magnets are pulled into the BT and the corresponding axial forces are transferred to the BT cryostats via special transfer points linking both magnet systems. The ECT cryostats rest on a rail system facilitating the movement and parking of the ECT magnets for access to the detector centre.

The magnets are indirectly cooled by forced flow of helium at 4.5 K through tubes welded on the casing of the windings. The cooling power is supplied by a central refrigeration plant located in the side cavern and the services are distributed among the four magnets. Electrically, the eight coils of the BT are connected in series, as are the 16 coils in the two ECTs. The toroid coil systems have a 21 kA power supply and are equipped with control systems for fast and slow energy dumps. The stored energy is 1.1 GJ in the BT and 0.5 GJ together in the two ECTs. A quench protection system has been designed to safely dissipate the total stored energy of 1.6 GJ without overheating the coil windings.

The full magnet system, including its fast safety discharge procedures, has been commissioned in its final configuration in the cavern in 2008. Since then it has worked very reliably during all data taking periods of the experiment. It is arguably the largest and most complex superconducting magnet system ever built for a particle physics experiment.

2.3. The Muon Spectrometer

One of the main requirements in the design of ATLAS was the capability to trigger and accurately measure leptons over the largest possible acceptance. For the case of muons (μ), ATLAS opted for an air-core toroid magnetic system to generate the external field used for momentum measurements. This configuration produces a magnetic field with maximal bending power that is perpendicular to the direction of the muon, even when close to the direction of the incoming protons.

To detect high-energy muons, one has to be able to trigger on them and identify the specific bunch crossing (every 25 ns, i.e., 40 million times a second) in which the particle was produced. This requires fast detectors that can provide an approximate position measurement along the muon trajectory with resolution in the timing at the level of a few ns. Two types of detectors with very good timing resolution (<4.5 ns) that can deal with the high background are used to trigger on high-momentum muons: Resistive Plate Chambers (RPC) in the barrel region and Thin Gap Chambers (TGC) in the

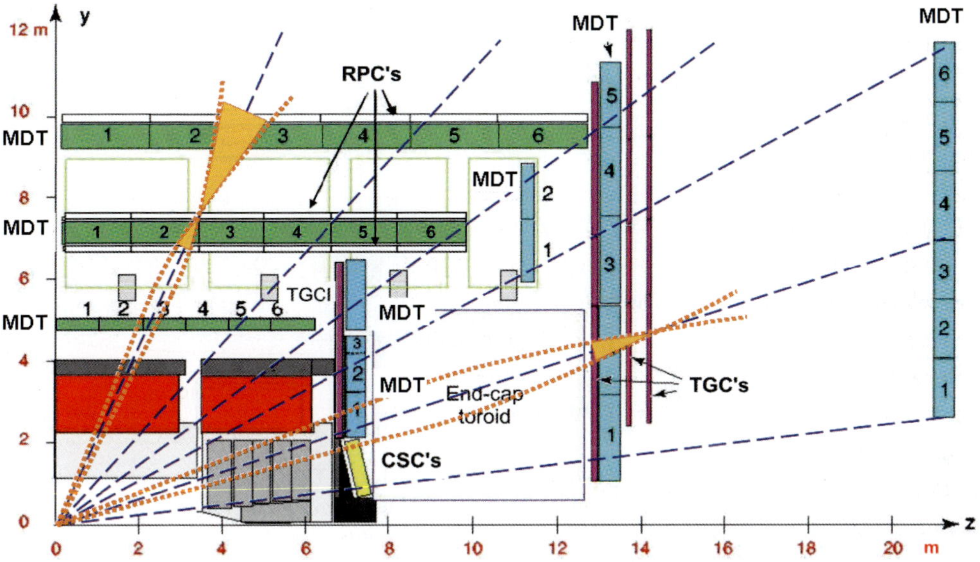

Fig. 2.7. Schematic of the trigger arrangement to determine muon momentum, profiting from the bending in the magnetic field. Infinite-momentum tracks are shown in blue and acceptance cones in yellow. Also includes positioning of the components (RPC, TGC, MDT and CSC).

endcap region, with both devices providing the two orthogonal coordinates due to the passage of the muons. The trigger detectors are arranged in three planes within the magnetic field, as shown in Fig. 2.7. The rejection of low-momentum muons is accomplished by defining acceptance cones on the three triggering planes, centred on the line that would correspond to an infinite momentum muon trajectory. The width of these cones defines the minimal muon momentum accepted due to the bending in the magnetic field.

Precise muon-momentum measurements to values exceeding 1 TeV are necessary for searches for new massive objects at the edge of the kinematic reach provided by the LHC collision energy. Although the momentum of muons is also measured in the Inner Detector (ID), the ID cannot measure with sufficient precision at very high momentum, and in particular for forward-going muons that traverse only part of the magnetic field provided by the central solenoid. Since the momentum resolution is inversely proportional to the magnetic field, B, times the squared length, L, of the measuring device (BL^2), the large toroid complex provides the desired lever arm to maximise L. This is done (shown schematically in Fig. 2.7) by comparing the curvature of the muon track to one of an infinite momentum (straight line)

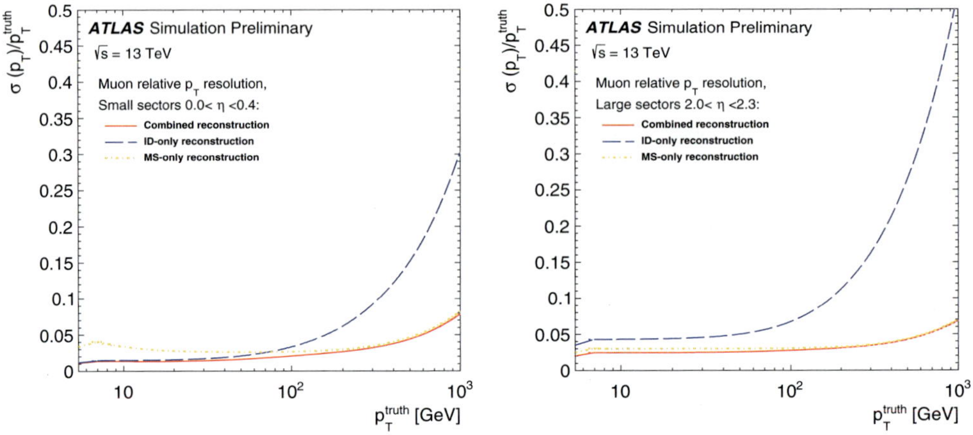

Fig. 2.8. Transverse momentum resolution using the Inner Detector measurement (blue), the Muon Spectrometer measurement (yellow) and their combination (red). (*Left*) For the barrel region ($0 < \eta < 0.4$). (*Right*) For the endcap region ($2.0 < \eta < 2.3$).

muon using three well-spaced measurements points. The maximal distance between the inner and outermost so-called MDT measuring stations varies between 5 m (barrel region) to 14 m (endcap region). To maximise the benefits of BL^2, the location of the muon detection elements must also be known to better than 0.05 mm inside a volume of $45 \times 25 \times 25\,\text{m}^3$, as well as the value of the local magnetic field to better than 1%. Figure 2.8 shows that by combining the momentum measurement in the Muon Spectrometer with that from the Inner Detector, a resolution of 2–3% is achieved for transverse momenta less than 100 GeV.

To accomplish such a momentum measurement performance with the Muon Spectrometer, a technology had to be developed that would monitor and correct deformations and provide a precision of better than 0.05 mm in the bending coordinate. It also had to be easily produced and tested at 30 laboratories around the world. Such a device is based on Monitored Drift Tubes (MDT), made of 30 mm diameter Al tubes with a precisely positioned 0.05 mm gold-plated tungsten wire in its centre. The tubes are precisely glued to each other, while their deformations are followed by an alignment system. The structure of the MDT detectors is shown in Fig. 2.9. In the high-background forward region, the MDT detectors are complemented by multi-wire proportional chambers, with high-precision cathode strips (CSC). The distribution of the various detector elements that constitute the Muon Spectrometer is shown schematically in Fig. 2.7.

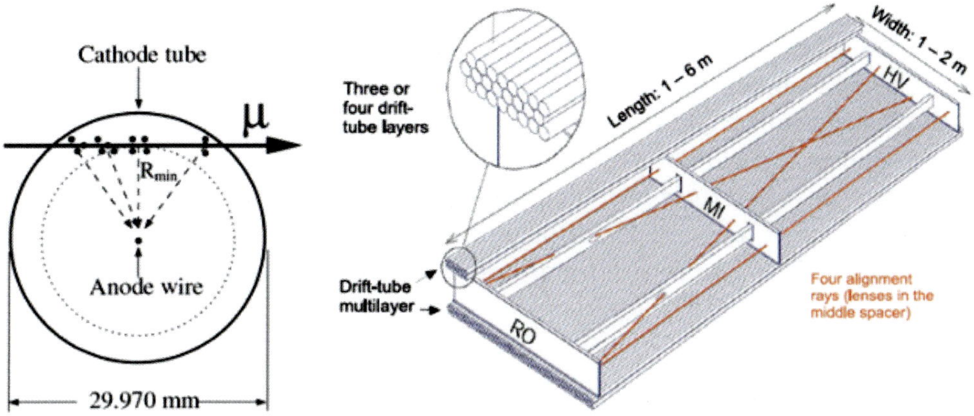

Fig. 2.9. Schematic of the MDT detector, showing alignment rays, which follow mechanical deformations.

Having precise detectors whose deformations are tracked continuously is not sufficient to precisely measure the muon momentum. It also requires a precise knowledge of the relative position of each layer of detector. This is accomplished by a very complex projective optical alignment system that measures the relative position of each layer of detector to within 0.05 mm. The alignment system together with the monitoring of the MDTs provide a very precise self-correcting reference system on which magnetic sensors can be mounted. These sensors provide a few thousand measurements, enabling a precise description of the very complex magnetic field of the toroids and their local interference. The measurements provided by these sensors permit to reconstruct the position of the very large coils ($25 \times 5\,\mathrm{m}^2$) with a precision of 1 mm in space.

A very important aspect in being able to achieve the required performance in such a large system as the ATLAS Muon Spectrometer has been the availability of very precise simulations of the detector performance, combined with the availability of reliable finite-element analysis programs to predict the mechanical deformations of large mechanical structures. In particular, the 25 m diameter barrel toroid structure deformations were simulated at the few mm level. The deformations in the endcaps (large wheels that are 25 m in diameter but only 30 cm in thickness) were predicted at the same level. Positioning the four endcap wheels to within a few mm and controlling their distortions at the same level was a key ingredient to muon triggering in the endcaps, where the distances between the trigger planes have to be controlled at the mm level, as well as knowing the position of the MDT tubes at the 0.05 mm level.

The above-mentioned goals have been achieved thanks to a very dedicated group of physicists and engineers; the individual detectors have operated over a period of 10 years with less than 1% failure, the deformation of the detectors and their positions have been monitored at the 0.05 mm level, and the complex magnetic field has been kept under control with a minimal failure rate. This has contributed to the determination of the mass of the Higgs boson at the 0.3% level.

2.4. The calorimeters

The low cross-sections and complexity of interesting hard-scatter collisions, furthermore obscured by the pile-up of additional events occurring during the same, or even neighbouring, bunch crossings, raised unprecedented challenges for calorimetry. This resulted in the quest for detectors capable of providing high-precision measurements and high background rejection (in particular for electromagnetic objects such as electrons and photons), with fast response, excellent time stability and high radiation resistance. After several dedicated R&D efforts, the choice of ATLAS was to use a liquid argon sampling calorimeter with a new "accordion" geometry for the absorbers and read-out electrodes in the electromagnetic (EM) part, iron-scintillator tiles with a new orientation of the scintillators for the barrel and extended-barrel parts of the hadronic calorimeter, both at a radius larger than 2.2 m, and again liquid argon sampling calorimeters for the hadronic parts more exposed to radiation, in the endcaps (HEC) and forward parts (FCal), at a radius lower than 2.2 m. The solenoid is located in front of the EM calorimeter, inside the insulating vacuum of the barrel calorimeter cryostat. The worsening of the energy resolution for electron and photons resulting from material in front of the calorimeter is mitigated by using a liquid argon presampler in the barrel and endcaps for $|\eta| < 1.8$. The overall layout is shown in Fig. 2.10 and the main parameters are summarised in Table 2.1.

The pseudorapidity coverage up to $|\eta| = 4.9$ is necessary for good missing transverse-energy measurements. Over the η region matched to the Inner Detector, the fine granularity of the EM calorimeter is suited for high jet-background rejection (over a factor 1000) and precision measurements of electrons and photons. The total thickness of the calorimeters is <22 radiation lengths (X_0) for the EM calorimeter and <9.7 interaction lengths (λ) of active electromagnetic + hadronic calorimetry, which is important

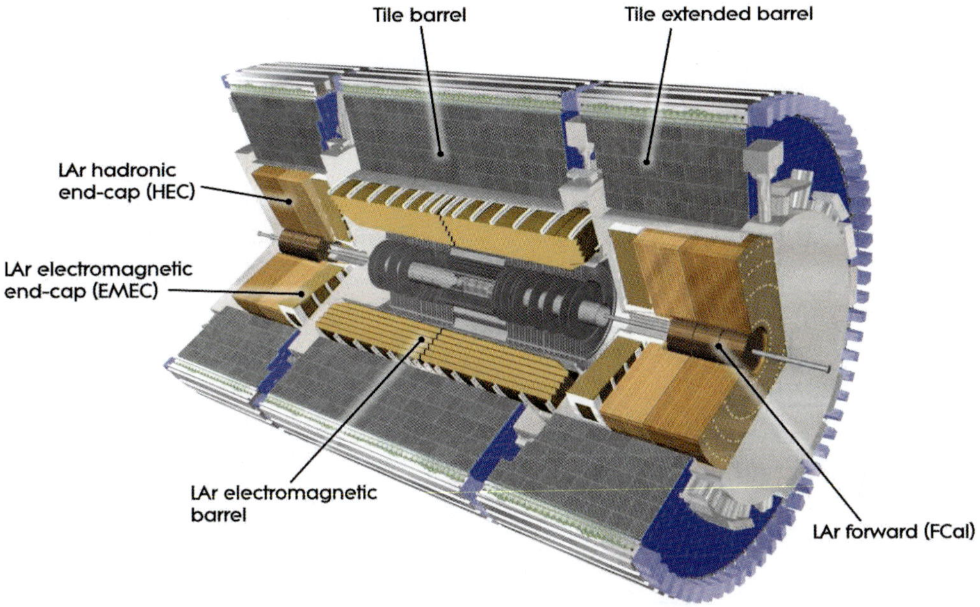

Fig. 2.10. View of the ATLAS calorimeter system.

to provide good resolution for the highest energy jets and to limit punch-through to the muon system.

Electromagnetic calorimeter. The accordion geometry, as sketched on the left hand side of Fig. 2.11 for the barrel part is rotationally symmetric in azimuth. The barrel part is split in two identical half-barrels, for $|\eta| > 0$ and $|\eta| < 0$. Each of them features 1024 lead-stainless steel absorbers and 1024 interleaved electrodes, with 2 gaps of 2 mm of active liquid argon gaps in between. The electrodes provide the read-out and the high-voltage distribution, nominally at 2 kV. Four read-out layers ganged in azimuth define cells of 0.025 in azimuth, while the granularity of pointing cells in η, and the depth in samplings are defined by the electrodes etching pattern. The structure of the endcaps is similar. The presampler, made of thin G10 copper-clad electrodes, samples the early part of the shower. Each cell is connected to the cold-to-warm feedthrough by a coaxial cable. The front-end electronics, with an input impedance matching the cables, is located directly on the feedthroughs. Downstream of the preamplifier, a three-gain shaping amplifier delivers a signal proportional to the derivative of the instantaneous current in the corresponding cell. The signal peaking time, of about 45 ns, was adjusted to minimise the total noise (electronics and

Table 2.1. Main ATLAS calorimeter parameters in layers s1, s2 and s3. Photon- and jet-energy resolutions are given in the last two lines. The symbol $\oplus$ implies addition in quadrature.

Params	Presamp.	EM barrel	Hadronic Tile	EM endcaps	HEC	FCal
Material	LAr	Pb / LAr	Fe/scint/fibres	Pb / LAr	Cu / LAr	Cu / LAr W / LAr
$\lvert\eta\rvert$ range	<1.8	<1.45	<1.7	1.45–3.2	1.5–3.2	3.0–4.9
Depth		>22 X_0	>9.7 λ	>22 X_0		
# Layers	1	3	3	3	4	3
Granularity ($\Delta\eta \times \Delta\varphi$)	0.025 × 0.1	0.003 × 0.1 s1 0.025 × 0.025 s2 0.05 × 0.025 s3	0.1 × 0.1 s1 0.1 × 0.1 s2 0.2 × 0.1 s3	(0.003–0.1)× 0.1 s1 0.025 × 0.025 s2 0.05 × 0.025 s3 0.1 × 0.1, $\lvert\eta\rvert$ = 2.5–3.2	0.1 × 0.1 0.2 × 0.2, $\lvert\eta\rvert$ >2.5	fixed size: (X × Y) = 30 × 26 mm^2
Channels	9344	101760	10000	62208	5632	3524
$\sigma_E/E(\gamma)$	$10\%/\sqrt{E} \oplus 0.8\%$ (barrel) or 1.2% (endcaps), E in GeV; noise ~0.45 GeV @ E_T = 50 GeV					
σ_E/E(jets)	$71\%/\sqrt{E_T} \oplus 3\%$, E_T in GeV					

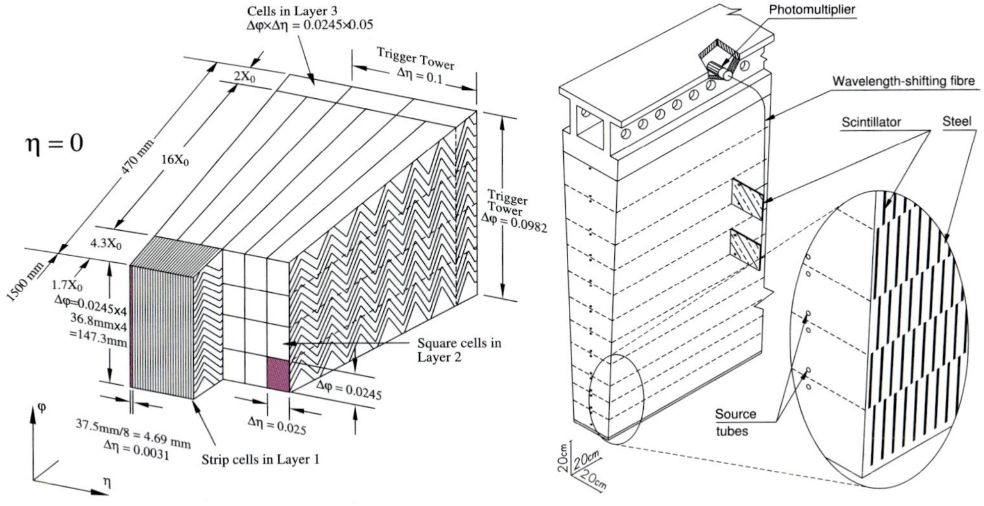

Fig. 2.11. Sketch of the electromagnetic LAr (left) and hadronic Tile Calorimeter (right).

pile-up) for the nominal luminosity. The signal, sampled and digitised every 25 ns, is sent to an analogue pipeline with a depth matched to the trigger latency. At the shaper level, signals from cells within towers of 0.1×0.1 in $\Delta\eta \times \Delta\varphi$ are added and sent to the level-one trigger system. Precisely controlled currents can be injected in precision resistances, in the liquid, for cell-by-cell calibration.

Hadronic calorimeter (Tile). The Tile hadronic calorimeter is a sampling calorimeter using steel and scintillator tiles, with a ratio of 4.7:1. The central barrel and the two extended barrels are divided azimuthally into 64 modules each and 3 longitudinal layers. The geometry is sketched on the right hand side of Fig. 2.11 and a summary of the main parameters is given in Table 2.1. The two sides of the scintillating tiles are read out by wavelength-shifting fibres into two separate photomultipliers to ensure good response uniformity in the cell and redundancy. In η, the read-out cells are built by grouping fibres into bundles connected to the photomultipliers. The orientation of the scintillator tiles radially and normal to the beam line, in combination with wavelength-shifting fibre read-out, allows for almost seamless azimuthal calorimeter coverage. The gaps between the barrel and extended barrels are instrumented with thin scintillator counters to partially recover the performance in a region occupied by services. The electronics are located at the outer radius in sliding "drawers" allowing for maintenance during access periods.

Hadronic Endcap Calorimeter (HEC). The HEC consists of two independent wheels per endcap, located behind the endcap EM calorimeter and sharing the same cryostats. Each wheel is built from 32 identical wedge-shaped modules and divided into two segments in depth, for a total of four samplings per endcap. The wheels closest to the interaction point are built from 25 mm parallel copper plates, while those further away use 50 mm plates. These plates are interleaved with 8.5 mm LAr gaps, in which three read-out planes define four gaps connected as an electrostatic transformer (EST). The signal from each pad of the central board of the EST structure is sent to GaAs preamplifiers located at the outer radius of the wheel, in the liquid. The preamplifier outputs are then ganged in depth (they are actively summed in a summing stage), within each cell of each of the four samplings, and connected by a coaxial cable to the feedthroughs. The downstream part of the electronics is basically identical to the EM part.

Forward Calorimeter (FCal). The FCal is integrated into the endcap cryostats. In order to reduce the amount of neutron albedo in the Inner Detector cavity, the front face of the FCal is recessed by about 1.2 m with respect to the EM calorimeter front face. This limits the depth of the calorimeter and therefore calls for a high-density design. The FCal is approximately 10 λ deep and consists of three modules in each endcap: the first, made of copper, is suited for EM measurements, while the other two, made of tungsten, measure predominantly the hadronic component of jets. Each module consists of a metal matrix, with regularly spaced longitudinal channels filled with the electrode structure consisting of concentric rods and tubes parallel to the beam axis. The LAr in the gap between the rod and the tube is the sensitive medium. This geometry allows for excellent control of the gaps, which are as small as 0.25 mm in the first section, in order to prevent ion build-up at high luminosity. The electronic chain is essentially identical to that of the EM calorimeter.

2.5. The Inner Detector

The Inner Detector (ID) aims to precisely measure the point of origin and trajectory of charged particles in the 2 T magnetic field of the central solenoid. This is achieved by combining very precise measurements from silicon detectors and continuous tracking from gaseous detectors, resulting in excellent pattern recognition and precise momentum measurements. The detector is as thin as possible, given the power, mechanical stability and cooling requirements, to minimise particle interactions before the calorimeter system.

To better match the topology of the tracks emerging from the LHC proton-proton collisions, the ID is separated in a barrel part (measuring particles below $|\eta| \approx 1$) and two endcaps (covering the range from $|\eta| \approx 1$ to 2.5). The barrel region is made of concentric cylinders around the beam axis and the two endcap regions are made of disks perpendicular to the beam axis. Different technologies are used at different radii to optimise the cost/performance ratio. The innermost layers, immediately surrounding the beam pipe and up to 15 cm, comprise the Pixel Detector that uses hybrid silicon pixel sensors for unambiguous pattern recognition, high accuracy and maximal radiation resistance. The intermediate region (radii from 30 to 60 cm) is covered by the SemiConductor Tracker (SCT), that uses microstrip detectors to provide excellent space resolution over a large area. The outer part (radii from 60 to 95 cm) is made of a large number of small diameter proportional drift tubes (straws) that provide good space resolution in the track bending plane and strongly contribute to the momentum resolution and the pattern recognition with multiple measurements. This detector has also transition radiation detection capability (which helps in electron identification) and hence is called the Transition Radiation Tracker (TRT).

The layout of the ID is shown in Fig. 2.12 together with its dimensions. More details of the barrel part are also shown in Fig. 2.2.

The typical spatial precision in the bending plane of tracks is of order 15 μm in the silicon part and 150 μm in the TRT. The ID was designed to perform efficient pattern recognition of all particle tracks with transverse momentum above 0.3 GeV up to a luminosity of $10^{34}\,\mathrm{cm}^{-2}\mathrm{s}^{-1}$, i.e., up to ~25 simultaneous proton-proton interactions. The detector in fact works efficiently with more than 50 simultaneous collisions. Both the radius of curvature of the tracks and their extrapolation to the interaction vertex are measured with high precision. This results in a transverse-momentum resolution $\Delta p_T/p_T$ of ~1% at 20 GeV and better than 50% at 1 TeV (important to measure high-mass particles) and transverse impact-parameter resolution better than 20 μm (important for the tagging of the short-lived heavy quarks and leptons, which come from a displaced or "secondary" vertex).

The choice of different detector technologies was driven by the role that each part plays in the measurements. A consequence of these choices is that the silicon part must operate below 0°C in a dry nitrogen environment to limit radiation damage, while the TRT works at room temperature with specialised gases.

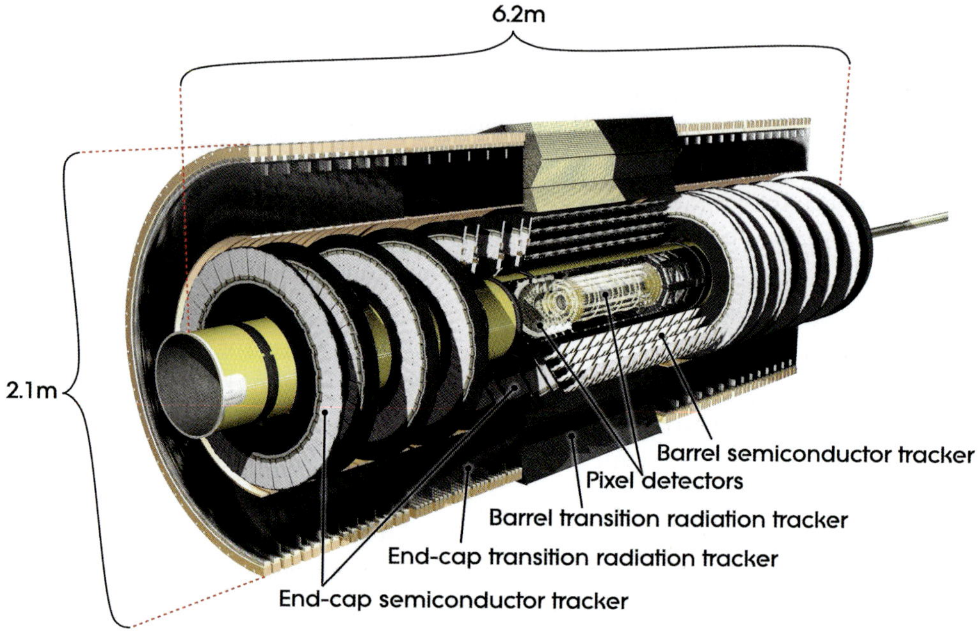

Fig. 2.12. View of the Inner Detector, which consists of a set of precision silicon tracking detectors (Pixel and SCT) extending up to a radius of 60 cm surrounded by the TRT gaseous detector up to a radius of 95 cm.

The Pixel Detector is crucial for the precise extrapolation of the tracks to the pp collision point or to secondary vertices. It was initially made of three barrel-layers (at radii 5, 9.8 and 12.2 cm) and three disks (of 8.9 cm inner radius and 15.0 cm outer radius) per side. Each layer is a mosaic of identical modules arranged so as to cover the sensitive surface without any acceptance loss. Each module is built with one sensor and sixteen read-out chips, and has a $6.04 \times 1.64\,\text{cm}^2$ active area. The $1.7\,\text{m}^2$ surface of the Pixel Detector is covered with 1744 modules giving a total of 80 million individual pixels. Each pixel is a reversely polarised diode of $50 \times 400\,\mu\text{m}^2$, in which a current pulse is released by ionisation on the passage of a charged particle. The pixel is created by an n+ implant on an n-doped silicon substrate. This arrangement partially allows depleted operation after type inversion due to radiation damage, should this be needed at the end of the detector's lifetime. Every pixel is directly connected to its read-out chain through high-density bump-bonding. The passage of a particle is recorded if the current pulse exceeds an adjustable threshold. The pixel efficiency is above 99% and its time resolution is below 25 ns, as needed to unambiguously associate the Pixel hits with a given LHC beam-beam interaction.

During the shutdown of the LHC in 2013–2014, an additional Pixel barrel layer was installed between the Pixel Detector described above and a new, smaller radius beam pipe. This new Pixel layer, the Insertable B-Layer (IBL), has a radius of 33.5 mm, covers up to $|\eta| = 2.7$ and improves the robustness and performance of the tracking system at high luminosity. Smaller pixels ($50 \times 250\,\mu\text{m}^2$), better front-end electronics, realised in 130 nm gate length technology, and minimisation of passive material, with the adoption of CO_2 evaporative cooling and Al power supply cables, are the most important improvements realised in this additional Pixel layer.

The SCT is crucial for the precise momentum measurement. It consists of 4088 modules — 2112 located over four equally spaced barrel layers and 1976 over eighteen disks (nine per side). The total number of silicon strips is about 6 million.

Nearly all the modules are made up of four strip sensors. Two are daisy-chained together to make effectively 12 cm long strips. The two pairs are mounted back-to-back on a thermally conductive pyrolytic graphite baseboard, with a 40 mrad angle between the strips on opposite sides to give a measurement in the second coordinate. The strips are wire-bonded to a double-sided read-out hybrid.

The sensors are single-sided p-in-n microstrip detectors, implanted on 285 μm thick high resistivity silicon wafers. Each sensor has 768 AC-coupled read-out strips resulting in 1536 read-out channels per module. Modules are all the same in the barrel and feature an 80 μm pitch. Four different shapes at different radial distances, with pitch ranging from 57 to 90 μm, are instead needed to match the endcap geometry. (The endcaps also have some shorter two-sensor modules.) The strip pitch was chosen to optimise two-track separation and to provide an occupancy below 1.0% at full luminosity. The 40 mrad angle is chosen to minimise the number of ghost hits (accidental additional strip pairings) and to keep a longitudinal resolution well below a millimetre. The fast front-end electronics associate all hits to a unique bunch crossing. The only information stored is a 1 if the signal is above a threshold, which is adjustable channel by channel. Data are read out through an optical link as soon as a trigger is received. To increase fault tolerance, a redundancy scheme is implemented to reroute signals in case a single chip or a read-out link stops functioning.

The main role of the TRT is to contribute to the momentum measurement and to pattern recognition (photon conversions and long-lived particles), and to provide electron identification in addition to that from the calorimeter. It does not cover the full pseudorapidity acceptance but ends at $|\eta| \sim 2.0$. The TRT is made up of 372,032 straws with a 4 mm inner diameter filled with a Xe-rich gas mixture. The straw diameter comes as a compromise between speed of response, number of ionisation clusters and operational stability. The length of the straws is different if they are used in the barrel (144 cm) or in the endcap (37 cm), but all other characteristics are the same.

The polypropylene transition radiator material is interspersed in the TRT in the spaces between straws. Transition radiation photons (TR) produced in the radiator are absorbed in the active gas of the straws, resulting in a significant higher probability (20–30%) to increase of signal, and hence exceed a higher-level threshold set for the detection high-momentum electrons. As the energy loss by ionisation is typically 2 keV (versus the 5 to 15 keV of the TR photons), the TR signal can easily be discriminated using double threshold electronics. The read-out requires recording the drift time with a 1 ns precision, corresponding to a 130 μm drift length.

Gas proportional detectors (like the TRT) require careful monitoring of the amplification factor in the gas mixture. This factor depends on the electric field, temperature and gas mixture. All of those parameters are continuously checked and the high voltage is adjusted to avoid gain variations that exceed 20%.

2.6. The forward detectors

The main part of the ATLAS detector is designed to detect the products from head-on collisions in which the created particles are produced with large angles relative to the incoming protons. However there exists a certain probability that the protons do not collide in a head-on manner but in a gentler way with a certain transverse distance between them. Such collisions lead to products with small angles relative to the beam line, i.e., in the forward direction. A large fraction of the interactions at the LHC are of this kind and are interesting themselves. They also provide information on the overall probability for interactions, also expressed as the total cross section (σ_{tot}) and on the luminosity delivered by the LHC. The latter quantity is of utmost importance for all absolute rate measurements made by ATLAS. The forward detectors are complementary to the central ATLAS detector and conceived to study interactions that produce particles in the very forward direction. The layout of this system is shown in Fig. 2.3.

ALFA is a scintillating fibre tracker, housed in so-called Roman Pots, located 240 m downstream of the interaction point, IP, on both sides. The detectors can be moved in the vertical plane and may approach the beam to distances of order of a few millimetres. ALFA measures protons that are intact after the collision and that originate mainly from elastic or diffractive scattering. σ_{tot} can be inferred from measurements of elastic scattering. Elastic scattering also provides a possibility to calibrate luminosity measurements.

AFP, like ALFA, is designed using Roman Pots to measure protons that emerge intact from the interaction region. A silicon tracker localised at 220 m from the IP can approach the beam from one side in the horizontal plane. The information from the central detector is combined with information from the AFP detector for measurements in a very clean environment without proton remnants. The corresponding physics processes go beyond elastic and diffractive scattering. Rare processes with particle production in the central ATLAS detector and still with intact protons can be measured.

LUCID consists of several small Cherenkov detectors. It is the only forward detector that is integrated within the central ATLAS detector and it is located only 17 m from the IP with the photomultipliers at a distance of 12 cm from the beamline. The novelty of this detector is that it uses the thin quartz windows of the photomultipliers as a Cherenkov medium and a radioactive ^{207}Bi source applied directly to the windows to continuously monitor the gain. It is used for luminosity measurements and luminosity monitoring.

The ZDC is a tungsten quartz sampling calorimeter installed between the two LHC beam pipes, in a configuration such that only neutral particles produced at the interaction point can reach the calorimeter. The ZDC is installed 140 m downstream of the IP. It is mainly used in the context of heavy-ion collisions and can be used as a trigger for ultra-peripheral ion-ion interactions. It is also used to measure how much energy was dissipated in the collision.

2.7. The trigger and data-acquisition system

The trigger is responsible for very quickly identifying which bunch crossings contain potentially interesting physics, based on distinctive features, or signatures, such as electrons, muons or photons emerging with high momentum transverse to the beam direction. The data-acquisition (DAQ) system is responsible for collecting the data read out from the detectors, putting them into a coherent format and structure, and transporting them, for the selected events, to permanent storage that is provided by the offline computing system and infrastructure. Once stored, they are made available worldwide for further processing and analysis in ATLAS institutes.

In practice, the trigger and DAQ components are intertwined as illustrated in Fig. 2.13. The trigger selection is performed in steps, first using a system of custom electronics processors to reduce the event rate (i.e., the rate of bunch crossings to be recorded) to less than 100 kHz — a massive reduction compared to the interaction rate of ∼2 GHz. The data for events that are retained after the first stage of selection, which is called the first-level trigger, are transferred to the DAQ system. The subsequent online selection, called the high-level trigger (HLT), is performed using a large number of rack-based commercial computers that can selectively access the data, guided by the results of the first-level trigger. High-speed commercial networks support data movement, and data storage and staging possibilities exist at various points in the system (in the read-out systems connected to the detector electronics, within the HLT processors, and at the output towards the offline computing system).

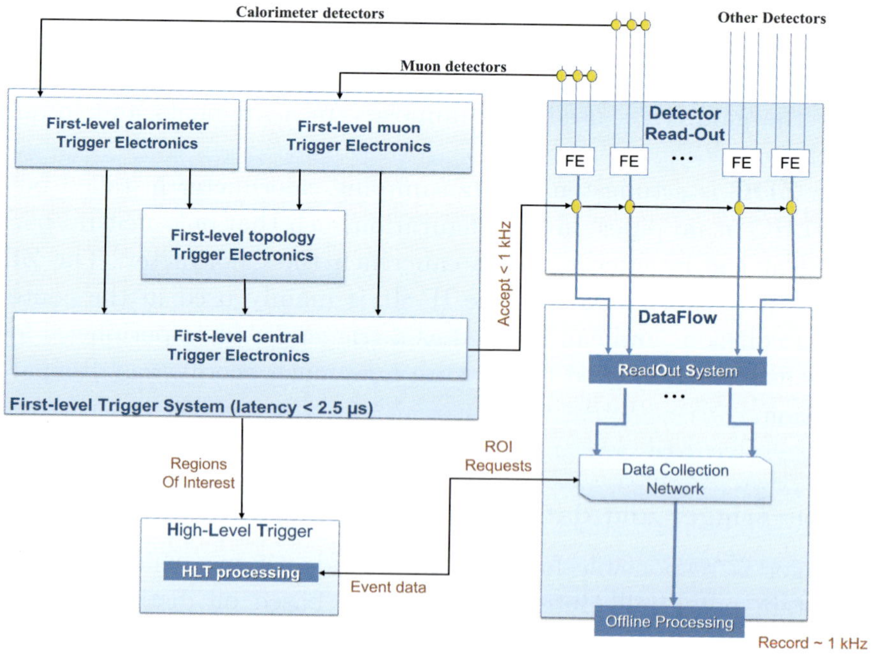

Fig. 2.13. Diagram of principal components of the Trigger/DAQ system and interconnections.

The first-level trigger is composed of subsystems searching for features in the calorimeters and in the muon detectors, of a topology subsystem that combines features, e.g., to search for pairs of particles resulting from the decays of heavy objects such as Higgs bosons, and the central trigger that assesses the event with respect to a list of physics-driven selection criteria, the so-called trigger menu.

The stages in the selection can be illustrated with a simplified example. Consider an event containing a high-energy electron, e.g., arising from a Higgs boson decay to a pair of W bosons, of which one decays to an electron and a neutrino. The electron traverses the inner-tracking detectors and produces a distinctive narrow deposit of energy in the calorimeters. The first-level trigger, using relatively coarse granularity calorimeter information, selects the event based on this signature, initiates the read-out, and indicates to the HLT where the electron was seen in the detector. The HLT accesses more detailed calorimeter information in the indicated region and retains the event if the electron hypothesis is confirmed. The HLT then accesses information from the inner-tracking detectors, only in the region of interest (ROI) identified by the previous step, and searches for a track that matches the energy cluster in the calorimeters. After this selection based on ROIs, the HLT may access and process the full event data before making a final selection, which can require additional signatures such as missing transverse energy due to the neutrino that escapes the detector without interacting. This final selection is then marked for full-scale data read-out from the detector and transferred to the offline system.

The trigger and DAQ system as implemented at the start of Run-1 contained some novel features, such as the ROI mechanism in the first stages of the HLT processing, and the use of offline software in the later stages. At that time, the HLT was composed of two separate computer systems — the second-level trigger and the event filter; these were merged for Run-2 starting in 2015. Other changes for Run-2 included the addition of a topology processor in the first-level trigger that could combine signatures, e.g., considering the angular separation between pairs of electrons, along with simple multiplicity requirements already available in Run-1.

With up to about 60 interactions per bunch crossing, the events are intrinsically very complex. It is therefore essential to associate detector activity with the correct bunch crossing and for the first-level trigger to uniquely identify the bunch crossings containing physics of interest. Otherwise one would need to aggregate the activity over two or more bunch crossings, further complicating the picture. The first-level trigger must start processing a new event every 25 ns, and it must make a decision every 25 ns. However, allowing for propagation delays on cables and fibres ~100 m long each way (corresponding to a transmission time of ~500 ns) from the detector to the electronics systems and back, as well as processing and data movement within the trigger system, there is a latency of about 2.5 μs. A technique called pipelined processing is used whereby the calculations are broken down into steps that can be completed within 25 ns. In each step, data from different parts of the detector may be processed in parallel. The information pertaining to a given event moves along the processing pipeline, one step every 25 ns, with the final step giving the result of the selection, i.e., to discard the event or to retain it for further examination in the HLT. Pending the first-level trigger decision, the information from all bunch crossings is retained in the detector electronics ("FE" in Fig. 2.13) that is mounted on or close to the detector. In many cases, the information (either digital data or analogue signals) is retained, at least logically, in a pipeline memory, i.e., in a first-in-first-out memory.

The great challenges of implementing the first-level trigger can be illustrated with the example of the calorimeter trigger. On-detector analogue electronics sum the calorimeter signals from ~200,000 cells to ~7000 trigger towers that are then digitised with 10-bits precision every 25 ns. This corresponds to 7×10^4 bits at 4×10^7 Hz, i.e., 2.8×10^{12} bits per second. A relevant comparison when ATLAS was being designed in the late 1990s, before the advent of smartphones and broadband home internet access, was the equivalent number of telephone calls, counting ~10 kbps per voice call. Using this analogy, the calorimeter trigger alone required transferring data equivalent to ~300 million phone calls into a system that occupies just a few electronic racks (see Fig. 2.14).

The ROI mechanism allows the HLT to operate without moving the bulk of the event data to the processing units. However, for the selected events, the full event data must be aggregated. Considering the ~1 kHz rate of events selected by the HLT and an event size of ~2 Mbytes (MB), ~2 GB of data must be sent for offline data storage per second. This is comparatively modest by today's standards. However, in the late 1990s, a relevant benchmark would have been CDs (compact discs) that had a capacity of only 700 MB per disk.

Technologies that made the first-level trigger implementation possible included application-specific integrated circuits (ASICs) where needed, e.g., in the on-detector's muon trigger system, large and affordable field-programmable gate arrays (FPGAs) that came onto the market at just the right time for several key applications, and very high-speed electrical and

Fig. 2.14. Racks of the first-level calorimeter trigger system installed close to the detector.

optical connections for inter-module data-transfer. The emphasis in the DAQ and HLT systems has been on exploiting commercial off-the-shelf products, in particular high-speed computer networks (e.g., Ethernet) and multi-core processors. Improvements in the speed of data links and in processing power per unit cost, driven by applications of these technologies in other areas, have allowed ATLAS to move and process more data, with more sophisticated algorithms, giving a trigger event selection that approaches the quality that can be reached in offline data analysis. Nevertheless, the DAQ and HLT system is quite large and demanding, e.g., the HLT requires about 2000 servers, each with two 12-core or 14-core processors.

ATLAS operations today and acting on the unforeseen

Since first data, ten years of experience have been accumulated. Much effort has gone into improving the stability and reliability of systems — from the control and monitoring of the components to the training of the shift crew. What hasn't changed much is the operation model, i.e., how ATLAS is run during collisions.

During the data taking period, typically eight months a year, a shift crew of eight operates the detector control, safety, and data acquisition system in three eight-hour shifts, 24 hours a day, seven days a week. During this period, the operations team comes together every morning for a short meeting to discuss any issues that occurred in the last day and to define the plan for the next 24 hours.

Collecting data in a big experiment such as ATLAS is often — but not always — uneventful. Error messages and alarm systems are there for a reason, and they inevitably do go off. Often the problem has a routine solution that the shift crew can apply. In case of problems, on-call experts, each a specialist for a certain component, will provide a first level of support. Depending on the nature of the problem, more experts may get called in. ATLAS rarely loses more than a few minutes of good data for these issues. When things get serious and data taking or data quality get compromised, subdetector experts or the Run Manager can get called in. At the end of this chain sits the Run Coordinator who is ultimately responsible for the efficient data taking of the experiment. The pressure can quickly rise, and everyone tries quickly and efficiently to do their job. The Run Coordinator may try to make the best use of compromised data or detector conditions. Situations, such as a temporarily unrecoverable LHC injection, or data collected with a given subsystem in a "bad" state are turned into opportunities. Reference samples for trigger studies or noise data, read-out scans for a given subdetector or alignment data are then collected.

Occasionally, something more serious may happen: a significant part of a detector needs to be turned off, a fluid leak develops, a hard to reach cooling manifold or a power supply critical to the operation of a given subdetector fails. The Technical Coordination team then inevitably jumps in. With the help of detector experts and Run Coordinators, the problem is discussed, accesses are coordinated when needed, interventions are planned, ad hoc tools and procedures prepared. Brainstorming sessions often ensue, where everyone's experience is precious. How clothes are dried in a Siberian winter, gluing practices from boat maintenance work, plumbing techniques, the ability of the thinner and more compact person in the room to squeeze in confined spaces: every kind of experience can suddenly become useful input in creatively saving the day. Everyone works as efficiently and quickly as possible, giving their time (and sometimes taking it from their family or a deserved break). Without detracting from the serious technical background and the professionalism of everyone involved, it can be definitely concluded that ATLAS has been quite fortunate through the years with the people facing these problems, their rich and disparate background, and their availability above and beyond their formal duties!

Interactions with the LHC and with the other experiments

The ATLAS Run Coordinator is part of a restricted LHC fauna, a bit like the elusive deer of the Prevessin site. The Run Coordinators of all experiments at the LHC usually meet daily at 8:30AM for the LHC operations meeting. These meetings happen as soon as the LHC comes back from the customary year-end-break's winter sleep. Formally, these meetings consist of executive 15–30 minutes LHC operation reports where the "plan of the day" is also discussed. No time goes wasted and every exchange is as frank and direct as it can be. This is often the first place where anything LHC-operations related will come up and will be discussed: be it about detector problems, broken machine parts, ferrets or birds being sadly electrocuted by the LHC supply system, or earthquakes across the globe disrupting the LHC beams. The atmosphere in these meetings is often tense, with everyone struggling to stretch every little bit of additional luminosity from the machine. Sometimes the CERN Director-General is present, to stay up-to-date on the latest developments. Sometimes the experiments' spokespeople and/or technical coordinators are also there (and then you know something serious may be going on...).

The constant presence is however that of the experiment and machine Run Coordinators, living and sharing the actual life in the trench. Everyone acts on behalf of their experiment, but the spirit is that of sympathetic mutual support, sometimes convivial, always friendly and refreshingly free of any experiment politics. Everyone bound by the same lifestyle and frantic pace, devoted for the duration of their charge to the cause of data collection efficiency and the exploitation of "the machine" to its best, genuinely cares about the mutual well-being, sharing — in the little time pockets surrounding this meeting — new-born pictures, stories of broken legs, empathically inquiring about the latest problems and difficult nights their peers had been facing.

ATLAS always thanks the LHC team for great beam delivery.

Chapter 3

Building up the Collaboration

3.1. The initial years and the global international context

The enthusiasm and motivation to explore particle physics at the high-energy frontier knows no borders between nations or regions on this planet; it is shared between physicists of widely different cultures and origins. This is evident nowadays when looking around large but still overcrowded auditoria where the latest results from the LHC are presented, as with the announcements of the Higgs boson discovery. Such results in turn are presented by speakers on behalf of LHC Collaborations that span the globe with physicists from all inhabited continents.

Today we take this for granted, but it is worth remembering that it took about two decades to grow and consolidate these worldwide scientific and human projects into the collaborative, creative and efficient partnerships that are now exploring LHC physics. CERN and its experimental collaborations at the LHC's predecessors — the Intersecting Storage Rings (ISR), the Super Proton Synchrotron (SPS) ppbar collider and the Large Electron-Positron collider — have long been a fertile cradle for physicists teaming up from different regions, but with the LHC Collaborations globalisation for the experiments has reached a new scale. This process of collaboration building is of course not yet finished and many challenges remain.

A formidable and fascinating evolution to observe took place over the past 25 years. Presenting the LHC and ATLAS in the early years was sometimes quite an adventure. There were places where electricity for the slides was not always guaranteed, many colleagues from potential new collaboration partners barely spoke English, and the local custom could be that only the most senior professor would be expected to speak up. Today, one may find at the same places the most modern conference installations and — even more enjoyable to see — confident, clever young students and postdoctoral researchers speaking up and expressing their curiosity and opinions.

What was also striking then was the great motivation to be part of the Collaborations and to contribute, sometimes also under difficult conditions, to the building up of the experiments. Often one had the impression that colleagues in less privileged countries made really extraordinary efforts, with many personal sacrifices, to fulfil their promises for the construction of the ATLAS detector. Those of us from the richer countries should not forget that!

Of course an experiment like ATLAS could not have been built without the massive and leading contributions from CERN member states and large and highly industrialised countries elsewhere on the globe, and we must be grateful for all of their support in the first place. They are the backbone that made it possible to be open to other countries that had great human talents but little in the way of material resources.

The building up and evolution of the ATLAS Collaboration has also to be seen in the context of the overall scene of large-scale worldwide collider projects in the 1980s and early 1990s. The United States reacted on the success of the SPS ppbar collider with its W and Z discoveries in 1982–83 with the recommendation from a Department of Energy High Energy Advisory Panel (HEPAP) to abandon an ongoing collider construction project (Isabelle, up to 400 GeV beams) at the Brookhaven National Laboratory. Instead, it recommended to go directly for a Superconducting Super Collider (SSC) with 20 TeV beams. The SSC was approved in 1987, with some European and Japanese groups engaging in it. However, very significant cost escalations became evident around 1990, leading finally to a cancellation of the project in October 1993. Shortly thereafter, another HEPAP Panel in summer 1994 recommended that the United States should join other countries in building the LHC, albeit within a very strictly limited budget envelope.

Since the late 1960s, CERN had established contacts and cooperation with the former Soviet Union and the Joint Institute of Nuclear Research (JINR) in Dubna. Thus, there was already a tradition of collaboration when it came to LHC partnerships in the late 1980s and early 1990s. In fact, Russia and CERN signed an update to their Cooperation Agreement in June 1993.

In the remainder of this section, and in the following one, we illustrate with some examples how the Collaboration has evolved to its current strength, 25 years after the Letter of Intent (LoI).

The Russian Federation and Joint Institute for Nuclear Research in ATLAS

The history of cooperation between physicists from CERN, the former Soviet Union (SU) and JINR already spans about 50 years. So it was natural that common work had already started on the concept of the LHC and its major detectors since the late 1980s.

However, one has to also remember that during these years the construction of a 3 TeV proton-proton collider (UNK) at the Institute for High Energy Physics (IHEP) at Protvino was advancing at full speed. In addition, significant parts of the SU and JINR particle physicists had been involved in the preparations for two detectors at the SSC — GEM and SDC. Unfortunately, both the UNK and SSC projects were cancelled in the early 1990s. Naturally, strong teams from Russia and JINR then directed and concentrated their efforts on the LHC project.

Financially, this cooperation was supported by a dedicated Protocol to the already mentioned 1993 CERN-Russia Cooperation Agreement, specifically about participation in the LHC project, signed in October 1993 between the Government of the Russian Federation and CERN. This resulted in the allocation of deliverables with values of 100 MCHF for the Russian contribution to the LHC accelerator construction, and another 100 MCHF for the detectors. In addition to the Russian Institutions, the Protocol included JINR as one of its major partners and contributors. JINR, from its own resources, promised to allocate at least 10 MCHF for some LHC machine elements and for detector constructions.

Together with JINR, seven Institutions from Russia have participated in the ATLAS project from the very beginning, with IHEP (Protvino) taking the role of the general coordinator. More recently a further Institution joined the Collaboration. The overall number of ATLAS collaborators from Russia and JINR is about 200. The total direct contribution to the ATLAS construction amounts to about 25.5 MCHF, or about 5% of the detector cost, while the overall value of detector parts actually produced for ATLAS in Russia and JINR Member States amounts to about 10%.

The specific feature of JINR as an international inter-governmental organisation motivated a number of Institutions from countries of the post-Soviet area to join ATLAS: Armenia, Azerbaijan, Belarus and Georgia. This role also facilitated contacts within the industry in these countries and resulted, for example, in fabrication of major components of the ATLAS Barrel Toroid warm structure and of parts for the Tile and Liquid Argon Calorimeters in Belarus, under the supervision of JINR.

At the LoI stage, researchers and engineers from Russia and JINR took part in the conceptual design and technological developments of ATLAS subsystems like, for example, for Monitored Drift Tubes, Tile and Liquid Argon Calorimeters and Transition Radiation Tracker. Valuable contributions have been further done to magnetic field calculations, DAQ architecture, and later on, to general software. It was also of great mutual benefit to the Collaboration and the Institutes concerned that ATLAS could profit from special inter-governmental funding programmes like ISTC and INTAS, to revitalise industries for peaceful applications in these countries.

Last but not least, highly qualified teams of engineers and technicians from Russia and JINR heavily contributed to the integration of the magnet system components at CERN. This activity was crucial for the successful realisation of the toroid magnet system (see Sec. 5.1). And when the mass production of detector components in Russia and JINR was nearing its end, such teams were also involved heavily in the underground detector assembly, integration and installation, thereby making a critical contribution to the excellent detector performance now on the LHC beams.

As an illustration of the excellent partnership with Russia and JINR, one can note that the first ATLAS Collaboration Week meeting outside CERN (see Chap. 11) took place in Dubna in June 2000 (see Fig. 3.1).

Migration of the US groups from the SSC to ATLAS

The cancellation of the SSC was a great shock to the US physicists. However, a High-Energy Physics Advisory Panel (HEPAP) subpanel on *Vision for the future of High-Energy Physics* was soon asked for recommendations on the direction of the US HEP Program. One recommendation was that the United States "should continue to be among the leaders in the worldwide pursuit of the fundamental questions of particle physics". In spite of some initial scepticism in parts of the community, it pointed to the LHC project. It stated that the United States should continue its tradition of success in this field through a strong programme that would include "significant participation in the LHC accelerator and detectors, both to provide research opportunities at the energy frontier and to ensure that the US physicists remain integrated in the international high-energy physics community". The envisaged participation would go well beyond previous cooperation on experiments at CERN's earlier colliders.

Although there were two experiments moving ahead when the SSC was cancelled, GEM (Gammas, Electrons and Muons) and SDC (Solenoidal

Fig. 3.1. First ATLAS Collaboration Week meeting outside CERN in Dubna, June 2000.

Detector Collaboration), the groups interested in joining ATLAS were more attracted by the common technology or detector R&D than their former SSC collaborators. Farsightedness had been exercised in the closeout of the SSC project to support the ongoing detector R&D activities judged relevant for the LHC's future. Many US scientists were already collaborating on CERN R&D projects such as the RD3 Collaboration (Liquid Argon Calorimetry with LHC-Performance Specifications). Thus, it was no surprise that the scientists proposing a noble liquid calorimeter for GEM would join ATLAS with its liquid argon calorimeters.

In addition, there were groups in the United States from SDC who were experts in pixels, silicon strips, straw-tube tracking detectors and scintillating tile calorimeters who also wanted to join ATLAS. The US groups were welcomed to ATLAS as collaborators, but were told that they could not change the basic design. However, some innovations from the SSC experiments found application in ATLAS, such as the fine strips in the first depth segment of the electromagnetic calorimeter, the forward calorimeter and the cathode strip chamber for the high rate section of the endcap muon chambers.

Agreements were needed between the two US funding agencies, the Department of Energy (DOE) and the National Science Foundation (NSF), and CERN in order for US high-energy physicists to fully participate in the LHC scientific programme. DOE and NSF jointly negotiated and signed a 20-year period agreement and protocols with CERN in 1997; this agreement was renewed and expanded in 2017. Before this, a famous footnote to the collaboration lists referred to the need of conclusions of bilateral agreements with CERN. This resulted in the allocation of \$164 M (\$103 M DOE, \$61 M NSF) for the US contribution to the ATLAS experiment's construction, as well as an equal amount to CMS and \$200 M from DOE to the LHC accelerator's construction.

In order to manage the US deliverables, an entity called "US ATLAS" was formed with a Project Management Plan and its host laboratory was at Brookhaven National Laboratory. The initial job of the Project Manager was to define the scope of the US deliverables, which could be promised from the available funds. The US was organised into six subsystems, each with a manager who negotiated between the ATLAS System Project Leaders, the US groups and the US ATLAS Management for the US deliverables for that subsystem.

Since one could not guarantee both the cost and scope of a project, and the total Project cost was absolutely fixed, US ATLAS was forced to not only allocate some of the funding to a cash contingency account to be used for approved cost increases, but to also develop "scope contingency". The scope contingency was called the "Management Contingency" and consisted of ~10% in value of the deliverables in that subsystem that would be not delivered if funding was not available. In the ATLAS Memorandum of Understanding (MoU) for the deliverables, 90% of the deliverables were promised while 10% was an additional goal that depended on the actual cost performance achieved during the construction. There were several meetings every year between ATLAS and the US ATLAS management to decide which of the Management Contingency items could be funded as the Project progressed and accumulated some cash contingency (by having more favourable cost performance on some items than estimated). In 2008, the US Construction Project concluded, delivering most of the Management Contingency items and all of the promised deliverables.

During the last decades, there has been a natural but modest growth in the number of US Institutions, with the full integration of the United States in all aspects of the ATLAS community.

3.2. Reaching out to the world

It was natural that many of the 88 Institutions signing the LoI came from CERN member states. It was also remarkable that at this early stage there were already several teams joining from distant continents, places which had previous or ongoing collaborations with the UA1, UA2 or LEP experiments. Very often, these groups were the nuclei that encouraged other groups from their region to join as well. Here are some of these stories, complementing the ones from the previous section. Others could be added, for example, the long-time Caucasian partners from Armenia, Azerbaijan and Georgia, as well as those from Turkey; all of them having joined ATLAS in the years preceding the Technical Proposal. Figure 3.2 shows the ATLAS Collaboration map as of 2018.

Israel

The Israeli participation in CERN as a full partner started with the OPAL experiment at LEP. The main Israeli experimental groups from Technion, Tel

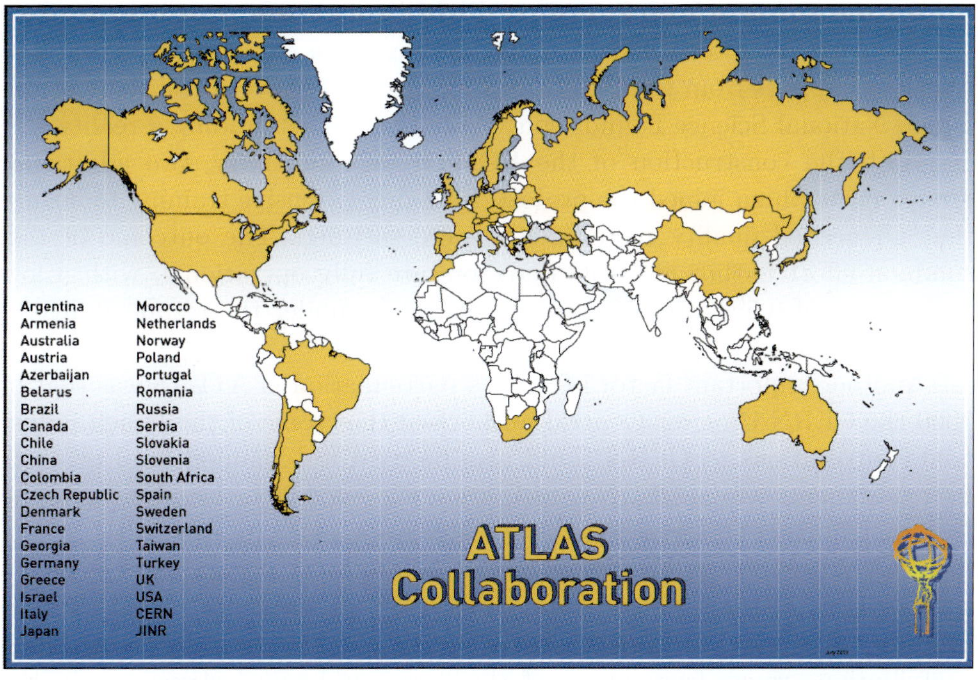

Fig. 3.2. The ATLAS Collaboration world map as of 2018.

Aviv University and the Weizmann Institute joined forces to obtain common funding, and also to participate in a common hardware effort for which they had full responsibility. This subsystem was the Pole-Tip Hadron Calorimeter of OPAL, based on the Thin Gap Chamber (TGC) technology, which at that point (1983) was considered a new and risky technology, due to the high gain operation of the detectors. This common effort proved to be a good procedure to create an impact in the experiment, both in terms of showing responsibility for the delivered hardware and its continuous maintenance throughout the 11 years of operation, and also by its impact in data analysis.

The responsible attitude towards the functioning of the subsystem was also shared with the Japanese groups, who were responsible for the barrel Electromagnetic Calorimeter. This led to a common respect and appreciation, which was fundamental to them deciding to participate together in the construction of a large system for the ATLAS adventure. Such a decision was not obvious because the political pressure was very strong on both sides to join the SSC Project. The common decision was then taken to construct together the Endcap Muon Trigger System (ECMTS) of ATLAS based on the TGC detector technology. With the end of the SSC, other Japanese groups joined the ECMTS as well as other systems.

On the Israeli side, the funding of the project was not obvious, since various parties opposed such a large funding for a single project, but the excellent support from the late Prof. Paul Singer, who was the chair of the Israeli National Science Foundation, made the needed funding a reality.

With the construction of the ECMTS well underway, two additional groups from China joined the project, following extensive training to ensure that all needed quality requirements were satisfied. The outcome of this common effort is that all TGC detectors are fully operational, with a low failure rate of 0.5% after 10 years of operation under the responsibility of the joined team.

Another important factor in Israel's participation in ATLAS was that in 2000 the CERN Director-General authorised that some of the Israeli industrial contributions to CERN could also be used for common fund projects for experiments where CERN contribution exceeded 20%. Israel became the first contributing Observer State to CERN from 1991 onwards, with 70% of its contribution being provided by industrial products. A number of elements that were crucial for ATLAS (and other LHC experiments) could thus be produced by the Israeli industry, following the usual tendering process. These contributions range from many of the large mechanical structures in the muon system to fibre-optics and various electronic components.

The combination of putting all the ATLAS Israeli effort into a single system, with excellent collaborators from Japan and China, important contributions in the analysis of the ATLAS data, as well as demonstrating that the Israeli industry could compete well within CERN's industrial requirements, allowed Israel to become the first non-European CERN Member State in December 2013.

Japan

On 10 May 1995, the government of Japan announced that it would help to finance the construction of LHC. At the 102nd CERN Council meeting on 23 June 1995, the Japanese Minister of Monbusho, Mr K. Yosano (who was responsible for science), stated during the signature of the agreement: "We decided to make a contribution to the LHC in advance of other non-member countries. I hope this decision will accelerate international cooperation in the project." The CERN Director-General C. Llewellyn Smith declared: "The scale of Japan's contribution to the LHC sets a new precedent in inter-regional scientific collaboration." A traditional Japanese Daruma doll with only one eye painted for the start of the project was handed over to Llewellyn Smith during the ceremony (Fig. 3.3). The second eye was dutifully painted in October 2008 at the LHC's inauguration.

Fig. 3.3. Signature ceremony of the CERN–Japan agreement.

Japan's bold action opened the path to make the LHC project truly international. One reason behind Japan's decision was the result of an outstanding achievement by a team from the Universities of Tokyo and Kobe in the OPAL Collaboration at LEP. Joined by physicists from KEK, the Japanese Accelerator Research Organisation in Tsukuba, and other universities, who once aimed for the SSC project, a Japanese team of about 40 physicists was formed in 1994 to participate in ATLAS. Today 16 Japanese Institutions are members of ATLAS.

Japan's major contributions in the ATLAS detector construction were the superconducting central solenoid and significant parts of the endcap muon triggering system and silicon micro-strip modules (SCT) modules. The solenoid (see Secs. 2.2 and 5.1) was designed by KEK, and the entire system arrived at CERN on 3 September 2001 via the Indian Sea, Rotterdam and the Rhine. The solenoid was the first major complete component of ATLAS.

The construction of the endcap muon triggering system, together with Israel and China, took more than five years for beam tests, R&D and pre-production before mass production started in January 2001. One-third of the TGC chambers were constructed at KEK and tested at Kobe before shipping to CERN. The entire front-end read-out was designed and fabricated in Japan. The preamplifier boards were tested in China in 2000. Special customised electronics were designed by graduate students. With the participation of a Japanese team, it took another two years to assemble all TGCs and electronics at CERN in Hall 180.

A quarter of the SCT barrel modules were constructed in Japan from 2002 to 2004. Flexible hybrid boards for barrel modules were designed by KEK, fabricated locally and delivered to other assembly centres in the United Kingdom, the United States and Scandinavia. A special robot used at Oxford University for auto-mounting all barrel modules was designed and made in Japan.

Many Japanese industries contributed to the ATLAS construction. Hamamatsu Photonics supplied 88% of the SCT sensors and all photomultipliers for the Tile Calorimeter, while Toshiba fabricated the superconducting solenoid as well as the CMOS timing chips for the MDT read-out. Kawasaki Heavy Industries constructed the aluminium cryostat for the LAr barrel calorimeter under a US-DOE contract. In addition, Kuraray supplied the fibres of the Tile Calorimeter read-out. Arisawa provided Cu-polyimide electrode sheets for the barrel LAr Calorimeter and Fujikura supplied radiation-hard optical fibres for the Inner Detectors.

China

Nanjing University (NJU), Shandong University (SDU) and the University of Science and Technology of China (USTC) initiated the participation of Chinese groups in ATLAS in the mid-1990s. Their request to join ATLAS was first put forward to the National Natural Science Foundation of China (NSFC). China had already joined the CMS experiment and there was an opinion that with the limited size of the Chinese high-energy physics community in the 1990s, China should only join one major LHC experiment. However, with the strong support from Prof. Naiyan Wang, the Deputy Director of NSFC, accompanied by the request from NSFC that the Institute of High Energy Physics (IHEP) should also play a major role, a Chinese cluster, consisting of IHEP, NJU, SDU and USTC, was approved and accepted as a member of the ATLAS Collaboration in September 1998.

From the start till now, the ATLAS project in China is jointly supported by NSFC, the Ministry of Science and Technology (MOST) and the Chinese Academy of Sciences (CAS). In November 1999, representatives of the Chinese Funding Agencies, NSFC, CERN and ATLAS signed an Addendum to the Construction MoU, for Chinese participation in the ATLAS detector construction of the muon system and of the endcap hadronic calorimeter system.

In order to better understand the constantly growing interest of physicists from various institutions in China to join the different LHC experiments, the NSFC organised an NSFC-CERN Workshop in Weihai, a beautiful city in the Shandong province, in autumn 2003. All four spokespersons from ALICE, ATLAS, CMS and LHCb were invited to attend that meeting. The Weihai Workshop became a milestone in China's participation in LHC experiments. During their visit, ATLAS colleagues also toured several collaborating Institutes.

According to the MoU, IHEP would construct monitored drift tube (MDT) chambers for the ATLAS muon spectrometer. IHEP's task was the production of 32 so-called BEE chambers and 16 so-called BIS.8 chambers, consisting of about 7000 drift tubes. A full-size BEE prototype was built in 2003. It was successfully tested at CERN's X-ray tomography laboratory, thereby triggering approval of IHEP as being a qualified muon MDT production site. At IHEP, all chambers were assembled and tested during 2005 and 2006. In January 2008, the installation and commissioning of the chambers were finished, and these chambers passed all the quality verification tests.

In the meantime, the NJU joined the R&D and mass production of the Liquid Argon Calorimeter system. For the Hadronic Endcap Calorimeter (HEC), they completed the production of 528 pieces, some 185 tonnes of copper absorber plates. NJU also took part in the assembly work at TRIUMF, Canada, and joined the beam tests over several years at CERN. For the forward calorimeters (FCal), NJU participated in developing and optimising the production technology and quality control programme, which guaranteed stable mass production of 8400 high-quality tungsten rods. Furthermore, SDU and USTC were involved in the muon endcap trigger system construction, which uses TGCs. From 2000 to 2005, SDU, in cooperation with Israel, produced 400 TGC detectors, whose failure rate is less than 0.5% even after several years of ATLAS running. Later, SDU took part in the installation and calibration of the trigger system. The USTC group tested more than ten thousand front-end electronics boards, collaborating with the Japanese TGC groups.

The Chinese ATLAS team has grown very quickly in recent years, becoming the largest LHC group in China. Shanghai Jiao Tong University (SJTU) and Tsinghua University (THU) were accepted as members in the Chinese ATLAS Cluster in 2012 and 2014, respectively. Given the continued growth, the Chinese Institutes were regrouped into two separate clusters with institutional membership in the ATLAS Collaboration. These two new Chinese clusters — the Chinese IHEP-NJU-THU Cluster and the Chinese USTC-SDU-SJTU Cluster — were officially approved (by the Collaboration Board) at the ATLAS Week held in New York in 2016. This paved the way to strong participation in the ATLAS high-luminosity upgrade phases.

The history of collaboration with physicists from Taiwan and Hong Kong followed similar patterns, albeit with an in-between gap of 15 years. First contacts were established when ATLAS spokespersons (and other senior colleagues) paid several visits to universities and their funding authorities. It typically took a couple of years to build up a mutually interesting partnership, resulting in concrete and excellent contributions of these teams. The Academia Sinica group from Taipei became an ATLAS member in 1999, followed in 2016 by a second group from Hsindu. In the case of Hong Kong, after the first contacts in 2011, three universities formed a common Hong Kong Cluster, which formally joined the Collaboration in 2014.

Canada

In the 1980s, the Canadian particle physics community was already engaged in international projects, for example, the SLD experiment at SLAC, the ARGUS and ZEUS experiments at DESY, and the ASTERIX, UA1 and OPAL experiments at CERN. Canada's nuclear and particle physics laboratory TRIUMF, operating a 500 MeV cyclotron in Vancouver since 1976, facilitated the construction of accelerator and detector components to be deployed abroad. By the time the first LHC R&D project was approved at CERN in September 1990, Canadian groups were also getting involved in the design of detectors for the SSC in Texas.

It is in this context that LHC R&D activities in calorimetry began in Canada in 1991, introduced by a returning member of the UA2 Collaboration. Two Canadian Institutions obtained funding in 1992, mainly to contribute to the RD3 Collaboration (see Sec. 1.5), concerned with the development of fast liquid argon calorimetry with an accordion design. The interest in LHC grew, and following the Evian meeting in March 1992 and the founding of the ATLAS Collaboration, four Canadian institutions signed the ATLAS LoI in October 1992.

The first attempt at securing Canadian funds for the ATLAS effort was of limited success; Canada was already involved in the SSC, which was at that time, unlike the LHC, an approved project. Following the cancellation of the SSC, the ATLAS Canada Collaboration quickly grew in size and expertise, which led to significant contributions to the ATLAS Technical Proposal. Capitalising on existing expertise, Canada pledged large contributions to the ATLAS liquid argon calorimeter system, namely the endcap hadronic calorimeter, the forward calorimeter, cryogenic signal feedthroughs and contributions to the read-out electronics. A large fraction of these components was built in Canada, following support from the national funding agency NSERC. In parallel, starting in 1994, the TRIUMF laboratory oversaw the construction and procurement of several magnet and power supply systems for the LHC and its injection systems.

The Canadian involvement in ATLAS has continued to grow, with a present total of ten Institutions and over 150 scientists involved in most aspects of ATLAS including detector operations, computing, trigger, detector upgrades, luminosity measurements and data analyses. Canadians are strongly engaged in the future of ATLAS, and with increasing involvement in the CERN programme comes the hope that Canada may explore a more formal relationship with CERN.

Australia

The Australian involvement in ATLAS evolved from years of participation by a one-faculty "group" at the University of Melbourne in the UA2 experiment at CERN. With an expansion of the faculty in the late 1980s, Melbourne and Sydney universities began participating in the RD2 Collaboration to develop silicon strip technology for the LHC environment. A strong and long-lasting collaboration emerged between Australian groups, CERN and the universities of Geneva and Cambridge, all becoming ATLAS founding members.

However, the Inner Detector of ATLAS, which was Australia's principle contribution, was not defined until its 1997 TDR. There were many interesting (sometimes rather heated) discussions leading to the final design. Layout, read-out chip architecture and SCT/TRT interactions were all important in these discussions. Australia, with Freiburg, Glasgow, Udine and others, carried out a long series of investigations into GaAs as a rad-hard alternative to Silicon. Silicon won out, of course, and the SCT was designed and built — a huge and difficult task that we revelled in.

The issue of a long development phase for an emerging field in Australia was both a benefit and a difficulty. We carved a good niche in detector development, but the groups needed some HEP data from other sources during the long ATLAS development. That was initially solved by working in the NOMAD experiment at CERN, and then, thanks to the silicon experience developed in RD2/ATLAS, by participation in the construction of the SVD vertex detector for the Belle experiment in Japan. During this time, however, the focus was on the SCT, a focus that led to the real stability of Australian funding through the ATLAS start-up and operation. From 2011, a 7-year funding for the Australian Research Council (ARC) Centre of Excellence for Particle Physics at the Terascale (or just CoEPP) was obtained. CoEPP was able to play a very strong role in ATLAS data analysis in the years that followed. It also led to the University of Adelaide starting an experimental group in the field.

Participation in the ATLAS/LHC programme from the early R&D phase, through ATLAS experiment development, design, construction, operation and data analysis has been a major success story for Australian science.

South America

The South American scientific community has had a long journey filled with tremendous effort to be able to join the ATLAS Experiment. The Brazilian experimental HEP community was the first to do so, in particular a group from UF Rio de Janeiro. After several decades of contributing physicists and technicians to CERN experiments and being part of LHC calorimeter R&D since the LAA and RD1 projects, it was only natural that they became founding members of ATLAS in the Tile Calorimeter system. Strong engineering contributions were also made in electronics and data base technologies. The Brazilian team then grew in the later years to include three other institutes, together forming the Brazilian Cluster as an Institutional Member.

However, other South American countries had only individual high-energy physicists participating in the experiments at CERN, hence they faced the challenge of consolidating a national participation. It was only in 2005, when Argentina hosted the 3rd CERN Latin American School of High-Energy Physics, that different possibilities for the participation of South American countries in the LHC were discussed. A month later, some South American physicists were invited to attend the ATLAS week, with a tangible enthusiasm for what might evolve. While ATLAS's construction had almost been finished, it was evident that there was still much to do before the start of the LHC. ATLAS presented, for the first time in its history, the opportunity for South America to be part of a unique scientific endeavour!

Without a doubt, ATLAS has had a high impact on the scientific communities of the Southern Cone, by giving them the opportunity of participating in forward thinking projects and extraordinary discoveries. These countries would also benefit from CERN's avant-garde technologies, potential technological developments, and mainly from highly qualified human resources with strong technical, theoretical and collaborative abilities. Small cores of lead scientists started elaborating proposals for their contributions with the full support and invaluable help of the ATLAS management. One of the hardest tasks was getting the support of the National Science Funding Agencies, but the relentless effort eventually paid off. In 2006, Argentina finally joined ATLAS, followed by both Chile and Colombia in 2007.

There was still a major obstacle with regard to the sustained participation of Latin American scientists taking part in an experiment such as ATLAS — a lack of regular travel funds. It was here that these countries received crucial help from two European Union projects: HELEN (High-Energy physics Latin american-European Network, 2006–9) and EPLANET (European Particle physics Latin American NETwork, 2011–15), which facilitated the scientific exchange between Europe and Latin America.

The first researchers and their PhD students from these countries created a vigorous and enthusiastic atmosphere in the ATLAS offices where they worked passionately contributing with studies for the High Level Trigger system and for several physics analyses for the imminent start of the LHC. Along with the arduous hard work, another noticeable trait was the sharing of *mate* and discussions of both physics and football with their new colleagues at ATLAS. At that time, there were barely a dozen physicists from Argentina, Chile and Colombia combined. Today, not only have the teams grown remarkably with young PhD students and researchers, they have also added engineers and computer scientists to help build hardware for the Phase-I upgrade in ATLAS, as well as to expand the contributions for Phase-II at the HL-LHC.

The participation in ATLAS provided the South American scientific communities with skilled human resources, new technologies and the exceptional possibility to be part of a global effort that pushes the frontiers of knowledge towards the understanding of nature.

Africa

Physicists from Morocco have had traditional links to France for a long time, in particular a small team from Casablanca and Rabat started working with the ATLAS Grenoble group in the mid-1990s. This participation in the LAr presampler design, development and testing led to them joining ATLAS formally as a joint team in September 1996, with active contributions throughout the construction and operation phases of this subdetector. Other Moroccan universities joined the "Morocco Cluster" in later years, which today has members from five Institutes.

Before the peacefully negotiated transition from the "apartheid era" to democracy in 1994, the scientific community in South Africa has had several pockets of excellence (for example, the first observation of the neutrino in cosmic rays in a deep gold mine in 1960), but these pockets were generally isolated, on a downward spiral and participation was not open to everyone. Towards the end of apartheid, South Africa was able to sign a Cooperation Agreement with CERN in 1992. This corresponded nicely with the beginning of the main LHC experiments, which sought global participation. Indeed, there was considerable investment by CERN and especially ATLAS in nurturing a broader South African participation.

The South African physics community undertook a major assessment of the state of Physics in South Africa to reverse the perceived decline of the discipline. The 2004 report from the "Shaping the Future of Physics" included a recommendation to prioritise the participation by South Africans in global science projects based at large-scale research infrastructures.

The SA-CERN Programme (ATLAS, ALICE, ISOLDE and Theory) was established on 15 December 2008 with the goal to grow and unify all CERN-based activities into one coherent network.

The SA-ATLAS group was a founding member of the national participation at CERN. This started with the University of Johannesburg, which had begun participating in ATLAS in early 2008 via an association with the Brookhaven National Laboratory. The new SA-CERN programme funding enabled universities to attract new HEP staff, reflecting both an international brain gain and some return from the African science diaspora. By 2010, the ATLAS participation spread to include the University of Witwatersrand. The same year, South Africa obtained its ATLAS membership. The University of Cape Town joined the SA Cluster in 2012. As of today, the SA Cluster has 5 authors, 4 scientific and technical staff, 5 postdoctoral researchers and 30 students. In addition, there are several scientific associates working on ATLAS projects from three other institutions. There has been a tremendous delivery to human capacity building and technology transfer, including high performance computing, accelerators, detectors and sensors, high throughput electronics and even an application in mineral sorting. There was also a significant number of students from Africa, who have received training via the SA-ATLAS cluster. In due course, we may see new direct participation from African countries.

3.3. The ATLAS organisation

Given the inherently cooperative and distributed nature of the Collaboration, a relatively lightweight organisation has been adopted in ATLAS. Nonetheless it is a significantly more complex structure than used in earlier generations of experiments, such as those at LEP, HERA or the Tevatron. Overall principles of the organisation include openness, democracy at the level of Institutions, and wide consultation within the Collaboration on key decisions and appointments. The high-level organisation of the Collaboration was defined in a lean constitution document in 1994, at the time of the detector construction Memorandum of Understanding. It is furthermore a fundamental principle of the Collaboration that every scientific result is circulated to all scientific authors for comment before it is made public, although further discussion on its implementation is beyond the scope of this chapter.

Two key organisational bodies of the Collaboration are the Collaboration Board (made up of one representative from each Institution in the

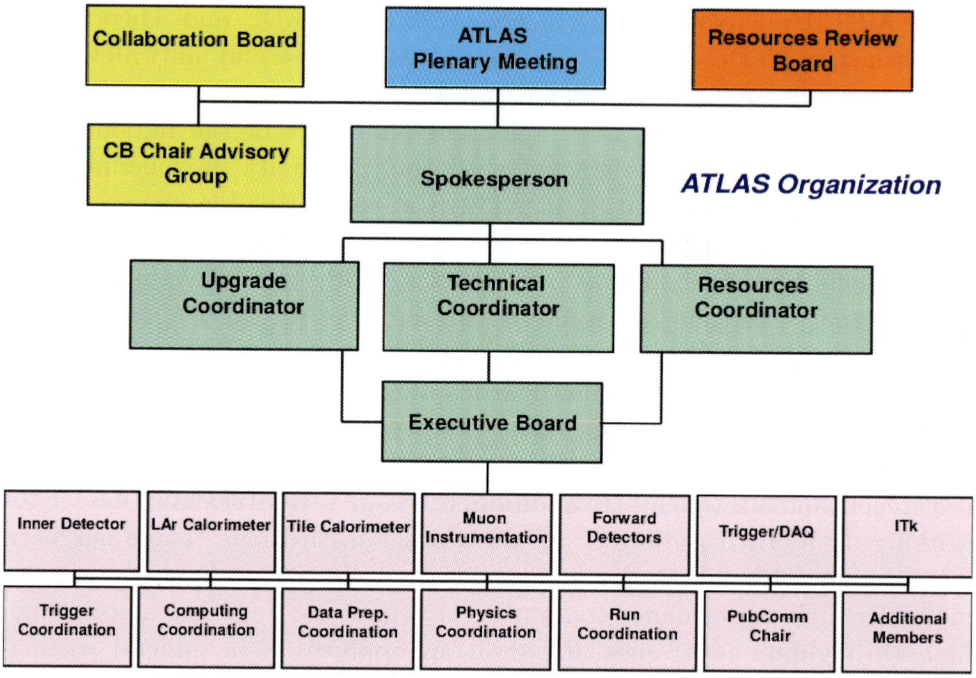

Fig. 3.4. Structure of the ATLAS Organisation at the time of writing. The Spokesperson typically has two deputies (not shown) who are also Executive Board members. The role of the Resources Review Board is discussed in Chap. 4.

Collaboration and chaired by an elected senior physicist among its members (the Collaboration Board Chair)) and the Executive Board, chaired by the Spokesperson. The inter-relationship of these bodies with each other, and with other important bodies discussed in the following, is illustrated in Figs. 3.4 and 4.6 for the current and construction periods (see Chap. 4) of ATLAS, respectively. The body of the Collaboration as a whole also plays a fundamental role: Plenary collaboration meetings in which key decisions, physics results and operational strategies are discussed, and to which all ATLAS members are invited, take place during ATLAS Weeks, which happen three times a year at present, and during an "ATLAS Weekly" plenary meeting.

The Collaboration Board forms the main voting body of the Collaboration, being responsible for general collaboration policy such as collaboration organisation, publication policies, etc., and for endorsing major decisions, such as those of detector changes and upgrades. The Collaboration Board is also responsible for admitting new member Institutions into the

Collaboration. The Collaboration Board presently meets three times a year, at the end of ATLAS Weeks. A key principle in ATLAS is that all holders of senior collaboration positions, including those who are members of the Executive Board, are either elected or endorsed by ballot in the Collaboration Board. In the case of the Spokesperson, the CERN laboratory's management is consulted regarding the candidates. For Technical and Resources Coordinators, the CERN's management approval is formally required. The selection of Spokesperson candidates for election is one of the tasks of the Collaboration Board Chair Advisory Group — a representative group of individuals from across the Collaboration whom the Collaboration Board Chair can seek advice on policy issues before bringing them to the full body. Further bodies elected by the Collaboration Board include the Publications Committee (responsible for overseeing ATLAS papers), the Speakers Committee (responsible for the quality of conference talks and managing the speaker allocation) and the Authorship Committee (responsible for managing the almost 3000-strong author list of the Collaboration).

The Executive Board is the primary steering body of the Collaboration, directing the execution of both the construction and operative aspects of the experiment. The Executive Board is chaired by the Spokesperson, with the Technical Coordinator being the deputy chair. The Spokesperson is able to rely on a small central "ATLAS Management" team for day-to-day central coordination, consisting of Deputy Spokesperson(s), Technical and Resources Coordinators, and since 2017, an Upgrade Coordinator. In addition, members of the Executive Board include Project Leaders of the individual detector subsystems, the coordinators of the "Activity Areas" (described below) and the Publications Committee chair. A small number of additional members help to ensure a balanced composition across regions of the world. The work of the Executive Board is wide-ranging — it spans from reviewing the operation and maintenance of the detector, trigger and computing systems, coordination of hardware and software changes, interactions with the main CERN review committee of the experiment, the Large Hadron Collider Committee (LHCC), to an internal review of M&O budgets and operation tasks (see Sec. 4.6) and preparation of major construction and upgrade hardware decisions for the Collaboration Board's endorsement. Significant challenges faced in any areas may be brought up to the Executive Board to be addressed. This Board typically meets ten times a year in full meetings, complemented by weekly meetings when needed. Major subsidiary bodies of the Executive Board include Technical Coordination, described in Chap. 6, the Upgrade activity headed by the Upgrade Coordinator and organised through the Upgrade Steering Committee, a sub-committee of the Executive Board, and the Offline Activity Coordination Board, also a sub-committee of the Executive Board, which coordinates in greater detail over-arching data processing, trigger and computing operations work.

The organisation below the Executive Board consists of two types of functional structures: the detector system (or "projects") and the "activity areas". The detector systems are each led by a Project Leader, elected by the Institutions which are part of the system, i.e., committed to the original construction of that subdetector or to its operation. There are six major operational systems in ATLAS today: the Inner Detector, Liquid-Argon and (separately) Tile Calorimeters, Muon Detectors, Forward Detectors and the Trigger-and-Data Acquisition. Within each subdetector system there are subsidiary responsibilities; typical roles could be project engineers, data taking ("run") coordinators, etc. Each detector system has its own System Steering Group and Institute Board, analogous to the Executive and Collaboration Boards, respectively. The subsystems have significant autonomy and important roles still in the operation phase, organising operations and maintenance activity, aligning and calibrating the detector, ensuring its data quality, ensuring coverage of project talks and publications, and so on.

The organisation of the upgrades to the detector, being prepared for installation in LS-2 (the 2019/20 LHC shutdown) or LS-3 (in 2025–2027), is generally also done through the subdetector systems, via upgrade projects within them, but also reporting to the Upgrade Coordinator and Technical Coordination via the Upgrade Steering Committee. An exception as regard to being contained within an existing system is the replacement inner tracker, ITk (see Chap. 10), which will replace the entire current Inner Detector. The ITk Project Leader is a member of the Executive Board.

The activity areas follow a different organisation, which provides some matrixing of aspects of work. These areas are related to the data taking coordination (led by the "Run Coordinator"), trigger operations and performance (led by the "Trigger Coordinator"), preparation of data for analysis in all aspects (led by the "Data Preparation Coordinator"), all aspects of the offline computing system and software (led by the "Computing Coordinator"), and combined-detector performance and physics analysis coordination (led by the "Physics Coordinator"). Each activity area has its own internal organisational structure, with some responsibility positions also appearing in the subdetector organisations. With the exception of the Physics and Computing activity areas, this part of the organisation was introduced in 2007–8, following extensive consultation (and a lively debate) within the Collaboration.

Perhaps the most organisationally complex of the activity areas (and of course the one which directly relates to the final physics results of the Collaboration) is the Physics Activity area. Both the performance of the detector for the reconstruction and identification of physics objects such as electrons, muons and hadronic jets, and the final data analyses leading to measurements of observed processes as well as searches for new ones, are covered by fifteen working groups divided into "Combined Performance" and "Physics Analysis" groups. The latter typically carrys out tens of analyses in parallel at any time, grouped in more topical subgroups, and then further into the individual analysis teams that often consist of a handful to tens of individuals.

In all areas of work of the Collaboration, individual team members, e.g., physicists, engineers, technicians, etc., are usually geographically separated and often spread across multiple continents and widely varying time zones. Multiple collaborative tools are essential to facilitate such collaboration, so that each team member can actively and effectively participate in discussions, decision forming, planning, etc. As an example, the CERN Indico product is used to schedule all meetings, allowing any ATLAS member to access meeting materials and talks in real and non-real times from a single web interface. Furthermore, all meetings in ATLAS are video-conference enabled, so as to allow ATLAS members to join in from all corners of the world, whether they are in their office, home, or in a meeting room at their institute or CERN. Ensuring reliable video-conference feeds from every meeting room at CERN — and further afield — was no small feat!

Run Nr	Task Name	User	taskID	Type	Status	Total	Done	Run.	Proc.	TBD	Abrt.	Failed	Events	Created (UTC)	Modified (UTC)
336852	data17_13TeV.00336852.physics_Main.recon.DESDM_ZMUMU.f871_m1...	tzprod	2220839	desdmerge	FINISHED	2328	2328	0	0	0	0	5	3119000	06/OCT 10:16	08/OCT 02:07
336852	data17_13TeV.00336852.physics_Main.recon.DAOD_ZMUMU.f871_m183...	tzprod	2220838	daodmerge	FINISHED	638	638	0	0	0	0	4	3119000	06/OCT 10:16	08/OCT 02:17
336852	data17_13TeV.00336852.physics_Main.recon.DESDM_EGZ.f871_m1831...	tzprod	2220830	desdmerge	FINISHED	612	612	0	0	0	0	4	988767	06/OCT 10:00	07/OCT 05:07
336852	data17_13TeV.00336852.physics_Main.recon.DAOD_EGZ.f871_m1831_f...	tzprod	2220829	daodmerge	FINISHED	165	165	0	0	0	0	1	988767	06/OCT 09:59	07/OCT 00:59
336852	data17_13TeV.00336852.physics_Main.merge.TAG.f871_m1879_m1818_...	tzprod	2220828	tagmerge	FINISHED	49	49	0	0	0	0	0	62769210	06/OCT 09:55	06/OCT 16:07
336852	data17_13TeV.00336852.physics_Main.recon.DAOD_TAUMUH.f871_m18...	tzprod	2220827	daodmerge	FINISHED	8	8	0	0	0	0	0	76324	06/OCT 09:55	06/OCT 17:18
336852	data17_13TeV.00336852.physics_Main.merge.DRAW_TAUMUH.f871_m1...	tzprod	2220788	recon	FINISHED	52	52	0	0	0	0	0	76324	06/OCT 07:36	06/OCT 16:17
336852	data17_13TeV.00336852.physics_Main.merge.DRAW_ZMUMU.f871_m18...	tzprod	2220787	recon	FINISHED	2329	2329	0	0	0	0	0	3119000	06/OCT 07:36	08/OCT 01:25
336852	data17_13TeV.00336852.physics_Main.merge.DRAW_EGZ.f871_m1831_...	tzprod	2220786	recon	FINISHED	612	612	0	0	0	0	0	988767	06/OCT 07:27	07/OCT 00:17
336852	data17_13TeV.00336852.physics_Main.recon.DRAW_RPVLL.f871_m183...	tzprod	2220781	drawmerge	FINISHED	2453	2453	0	0	0	0	0	3827365	06/OCT 06:39	06/OCT 07:38
336852	data17_13TeV.00336852.physics_Main.recon.DRAW_TAUMUH.f871_m18...	tzprod	2220780	drawmerge	FINISHED	52	52	0	0	0	0	0	76324	06/OCT 06:39	06/OCT 07:28
336852	data17_13TeV.00336852.physics_Main.recon.DRAW_ZMUMU.f871_m18...	tzprod	2220779	drawmerge	FINISHED	2329	2329	0	0	0	0	0	3119000	06/OCT 06:39	06/OCT 07:28
336852	data17_13TeV.00336852.physics_Main.recon.DRAW_EGZ.f871_m1831.d...	tzprod	2220778	drawmerge	FINISHED	612	612	0	0	0	0	0	988767	06/OCT 06:39	06/OCT 07:17

Screenshot insight into a PROC's life.

It's a PROC's life...

Most members of the Collaboration when asked to become the next PROC had to look it up to see what it meant. However, by the end of their term, they know that PROC not only stands for Prompt Reconstruction Operations Coordinator but it also defines a new way of living (at least for the duration of the appointment but also a bit beyond as it is very hard to leave this position).

What is it all about? It is best described by going through a typical working day. Our day starts by looking at the software release builds trying to assess if the updates that were put in the previous night broke everything or just a small part of the release. Then, we usually proceed to looking at the software comparison tests trying to understand why they don't look as they should even if the changes were "minor" and "couldn't possibly have an impact on the reconstruction". Then the real day starts with (depending on the time of the year and on the current state of the LHC) the daily operations meeting where it is important to follow all the activities from all the sub-detectors, just in case there is a special run taking place that needs special requirements and care (yes, it is a PROC who takes care of the special reconstruction requests). Then we continue with a day full of testing new software, preparing new settings for data reconstruction, babysitting all the jobs that the whole Collaboration is waiting for, replying to emails asking why the jobs are not finished, and answering all types of questions related to the state of the new data to be processed in search for new physics. There are of course a few other meetings per day but that is the case for everyone else, no matter what the position. The day ends with a rush against the deadline for accepting any new software patches that desperately need to be included in the next build release that will definitely be built the next day (spoiler alert: it won't!). But most importantly, the day ends with a feeling of accomplishment and success that everything is still running and working as it should. Then, take this day and multiply it by 365 to get a year out of PROC's life!

It is an extremely interesting, important, and rewarding job (not many people pride themselves in knowing by heart all the previous configurations of data reconstruction)! It gives us a direct handle on the data collected and to be analysed by the physics and combined performance groups. It gives us an overview of the operation of the detector and data preparation from raw collisions to final data format that can be downloaded on a laptop for analysis. But above all, it exposes us to a whole new group of people, so devoted to their job, so friendly and willing to help, so cheerful and fantastic! And these people later on become our friends, our family, for better and for worse! Someone wise said "once a PROC, always a PROC". And this couldn't be more true!

Data quality as seen by the liquid argon team

Preparing for a new year of data taking is a busy and exciting time. Operations teams are determined to ensure the detector is ready to collect data efficiently, and analysis teams are restless, eager to start sifting through the data, looking for new physics, or to further expand our understanding of the Standard Model. Meanwhile, behind the scenes, monitoring experts sit in wait. Once the data start coming in, it's up to them to ensure that before the data are used for physics results, they are ready — "clean" — with no software glitches or detector noise anomalies masquerading as real "physics" effects.

Towards the end of June 2017, the LHC reached new heights, with a record 2544 colliding proton bunches and a peak instantaneous luminosity of $1.58 \times 10^{34}\,\mathrm{cm}^{-2}\mathrm{s}^{-1}$. This marked the end of the first 2017 intensity ramp up, where the LHC machine gradually increased the number of bunches, step by step. For three weeks leading up to this, the LHC had been delivering over $2\,\mathrm{fb}^{-1}$ of data per week to the ATLAS experiment, meaning that by this stage the experiment had in hand more data than was collected during the whole of 2015!

Ensuring the quality of the data takes time, time during which the analysis teams would love to already have access to the data and are, understandably, impatient to start analysing it. The detector, however, is still waking up from the winter break. In the calorimeter, for example, new regions of the liquid argon system are producing energetic bursts of noise that must be removed from the data before physicists can use it. To make sure any new noise from the electronics is identified and understood quickly, the calorimeter experts spring into action, meeting daily to monitor their system and apply fixes where necessary. These new "hot" calorimeter cells, these bursts of energy, are flagged as noise and filtered out from the data before it reaches any physics analysis.

Further upstream, the energy deposited in the calorimeter cells is gathered to form "clusters", which are used to reconstruct traversing particles, such as electrons and photons. To ensure that the data have been sufficiently cleaned, other teams of data-quality experts look at these reconstructed objects. Should they spot something suspicious, like a recurring energetic jet in a localised region of the calorimeter, characteristic of high-voltage-induced electronic noise, it must be rejected.

This is a busy couple of weeks; the data-quality-checking procedure must be carried out on all of the data collected before they can be made available for analysis. Despite the heavy workload and time-pressure, it's all made worthwhile when the clean, certified data are unveiled. Finally, there's new data to sink our teeth into. It's exciting to think that somewhere buried deep within this new dataset, there could be answers lying in wait. All we have to do is ask the right questions.

Chapter 4

From the LoI to the Detector Construction

4.1. Early technology decisions

The newly formed ATLAS Collaboration was well aware, at the time of submitting the LoI, that making technology choices would not be easy in several cases where different approaches could meet the physics and budgetary requirements. It was therefore important to establish a transparent procedure for these situations. The decision was to establish the so-called Detector Review Panels, including (and chaired by) senior collaborators not directly involved in developing the technologies in question, together with the expert proponents. The technology evaluations consisted of performance requirements with aspects of schedule, costs, available resources and ability to handle construction and overall integration into the ATLAS detector design. The evaluation progress was regularly reported at all levels of the ATLAS organisation and project management, as well as in plenary meetings. This concluded with a recommendation to the Collaboration Board.

A major decision was made very early on the choice of technology for the toroid magnet system between a conventional iron-core magnet and superconducting coils in an air-core configuration. In April 1993, only half a year after the LoI, the Collaboration chose the air-core design, being fully aware that this meant optimising the overall use of funds. In fact, only a few months later in July 1993, the initial design of 12-fold symmetry with 12 coils around the beam axis was reduced to a final cheaper 8-coil design. This turned out to be a wise choice as it also increased the solid angle for muons that avoid passing through coil material, while the slightly increased magnetic field inhomogeneity was acceptable. More details about this central choice are given in Sec. 4.2.

Also within the first year after the LoI, the calorimeter technologies and layout for the barrel and endcap regions were settled from various options still considered in the LoI, following the intense evaluation work of the Calorimeter Review Panel. By the end of 1993, ATLAS had decided on a configuration with a barrel and two extended barrel cylinders, with fine-grained iron-scintillator tile modules, for the hadronic calorimeter. These formed an outer cylinder around the three-liquid argon cryostats, the barrel one housing the novel highly granular lead-LAr sampling electromagnetic calorimeter with "accordion" geometry and pre-sampler, also used in the endcaps together with hadronic calorimetry with copper absorbers. Somewhat more time (until September 1994) was required to converge on the choice of an integrated liquid argon forward calorimeter within the endcap cryostats.

A further major technology and layout choice needed to be made for the muon spectrometer instrumentation, which employs separate precision and trigger chambers. Here, the Muon Chamber Panel and the ATLAS Technical Coordinator helped the muon community to converge in 1994, well before the Technical Proposal (TP) submission date, on the baseline solution that is implemented in the ATLAS detector now.

The Inner Detector (ID) tracking system is particularly challenging in the harsh high-rate environment of the LHC. When the TP was submitted at the end of 1994, it had been too early to arrive at a definite layout decision that would be backed up by the necessary R&D. The Inner Detector Panel was active until September 1995 when ATLAS adopted its ID layout with pixel detectors in the innermost part, followed by Silicon strip detectors, with the two providing the discrete precision tracking points, followed at larger radii by a straw-tube tracker that allowed continuous tracking and transition radiation detection to enhance electron identification. Only much later, before the second LHC Run starting in 2015, was an additional layer of advanced pixel technology added to improve further secondary vertexing capabilities at high luminosities.

By the end of 1995, the basic detector concept was settled and the project was, in principle, ready to enter the construction phase. However, many detailed developments and decisions still had to be made over the decade of detector construction, especially in the areas of detector controls (DCS), read-out electronics, triggering and data acquisition (DAQ), which profited from the rapid developments in the commercial sector and were able to benefit from the longer timescales. As an indication, the Level-1 trigger

Technical Design Report (TDR), still with much custom made hardware, was written in 1998, and the DAQ, high-level trigger and DCS TDR, in 2003.

However, cost limitations imposed further major optimisation and redesign efforts. CERN had set a ceiling of 475 MCHF (1995 currency) on materials and industrial manpower, while the pledged resources from the Funding Agencies supporting the Collaboration partners were even more constrained. In a global redesign effort, judging each detector component against the most essential physics goals, the Collaboration undertook an optimisation exercise of the full detector (as described in Sec. 4.3) in 1995. This included slight reductions of some dimensions and granularities in order to gain a reduction of some 10% on the estimated project costs. An important side effect was that this made the detector fit more comfortably into the fixed dimensions of the underground cavern, which was nevertheless a tremendous challenge. The re-scoped detector was the basis of the recommendation by the LHCC to submit TDRs for the various detector components before entering the construction phase, with formal construction authorisation given on 1 July 1997, after the approval of the first series of TDRs (Sec. 4.4) on the calorimetry.

4.2. The example of the toroid magnet technology choice

As leptons and photons with high transverse momentum can be messengers of exciting new physics, the ATLAS Collaboration gave high priority to the precision measurements of muons and electrons, as well as photons. Toroid magnets were considered for the muon spectrometer, since their bending power increases as the inverse sine of the scattering angle in the forward direction, just as the momentum of a particle does for a given transverse momentum. Hence, a toroid magnet configuration in the central and forward region formed the basis of a large acceptance and robust stand-alone muon system; this was complemented by a superconducting inner solenoid magnet covering the inner tracker.

In the LoI three magnet configurations were considered:

1. Superconducting air-core toroids for both barrel and endcaps providing the best muon momentum resolution.
2. All iron-core toroids.
3. Superconducting air-core barrel and iron-core endcap toroids.

For the Technical Proposal, the three options had to be evaluated and a best choice made under the guiding principle that the muon spectrometer should not be optimised at the expense of other subdetectors. For that purpose, a Magnet Panel was appointed by the Collaboration to provide a recommendation based on the study of the physics performance, the technical feasibility, the construction and installation schedule and the overall costs. The panel was in frequent contact with the independent LHCC Magnet Panel, which was composed of well-known senior magnet experts from CERN and a few European Laboratories with experience in large-scale magnet projects, mandated to assess for CERN the technical aspects such as feasibility, safety, global risks and overall costs of the magnets for all experiments.

It was the consensus of the magnet experts that both the air-core and iron-core solutions were technically feasible and their operation would be quite safe. The air-core toroids employ a rather conservative design for the conductor that was considered to be quite stable and robust within the expected limits. However, the major challenge of this magnet system stems from its enormous size (26 m (length) by 22 m (diameter)). It determines the size of the underground experimental hall and the size of the access shaft necessary to lower the magnet components into the underground cavern. The electrical power consumption for the air-core magnets was estimated to be with 1.3 MW less than that for the iron-core magnets with 4 MW. However, this advantage would be essentially compensated by the higher maintenance costs of the refrigeration system.

These cost estimates were made by experts from CEA Saclay (for the air-core barrel), the Rutherford Appleton Laboratory (for the air-core endcaps) and CERN (for the iron-core magnets). The air-core magnets turned out to be 28% more expensive than the iron-core magnets. Although the ATLAS Magnet Panel strongly recommended the air-core solution for its greater discovery potential and better stand-alone capability, it also recognised that this recommendation might imply some staging of the air-core endcaps to meet financial constraints.

The recommendation of the ATLAS panel was explained in detail to, and finally adopted by, the Collaboration in a plenary meeting and then formally approved by the Collaboration Board on 4 April 1993. The decision statement, voted with a very large majority, also emphasises the need to contain the overall costs of the experiment, a constraint one was aware of constantly during the early design phase:

1. The all air-core toroid magnet is chosen as the baseline design.
2. Every effort shall be made to reduce the costs of the air-core magnets along the lines described in the staging/down-grading document.

3. The endcap magnets shall be staged if so required to match the spending profile.
4. In case of a serious financial shortfall, the Collaboration agrees that it should be met by all subdetectors in a concerted effort to minimise the impact on the ATLAS physics potential. This includes re-evaluation of the magnet concept as well as the design of all other subdetectors.

The choice was based on the results of many physics studies. Most importantly, the all air-core option provides approximately a factor three better di-muon and 4-muon mass resolutions than the all iron-core options, and would permit the observation of a resonant multi-muon Higgs boson signal in the stand-alone mode, i.e., without the use of the Inner Detector tracking. Owing to the better resolution, the 5σ discovery level for an anticipated Higgs boson could already be reached with a lower integrated luminosity. The stand-alone mode allows the independent comparison of signals in multi-muon channels with those obtained in electron channels, providing redundancy for the identification of weak signals. Compared to the iron-core option it permits the lowering of the transverse momentum trigger threshold by a factor two. This feature leads to a considerable event enhancement for B-physics. It also allows the study of the production and the suppression of Upsilon states in heavy ion collisions, providing a sensitive discrimination between hot hadronic matter and quark gluon plasma. The chosen magnet configuration for the muon spectrometer consists of a long barrel of superconducting coils arranged at equally spaced azimuthal locations, embedded in individual cryogenic coil casings. The endcap toroids are inserted in the barrel toroid at each end. Their superconducting coils are mounted in two huge cryostats. The barrel and endcap cryostats are fed by a common refrigerator. As a result of optimising the overall detector costs for the Technical Proposal the originally foreseen 12 coils in the barrel and the endcaps, see Fig. 4.1, had to be reduced to 8. On the positive side, this simplified the access and installation of the muon chambers and reduced the obstructed phase space for muons.

4.3. Descoping to meet the CORE budget

In 1995 the LHCC requested a list of detector components that could be part of a descoping or staging of the apparatus (in order to meet the 475 MCHF cost ceiling) for the meeting in mid-November. The ATLAS Collaboration Board met in September and there was a consensus that a global point of view, based on overall physics performance, was needed for any sensible

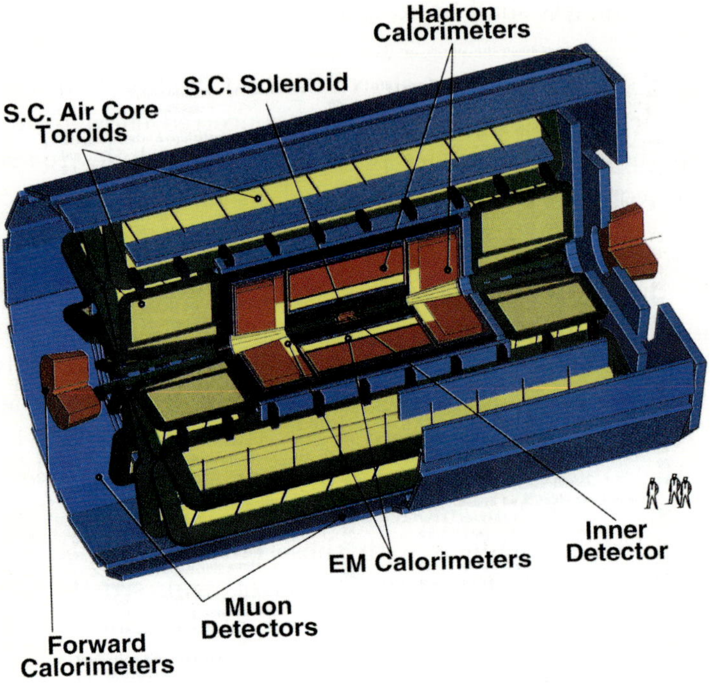

Fig. 4.1. Initial ATLAS design with a 12-coil toroid system as presented in the LoI.

response. It was indicated that even if such a plan could not be fully developed in time for the LHCC, one might gain a useful early perspective and that a modest voluntary descoping might be healthy for ATLAS. Such an exercise would be especially important for the large mechanical systems, whose dimensions were about to be frozen for detailed mechanical design and construction of full-size prototypes. In these cases descoping at a later time would have been very counter-productive. It was felt that ATLAS should not miss a last chance to create some safety margin in the detector costs relative to likely resources.

Members of a Global Descoping Task Force were appointed in early October 1995 and were tasked with recommending a modest, "physics-driven" descoping of the detector, which would retain good overall physics performance. Members were nominated by the ATLAS spokesperson, in consultation with the task force chair and the project leaders of the detector systems. Members were included from each of the major ATLAS systems, although they were charged with taking a global perspective rather than acting as a representative for their system. The final group consisted of seven regular members, plus the spokesperson and technical coordinator in ex-officio capacities.

The task force met in five closed sessions. Following the first meeting a brief summary of the intentions and planning was distributed to members of the Collaboration Board and input to the task force was solicited. During the review process, the chair presented two status reports to the Executive Board.

At its first meeting, the task force considered its mandate and global issues, such as overall detector dimensions and rapidity coverage. At subsequent meetings descoping options for each detector system were reviewed from the perspective of their cost savings and their impact on the performance of that system, and of the overall detector. Throughout the process, close contact was maintained between task force members from any given system and their respective system project leaders.

The final report contained two classes of recommendations. The first involved global aspects of the detector such as overall dimensions, rapidity coverage, and trigger and data acquisition rates. Specific recommendations were made on these issues. The second involved further descoping in individual detector systems. The task force believed it should not make detailed design recommendations to the subsystems since these groups themselves were in the best position to optimise their systems within clearly understood constraints. Instead, it tried to identify aspects of each system that could be legitimately descoped without major impact on the performance or safety margin. The recommendation was in the form of a reduced cost ceiling for each system, with one or more examples of how this target could be reached. It was expected that the subsystem designs would be altered in such a way, so as to fall within the lowered cost ceiling.

The recommendation called for reducing the cost ceiling of the muon toroids by 5 MCHF, to a total of 120 MCHF. About 50% of the savings could be achieved by reducing the overall detector diameter by 1.1 m and the length by 2.1 m to give a better margin with respect to the cavern size, which was taken as fixed. Better design optimisation of conductor placement and a possible current reduction would achieve the other 50%. This appeared to be possible with no more than a 10% reduction in bending strength. The size reduction of the toroids would also reduce the coverage needed by the muon chambers and hence their cost.

An effort was made to harmonise the rapidity coverage of all detector systems for precise measurements over the interval $|\eta| < 2.5$. To this end, it was recommended that the inner wheel of the electromagnetic endcap calorimeter, which covered the rapidity range $2.4 < |\eta| < 3.2$, be built with reduced transverse and longitudinal granularity. It was also proposed to limit the fiducial coverage of the muon system to $|\eta| < 2.6$. The radius of the inner bore of the endcap toroid would be increased by 25 cm, allowing additional shielding around the beam to reduce the rates in the muon detectors and the area for them to cover. Simulations of the revised shielding showed a 30% reduction of rates in the barrel and factors of three to five in the forward regions. It was suggested that the additional shielding be added in phases as the luminosity of the machine increased. Another important proposal was that the initial maximum LVL1 trigger rate be reduced to 75 kHz from 100 kHz, but in an upgradeable manner so that it could be extended once the machine reached high luminosity.

Within the detector systems, costs for the Inner Detector could be reduced by modifying the geometry of both the barrel and forward sections of the SCT, so as to reduce the area required. The granularity of the liquid argon calorimeters could be reduced in regions where the performance would be compromised by dead material and be outside the primary rapidity region. A series of ten suggestions were made for the muon system. The final total cost reduction proposed was 25 MCHF, with about 40% associated with detector instrumentation and the rest coming from more global changes.

Throughout the descoping exercise the power of a large Collaboration like ATLAS became apparent. It was possible to get detailed responses to complicated questions in a remarkably short time. Physics simulations were run for a number of topics to evaluate performance with alternative geometries. Detector shielding simulations were also rerun. The powerful engineering teams were also able to estimate expected cost changes. The whole process went smoothly but a substantial and constructive effort was required across the Collaboration. In the end it was clear that ATLAS was already well optimised for the level of performance proposed. One could carve out some savings but only levels of less than 10%, especially if one wanted to preserve the desired level of performance. At the time of the study the mass scale and details of the new physics to be studied were unknown. The only way to be confident of being able to explore the full energy range of the LHC was a general-purpose detector with a broad, powerful performance.

4.4. The Technical Design Reports

Following the Technical Proposal published on 15 December 1994, and its subsequent review by the LHCC, the ATLAS Collaboration was an entity asked by CERN to proceed towards a definition of the experiment that CERN could approve as being one of the experiments for the LHC. The experimental design was not really fully converged at this point, e.g., there were too many different detector technologies. ATLAS continued to develop the experiment design and to progress on the different detectors through prototypes and simulation studies, to be ready for formal approval. To arrive at this point, CERN decided that the experiment should be documented in a sequence of Technical Design Reports (TDRs). These reports were to be refereed by the LHCC and the dedicated referee teams that the LHCC appointed for this purpose. For some systems where the implementation could wait and benefit from further developments, like High-Level Trigger, Data Acquisition and Computing, substantial progress reports in the same format as the TDRs were submitted to the LHCC.

From 15 December 1996 to 18 March 2005, ATLAS produced seventeen TDRs (Table 4.1), with detailed descriptions of the experiment's different detector systems, trigger and data acquisition, and its overall performance.

Templates were made to have a common style, and Framemaker was chosen as the work processing system, based on its ability to handle large text volumes. All volumes have the same graphical design on their covers, with colours distinguishing them (Fig. 4.2). The TDRs have a harmonised disposition of the texts for the detector systems and for the trigger and data acquisition. The detector systems could be harmonised quite naturally,

Fig. 4.2. The series of ATLAS Technical Design Reports.

Table 4.1. The 17 Technical Design Reports for the ATLAS construction.

TDR	Pages	Titles	Date
1	178	*Calorimeter Performance*	1996-12-15
2	606	*Liquid Argon Calorimeter*	1996-12-15
3	330	*Tile Calorimeter*	1996-12-15
4	256	*Inner Detector Vol 1*	1997-04-30
5	898	*Inner Detector Vol 2*	1997-04-30
6	101	*Magnet System*	1997-04-30
7	208	*Barrel Toroid*	1997-04-30
8	282	*End-Cap Toroids*	1997-04-30
9	85	*Central Solenoid*	1997-04-30
10	513	*Muon Spectrometer*	1997-05-31
11	317	*Pixel Detector*	1998-05-31
12	500	*First-Level Trigger*	1998-06-30
13	598	*Technical Coordination*	1999-01-31
14	458	*Detector and Physics Performance Vol 1*	1999-05-25
15	506	*Detector and Physics Performance Vol 2*	1999-05-25
16	370	*High-Leve! Trigger Data Acquisition and Controls*	2003-06-30
17	234	*Computing*	2005-03-18
Total	**6440**	**pages**	

but there are also volumes describing performance, the organisation of the Technical Coordination and one for the Computing.

The first TDRs reviewed were of the Calorimeter system — the Liquid Argon and Tile Calorimeters. The conclusion of these reviews led C. Llewellyn-Smith, CERN's Director-General and Chairman of the Research Board, to approve these systems for construction. This first TDR approval became the de facto approval of the ATLAS experiment.

The process of writing these TDRs was a big effort for the Collaboration. It acted as a catalyser for convergence and decision-making, and helped the different systems to organise themselves. However, this didn't mean that everything was settled and decided. There were still a number of technical issues that concluded at a later time, for examples, practical manufacturing aspects resulted in significant changes in the calorimeter cryostat designs, manufacturing of components could reveal weaknesses that resulted in redesigns, or the adaption to the machines at a given manufacturer could result in design changes. Therefore, although these TDRs do not necessarily contain the exact technical descriptions of what was finally installed in the experiment, they are still close.

Nevertheless, the main impact of the TDRs was to coherently define all technical details at a given time, which were necessary to facilitate the approval of the experiment. The TDRs induced convergence and a cultural transition in the detector systems, from R&D projects to construction projects.

4.5. CORE costs, overcosts and costs to completion

In the Memorandum of Understanding (MoU), dating back to 1998 and predated by an Interim MoU in 1996, the ATLAS detector construction budget was set at 475 MCHF in 1995 value. This value followed a review by the LHC Cost Review committee and was thereafter referred to as the CORE value. This value referred to *direct production costs* and did not include the cost of R&D, home institute infrastructure, salaries or taxes. The figure for the total CORE costs resulted from careful scrutiny of expected expenditures and numerous discussions with collaborating institutes. After having been pared down to agreed values of deliverables from original requests for around 500 MCHF, the 475 MCHF formed the backbone of the MoU. This number, however, does not reflect all the tremendous efforts made in human and additional financial investments by all ATLAS Institutions.

The ATLAS MoU is a 7-page document (excluding the long list of annexes and the signature page) and lays out the principle of deliverables, common projects and related in-kind contributions. One often-used analogy to summarise these key concepts was to describe ATLAS as a "high-tech picnic party", where its members first subscribe to the shared list of deliverables associated with a CORE value and deliver them for that recognised value. This process often required lengthy negotiations in order to match the interests of Institutes with the needs of the Collaboration to cover all aspects of the detector. Then, what was left was grouped as Common Project items, to which members contributed either cash (Common Fund) or in-kind contributions. About 56% of the construction cost of 475 MCHF were allocated to deliverables with a recognised CORE value. The nature of deliverables reflected the core competencies of the institutes providing them. The remaining 44% were common items shared among the Funding Agencies (FAs) in proportion to the CORE values of their deliverables, and of which around 60% were provided as in-kind contributions. The mechanism for providing the Common Project items is shown in Fig. 4.3, the overall share of systems in Fig. 4.4.

It is important to note that by convention the 475 MCHF did not include any explicit contingency. No funds were included for example for unforeseen raw material fluctuations or production cost increases, inflation or currency

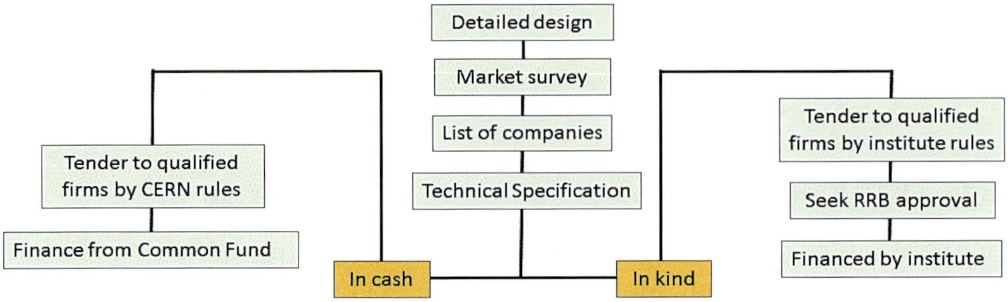

Fig. 4.3. Flow chart describing the provision of equipment via the Common Projects.

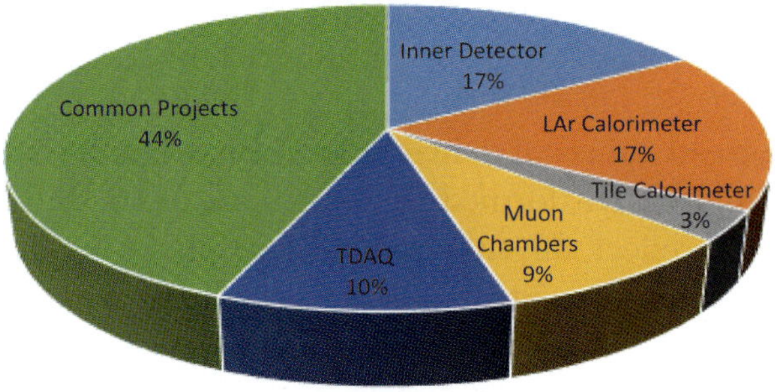

Fig. 4.4. Share of ATLAS systems as percentage of the CORE 475 MCHF.

exchange rate changes. This all meant that FAs would then need to handle such unforeseen costs separately when such events happened. Moreover, the 475 MCHF concentrated on producing the detector system components, but the commissioning and integration aspects were not included in the scope of the original CORE review. The role of the Technical Coordination was also not very clearly defined between CERN as the Host Laboratory and ATLAS.

By 2001, ATLAS was facing overcosts related to the construction budget as stated in the MoU. The above-mentioned factors eventually resulted in cost increases that surfaced at that time. A second CORE review took place where 72 MCHF worth of additional resources was reported to be needed in order to successfully complete the initial detector configuration, taking into account the staging of selected detector components. Some 70% of these so-called Cost-to-Completion (CtC) were directly associated with the construction and the remaining 30% to commissioning and installation in the ATLAS cavern as of 2003. About 70% of CtC were assigned for common items and 30% to other subsystems.

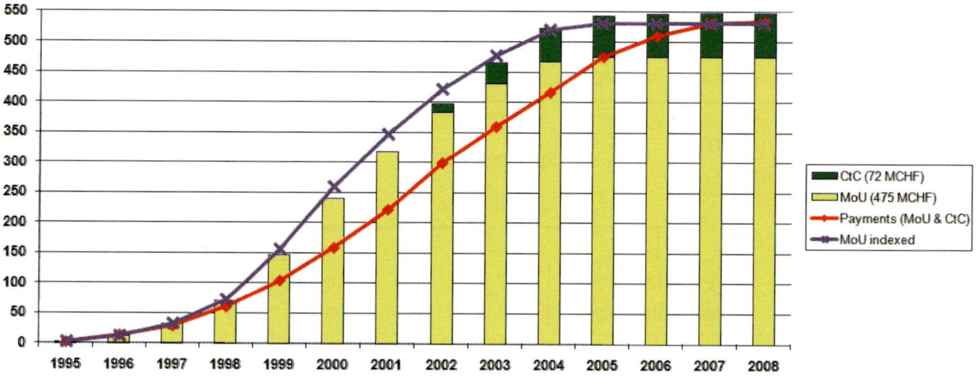

Fig. 4.5. Original CORE expenditures (475 MCHF) in yellow and CtC (72 MCHF) in green. The red line shows the actual spending and the violet line indicates the original CORE budget had it been indexed annually by 2%.

Was this overcost significant? Yes, though the additional CORE cost of 72 MCHF corresponded approximately to an intervening escalation of 2% per annum on the original CORE cost (corresponding to the figure applied for the LHC accelerator). The additional resources needed to cover the CtC were to be committed (as agreed upon by the FAs) on a best-effort basis in two funding stages. The annual distribution of the CtC, over and above the original construction CORE, is shown in the chart in Fig. 4.5.

The CORE and CtC expenditures were monitored and endorsed twice a year by the ATLAS Resources Review Board (RRB). The RRB consists of representatives of the FAs, as well as the ATLAS and CERN managements. Each year, the ATLAS management would present for approval, the expenditures for the previous year and the ones planned for the current and following years for information. The payments did not necessarily coincide with the pledged contributions. To manage this gap, the CERN management put in place a so-called deferral mechanism, whereby remaining original CORE (baseline) investments were to be delayed as long as possible (without harming the completion of the experiment) so as to liberate cash. The resulting cash flow balance became one of the important parameters followed by the RRB. The relationship between the RRB and the ATLAS Collaboration structures is shown in Fig. 4.6.

The handling of the above mentioned overcosts relied on the following key principles:

- Technical problems could be solved where the core competence resided;
- Institutes agreed to share *both* technical and financial risks;

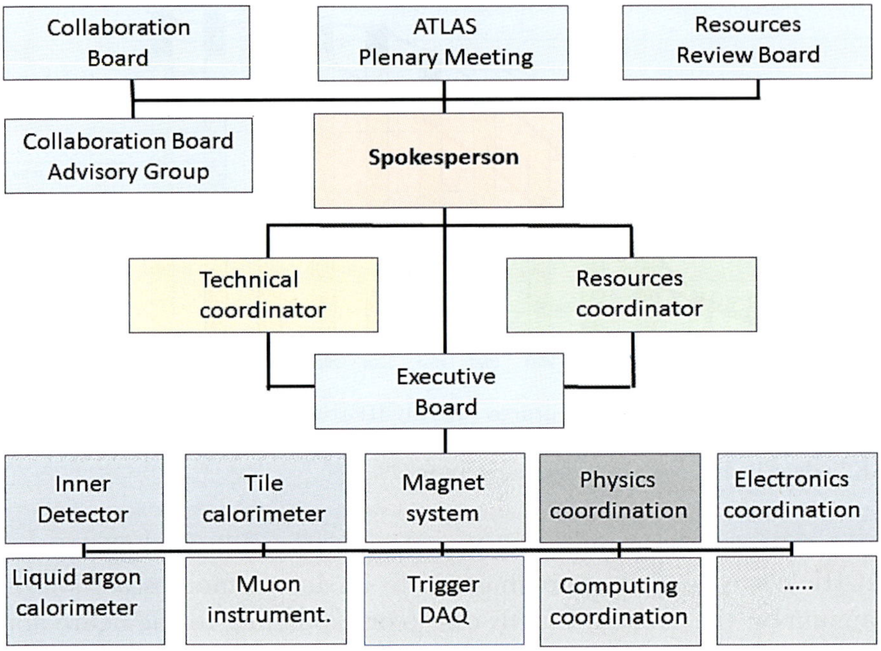

Fig. 4.6. The structure of the ATLAS Organisation during its construction period.

- FAs liberated required resources on a best-effort basis;
- While waiting for these resources to become available, ATLAS postponed or delayed baseline CORE investments wherever possible, without jeopardising the overall success of the experiment;
- A light financial reporting system was accepted (enabled by the CORE value arrangement).

It is worth noting that a special framework was put in place to handle the ATLAS overcosts. Implementing this required great flexibility from the FAs, ATLAS management and CERN. It also helped to pave the way towards addressing the maintenance and operation costs well ahead of time.

4.6. Sharing the operation costs

With the construction of many detector components in full swing in 1999, it became clear that there were rapidly growing costs relating to the testing, installation and operation of the new equipment. These costs were not included in the CORE construction budget, as mentioned before. In the construction MoU it was stated explicitly that these Maintenance and Operation (M&O) costs were still to be defined at a later stage.

Traditionally, the experiments at CERN had to deal with M&O costs only seriously once when the experiments were in a data taking mode. That this issue surfaced so early in the project meant that CERN, the FAs and its collaborators had to reluctantly address it. In fact it was not readily understood by some of the FAs, who while being fully dedicated to managing the investment costs, did not grasp the possibility that the CERN budget would not, in turn, cover the running expenses of the experiments.

In order to get the ball rolling, ATLAS set up an internal working group for M&O in early 2000, jointly by the Collaboration Board (CB) and the management. Its charge was to review M&O cost estimates and to propose a cost sharing strategy. It took several iterations of discussions with the various detector system communities and the CB to get a framework approved in October 2000 within ATLAS, which could then be brought forward to the CERN RRB level as a basis for formal discussions with the FAs. The working group forecasted an annual M&O cost during data taking of 25 MCHF, remarkably close to the reality today.

The M&O costs are categorised into three classes: Class A is for costs shared by the entire Collaboration, related to the overall operation of the experiment, in particular also covering the Common Project components; Class B relates to the operation of detector subsystems and is to be borne by the community responsible for the given component; and Class C denotes the general maintenance and operation provided to the Collaboration by CERN, acting in its role as Host Laboratory. Rather intense discussions and negotiations were held over the scope of Class C, as CERN's budget was very tight given the large LHC construction costs.

It took several iterations until ATLAS, CERN and the RRB finally agreed in May 2002 on the M&O framework (in common for both ATLAS and CMS), which was formally ratified by the RRB in the form of a separate Memorandum of Understanding for M&O. An important ingredient is the way of sharing Class A costs. Several sharing models were initially proposed; the model that was finally adopted is the one sharing these costs "amongst the FAs in proportion to the number of their scientific staff holding PhD or equivalent qualifications, who are entitled to be named as authors of scientific publications". This list is updated annually in September. Note that PhD students are not "taxed", a great thing for the Collaborations as they are doing so much for the experiment! The Class B sharing is left as an internal matter to ATLAS.

The RRB also installed a very severe procedure to check the yearly M&O budget requests, the so-called RRB M&O Scrutiny Group. The yearly cycle

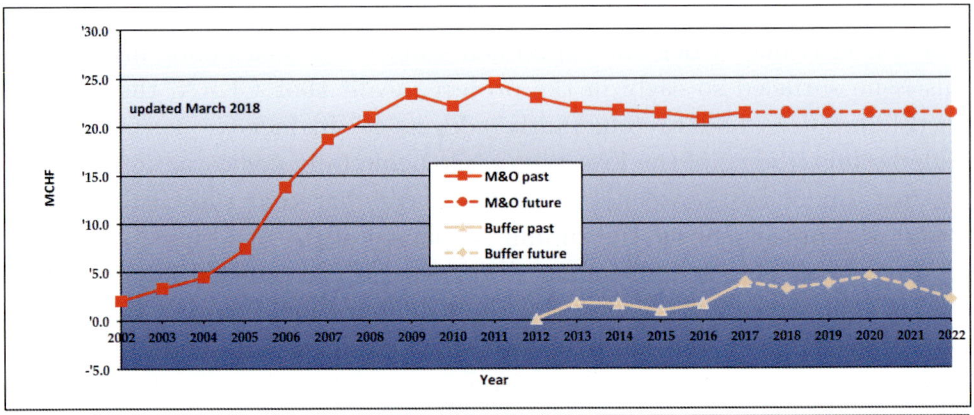

	2002	2003	2004	2005	2006	2007	2008	2009	2010	2011	2012	2013	2014	2015	2016	2017	2018	2019	2020	2021	2022
Category A	1.2	2.5	3.7	5.5	9.4	11.7	14.1	16.4	16.7	18.4	17.9	16.8	16.6	16.8	16.2	16.6	16.6	16.6	16.8	16.8	16.8
Category B	0.9	0.9	0.8	2.0	4.4	7.0	6.9	7.0	5.5	6.1	5.2	5.1	5.0	4.6	4.6	4.8	4.8	4.8	4.6	4.6	4.6
Total (A+B)	2.1	3.3	4.5	7.5	13.8	18.7	21.0	23.4	22.2	24.6	23.0	21.9	21.6	21.4	20.8	21.4	21.4	21.4	21.4	21.4	21.4
TDAQ buffer account cumulative cash balance											0.14	1.73	1.63	0.99	1.65	3.95	3.28	3.74	4.52	3.43	2.11

Fig. 4.7. Evolution and forecast of the total M&O costs in MCHF for ATLAS, including provisions for Trigger and DAQ computing equipment renewals. The breakdown in Class A and Class B is also indicated in the table in the lower part of the figure.

of M&O budgeting, meetings with the Scrutiny Group, and the presentation of preliminary and final budget requests to the RRB, as well as closing reports from the previous year, keeps the Resources Coordinator and his finance colleagues in the detector subsystems busy. Figure 4.7 shows the history and forecast of the M&O costs, now at a constant level of around 21.5 MCHF. Note that Class A are typically $\frac{3}{4}$ of the total (A+B).

The M&O procedure described above provides for the consumable material costs and services, including expert person-power, that needs to be paid for in order to operate and maintain the detector in best running conditions. In detail, the tasks and their budget needs can vary greatly between years with data taking and years within the long shutdown periods of the LHC.

However, a much larger effort is needed from the Collaboration members for this experiment, for example, from those in detector operation and data taking to those in the many technical service tasks along the data flow chain, ensuring for example smooth functioning of the complex computing infrastructure, providing calibration and alignment information, and the many other tasks necessary to "deliver" the data for the physics analyses. The Collaboration became aware of the need to define a framework for all the operation tasks in early 2006, well ahead of the actual data taking. A rather large working group was initiated with the internally famous name of OTSMOU, meaning "Operation Task Sharing and M&O Update". Its membership covered all concerned ATLAS communities, from the CB and management to representatives from the activity areas and detector systems, as well as from the National Contact Physicists. Their mandate was to work out a strategy to ensure work is fairly distributed amongst all Institutions. In February 2007, the CB approved an important document, which remains the basis for today's organisation of the operation tasks across the Collaboration. A sophisticated software tool called the Operation Task Planner (OTP) was developed to document and plan in a fair and transparent way all the work needed to operate, in the largest sense, the experiment.

Luminosity and a van der Meer scan

Whether one is searching for signs of new physics or measuring known processes, it is essential to know the total number of pp collisions delivered by the LHC. A typical example is the production of W and Z bosons: a highly precise measurement of the rate at which these processes take place is an important test of the Standard Model. *Luminosity* describes the number of collisions per second. The denser the proton bunches and the more they collide, the larger the chance that two protons interact, i.e., the larger the luminosity.

The absolute luminosity scale is calibrated by combining simultaneous precision measurements of the LHC bunch currents and of the convolved transverse size of the colliding bunches. Using a technique pioneered by S. van der Meer at the ISR 50 years ago, the inelastic pp collision rate is monitored as the beams are separated first in the horizontal, then in the vertical direction (see figure). This "vdM scan" provides a measurement of the beam-overlap area, which when combined with the numbers of protons in each bunch, determines the absolute luminosity produced in head-on collisions.

The vdM scans are performed in special fills with beam conditions carefully optimised for the accuracy of the luminosity calibration, while remaining within acceptable operational parameters for the accelerator complex. The precision requirements are exacting: the best absolute accuracy achieved at the LHC on the vdM calibration is 1.2%. The success of the procedure depends critically on the strong cooperation, in the accelerator control room, between the LHC operations crew, their instrumentation colleagues, and the ATLAS luminosity experts. The experts in the ATLAS control room are not idle either, juggling multiple trigger menus to maximise the usefulness of the vertex- and track-based luminosity data that are crucial to minimising the systematic uncertainties. To make optimal use of the severely constrained beam time as well as provide additional consistency checks, the ATLAS and CMS calibrations are carried out in the same LHC fill, in a carefully orchestrated dance in which various scan protocols alternate at the two IPs. This makes for quite intense, but also very long calibration sessions: after 16 straight hours, the players in all three control rooms are on their knees.

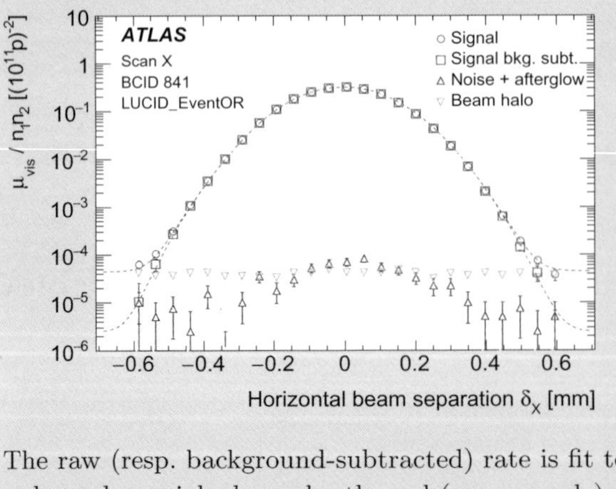

Visible interaction rate per bunch crossing and per unit bunch-population product, for the luminometer LUCID_EventOR, vs. horizontal beam separation. The total rate measured for a single colliding-bunch pair at position 841 in the fill pattern is displayed before (red circles) and after (purple squares) subtracting backgrounds from noise and afterglow (blue triangles) and from single-beam halo (green triangles). The raw (resp. background-subtracted) rate is fit to a Gaussian multiplied by a sixth-order polynomial, shown by the red (resp. purple) curve.

From data to big data in ATLAS

The evolution of ATLAS coincides with the remarkable growth of the internet and its global bloodstream of data. ATLAS and the web are about the same age; Tim Berners-Lee invented the web at CERN in 1989, and months after the creation of ATLAS, CERN in 1993 took the pivotal step of putting the World Wide Web software in the public domain. The design of the LHC's global distributed computing model began around 1998, the year Google was founded and housed in one room at Stanford. It was a visionary model driven by necessity: the computing resources available to the LHC experiments would necessarily be globally distributed, making the unprecedented approach of a global integrated scientific computing infrastructure essential. Nothing like it existed, and it was not clear it could be created; the early models foresaw tapes shipped in airplanes as possibly the most realistic means of moving LHC data across the Atlantic. But then the "internet bubble" of 2000–2001 happened, and with its subsequent collapse was a tremendous boon to LHC computing. The world was wired up with fibres that then became cheap with over-supply after the bubble burst, making a fully networked global LHC infrastructure affordable. With the creation of the Worldwide LHC Computing Grid (WLCG) in 2001, LHC computing was on its way.

And how far we have come! By the time "big data" became a buzz word in 2013, ATLAS was big data by any standard. The 150PB ATLAS data set after Run-1 was similar in size to the Google search index (100PB) and yearly Facebook uploads (180PB), with 2 million files moving across the ATLAS global grid every day. ATLAS processing has been "Exascale" since 2014 when the annual volume of processed ATLAS data first exceeded an exabyte; in 2017 it reached 1.6EB. The 300PB ATLAS data set today is ten times the size of the openly accessible web, estimating from the 30PB size of the Internet Archive's data set (https://archive.org/about/).

Today, physics analyses in ATLAS and the other LHC experiments rely on the largest distributed data intensive scientific computing infrastructure ever built. The phenomenal scale today compared to the late 1990s when LHC computing was first being designed can be seen by looking back to the precursor collider in the present LHC tunnel, LEP. The entire data sample of the LEP experiments was about a Terabyte, a data volume that the distributed computing system of ATLAS processes every three minutes.

ATLAS and CMS today are preparing for the next leap in scale that the HL-LHC will bring in 2026, when the annual raw data volume alone will be about 600PB, and the total annual sample about an Exabyte. Once again, vision and creativity and perhaps some good luck will be the ingredients enabling us to build a capability that we couldn't possibly build and pay for today. Fortunately for us, "big data" is moving from the exotic to the mundane, as data of all sorts grow exponentially in the wider world, making our needs comparatively less extreme.

Chapter 5

Detector Construction Around the World

5.1. The magnet system

The construction of the ATLAS magnet system was a great effort of many collaborators worldwide. In normal circumstances, such a magnet project would have included a tendering process resulting in a consortium of companies that would take care of production design, procurements, integration and testing. However this would require all financial resources to be present and collected in a bank account in Geneva, from which the deliveries could be paid. This was not so in ATLAS. Since cash was only available for some 15% of the project cost, the decentralised resources of the funding agencies interested in the magnet system had to be used, which ranged from a few hundred thousand to millions of Swiss francs.

The solution found was to split the magnet system up into smaller units, which could then be covered by the budgets available at the various funding agencies over the construction period. At first impression, this looked like a recipe for disaster, as many of the additional technical and managerial interfaces between collaborating companies and institutes and their workshops had to be controlled, which in principle could lead to more risk and potential delay due to inefficiency. But on the other hand, this manner of system procurement could be a brilliant solution, as it allows to deliver a 150 MCHF project at a lower cost, since financial reserves are decentralised and substantial centralised cash does not have to be available.

Let us illustrate the case for the Barrel Toroid, by showing the procurement circumstances for the main parts, as presenting the details in full is well beyond the scope of this book! As explained above, in a conventional project tender, a single leading contractor in a consortium would take responsibility and manage all procurements and deliver the system to CERN, ready for installation. Not in this case. In the project's set-up all manufacturers were only responsible for their own deliveries; the ATLAS magnet team at CERN handled the overall financial and technical responsibility. When going from inside out, from the conductor to assembled toroid, we see the following procurements, with the percentage of the system cost in brackets:

- Production of superconductor (18%) was a 50–50 effort of funding agencies from Germany and Italy, using many companies in the two countries for the delivery of main ingredients comprising superconducting wire and billets of pure aluminium, conductor cabling and co-extrusion at two sites, and quality control by yet other companies and the supervising magnet laboratories CEA, INFN and CERN.
- The manufacturing of 16 coil winding packs (17%) was fully funded by Italy, and make in an Italian company.
- The eight cold mass structures called coil casings (22%) were funded by Switzerland and Germany. The sixteen coil packs and eight cold masses were delivered to CERN for integration, which was performed by a company, thus delivering eight cold masses ready for integration with the eight cryostats.
- A cryostat comprises an outer vacuum vessel, a radiation screen and MLI (multi-layer) superinsulation, and cold supports locking the cold mass in the vacuum vessel.
- The cold mass supports have two families, 17 stops and 8 tie rods per cryostat. The stops (1%) were made at a small company in France, while the special and expensive titanium tie rods (2%) were delivered in-kind by Russia, benefitting from an ISTC project (see Sec. 3.1).
- The thermal radiation shields for all eight cryostats (3%) are in-kind from Italy and were made by an Italian company. The superinsulation blankets and installation (1%) were in-kind by an institute in Russia.
- The vacuum vessels (10%) were in-kind from Spain and manufactured there. However, the stainless steel used was delivered in-kind from Sweden by a Swedish company. Eight cryo-ring sections and a cryogenics valve box (3%), which guide bus bars for interlinking coils and helium cooling lines, interconnect the eight cryostats.
- When the eight coils were completed, bolting the parts together into a toroid took place in the cavern using the so-called warm structure comprising an inner and an outer ring of very special aluminium alloy struts (8%), delivered in-kind by Russia and Belarus through forging and

H-profile extrusion and welding of connection flanges at a company in the Netherlands.

The coil parts listed above were delivered to CERN where all integration and coil testing on the surface prior to installation (4%) in the cavern took place.

It was a deliberate and conscious choice to do all integration activities (12%) at CERN under the direct responsibility of the magnet team, in order to minimise technical risk and to control cost. Figure 5.1 shows the vast building at CERN used for the Barrel Toroid coil integration; the various parts can be identified. The team was supported in these works by the engineering laboratories involved — CEA, INFN and JINR providing most of the manpower and manufacture of all large scale tooling, as well as a few companies for mechanical integration services and others specialised in welding repairs and quality control.

Without going into detail here, a similar procurement organisation for the manufacturing of conductor, coils, cold masses, cryostat parts, integration and testing was set up for both Endcap Toroids. However, this was supported by other funding agencies, in particular from the Netherlands and Israel, and other companies, but again, leaving the integration of coils

Fig. 5.1. The Barrel Toroid cold mass and cryostat integration area in building 180 at CERN, showing several stages in coil integration from cold mass (bottom) to cryostats (upper right).

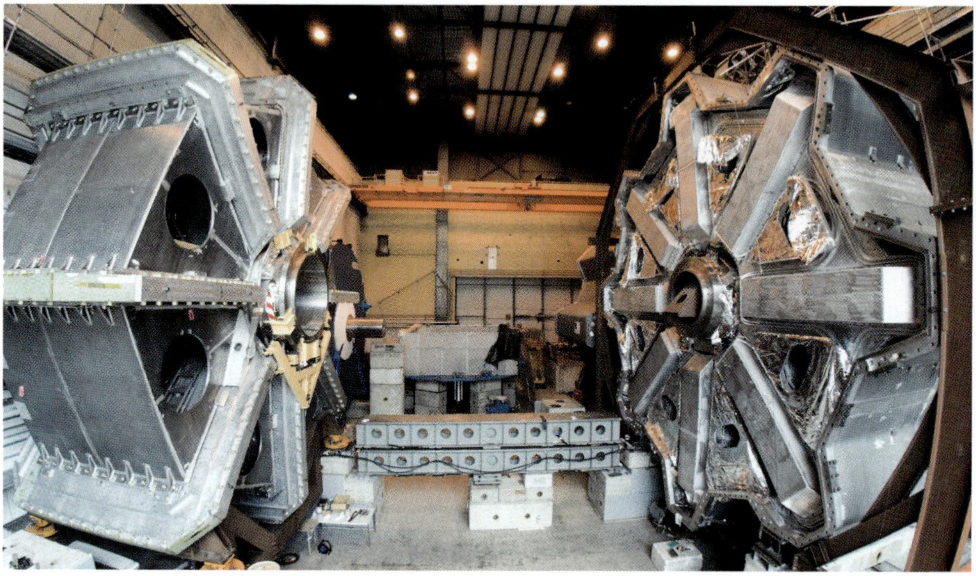

Fig. 5.2. The Endcap Toroid cold mass and cryostat integration area in building 191 at CERN, showing the assembled cold mass (left) and the cold mass in cryostat integration (right).

and cryostats at CERN for the same good reasons. Again, the integration activities were performed by the ATLAS magnet team, supported by the magnet engineering laboratories RAL in the United Kingdom and NIKHEF in the Netherlands, as well as JINR in Russia, which provided most of the tooling and integration manpower. A flavour of the integration activities for the Endcap Toroids is shown in Fig. 5.2.

An exception was the procurement of the central solenoid, shown in Fig. 2.5 of Chap. 2. Its complete cold mass and proximity cryogenics were an in-kind contribution of Japan, with engineering leadership coming from the KEK laboratory. The solenoid was manufactured almost entirely and pre-tested in Japan, whereafter it was shipped to CERN and integrated with the barrel calorimeter. In this configuration, an ultimate test was performed before its installation in the experimental cavern in 2005.

5.2. The Inner Detector and its subsystems

The initial Inner Detector, as discussed in Chap. 2, comprises three subsystems using different technologies at different radii: the Pixel Detector utilising silicon pixel sensors, the silicon-strip tracker ("SemiConductor Tracker" or "SCT"), and the Transition Radiation Tracker ("TRT"). While the design of the Inner Detector was optimised for its overall performance, the construction of the different subsystems was carried out by separate teams of institutes. Some of the services for the different systems are shared, for example, the off-detector ("back-end") read-out electronics use the same devices for the Pixel and SCT systems. Further discussion of these is beyond the scope of this chapter. A new, innermost, layer of pixel sensors was added during the first long shutdown of the LHC; the construction and installation of this upgrade is discussed in Chap. 10, this chapter reflects the original Pixel Detector work.

5.2.1. *The Pixel Detector*

An extensive R&D programme was pursued to design the pixel sensors and on-detector ("front-end") electronics. The sensors needed to satisfy physical size constraints on thickness and granularity, as well as to keep a high charge-collection efficiency for charged particles, even after exposure to a high dose of radiation. The former constrained the design of the pixel sensor structure while the latter dictated the choice of silicon-sensor material. A variety of different designs were extensively tested during the prototype phase, including different geometries, implant-isolation technology and dopant concentrations. Quality control of sensors consisted of mechanical and electrical inspections and tests, including checks of alignment, planarity and thickness of wafers, current and capacitance and leakage current measurements. Prototype sensors were tested during irradiation with particle fluences of up to 10^{15} neutrons per cm^2 and were later extensively characterised in laboratory set-ups and test beams. Oxygenated silicon was chosen for the bulk, and the so-called "moderated p-spray" technology was used for pixel isolation.

Sensors were fabricated by two vendors (CiS and ON Semiconductor); quality assurance procedures were deployed for the production of sensors, carried out as a collaborative effort at four different pixel sensor institutes spanning the Czech Republic, Germany, Italy and the United States. An extensive cross-calibration of mechanical and electrical measurements was performed before the production, to ensure uniform quality assurance procedures. More than 2500 sensors passed the quality assurance process required before they could be attached to the front-end electronics.

The design of the pixel front-end electronics also had to satisfy multiple requirements: operation at high radiation doses, a time resolution of 25 ns to separate hits from two consecutive LHC bunch crossings, low noise, a low charge threshold and dispersion, and satisfactory power consumption. The first front-end chips with the required analogue and digital architecture were developed in the second half of the 1990s. Several subsequent chips were developed using different technologies (BiCMOS and CMOS), first in a radiation-soft version, demonstrating all the basic performance goals and then in radiation-hard technology. The failure of two traditional vendors to produce radiation-hard chips pushed the Collaboration toward the deep sub-micron technology, based on a commercial $0.25\,\mu$m CMOS process with radiation-tolerant layout. The design for the final chip (known as the "FE-I3") was ready in late 2003.

The Pixel Detector contains a total of 27,904 front-end chips, which had been fabricated by IBM in six batches of 48 wafers each. The wafers were tested using semi-automatic probe stations in two institutes in Germany and the United States. Each chip was extensively characterised, including the measurement of a reference calibration capacitor value using a dedicated circuit that was only available during this initial wafer probing. The test time was approximately 30 hours per wafer. The selection of good chips was made based on the evaluation of 30 analogue and digital parameters. The average yield was about 80%. Each chip was also probed after dicing the wafers using custom vacuum chucks to hold up to 60 chips. About 2% of these chips failed after the dicing process. A similar R&D programme was carried out for the module controller chip ("MCC"), as well as other ancillary electronics.

The Pixel Detector consists of 1744 modules — 1456 in the three-barrel layers and 288 in the three disks in the two endcaps, for a total of around 80 million pixels, with a total sensitive area of $1.7\,\mathrm{m}^2$. Each module consists of a sensor attached "face-to-face" to sixteen read-out chips, using the so-called "flip-chip" technology, with an array of tiny metal "bumps" making the electrical connections between the sensor and read-out chip planes via a technique known as "bump-bonding". A fine-pitch, double-sided, flexible printed circuit (flex-hybrid) contains an MCC and routes signals and power. Two different bump-bonding techniques were used in nearly equal proportions: electroplated-solder (PbSn) bumping at IZM-Berlin and evaporative-indium bumping by SELEX-Rome, with bump defect rates of 10^{-5}–10^{-4} at the wafer level and 10^{-4}–10^{-3} after the flip-chip process.

The bump bonding and flip-chip operation results in a so-called bare module. All front-end chips of bare modules were electrically tested on a probe station. Rework on modules could be done at this point of the assembly sequence, with a success rate of more than 95%. Once a bare module passed the acceptance test, it was equipped with a flex-hybrid to provide the connections between the module controller chip and the front-end electronics and a micro-cable. Such assembled modules were tested, both electrically and mechanically, at room temperature and at the operation temperature of approximately –10 C, before and after thermal cycling. The assembly of a module and its subsequent characterisation took about seven days. Therefore the module assembly and testing capability was replicated at six production sites, spanning France, Germany, Italy and the United States. Out of the ~2500 bare modules, about 2300 were assembled and tested, of which 2100 were accepted and ranked for mounting on the detector — the best modules were loaded on the innermost layer, leaving the lowest-ranked modules to be set aside as spares. Accepted modules were shipped to specific sites, one in the United States (for the endcaps) and three in Europe, i.e., France, Germany and Italy (for the barrel), where they were mounted onto high precision support structures. In the barrel region, 13 pixel modules were mounted on a linear structure, the "stave". For the endcap, six pixel modules were directly mounted on each sector, the endcap equivalent of a stave, as shown in Fig. 5.3. These support structures were equipped with cooling and electrical services for the modules.

In the barrel region, the staves were mounted in a thin carbon fibre half-shell as illustrated in Fig. 5.4. Two fully loaded half-shells formed a barrel

Fig. 5.3. Pixel disk sector during the attachment of modules.

Fig. 5.4. A pixel barrel half-shell being loaded with barrel staves and services.

layer, as shown in Fig. 5.5. The disks of the endcaps were positioned onto a carbon composite support ring (Fig. 5.6). The final integration onto the pixel support tube was done at CERN, for insertion into the SCT in 2007.

5.2.2. *The SemiConductor Tracker (SCT)*

An extensive R&D programme had been required to establish sufficiently radiation-hard semiconductor strip sensors for operation in LHC conditions, building on the experience from previous generations of detectors at the LEP, Tevatron and HERA experiments. Based on this work, including extensive irradiation studies, the strip sensor technology chosen uses p-doped strips on an n-doped bulk silicon substrate, essentially because it was the lowest-cost option with the required radiation hardness. Another challenging technology choice for the strips was for the read-out where analogue, digital and binary options were all initially considered. Cost played a key role

Fig. 5.5. Outermost barrel pixel layer, loaded with staves, after the joining of the half-shells.

Fig. 5.6. A pixel endcap at the last stage of assembly, after connection of its cooling circuits.

in the selection of a binary read-out, despite there being rather limited operational experience with it in previous detectors. The DMILL process, a mixed analogue-digital radiation-hard BiCMOS technology, was chosen for the main on-detector read-out device, the so-called "ABCD" chip. Cooling, opto-electronics, module designs and supports all also proved to be highly demanding areas.

The SCT overall structure consists of four concentric barrels, together with nine ring-shaped discs forming each endcap. Each layer (barrel or disc) is composed of a large number of individual detector modules: the modules in the barrel are all of the same design, while those in the endcap fall into three types, the so-called "Inner", "Middle" and "Outer" forward modules. There are a total of 4088 modules, most including four silicon sensors. The modules are typically $12 \times 6\,\text{cm}^2$ in area. The module construction required very tight tolerances in terms of thermal properties and robustness against thermal cycling, sensor front-back alignment to $5\,\mu$m (the most difficult aspect) and electrical properties, including projected performance after irradiation. Extensive quality assurance procedures were put in place, which were common across production sites.

For the barrel modules, the design is shown in Fig. 5.7: the four sensors are mounted on a thermally conducting thermal pyrolytic graphite ("TPG") baseboard, two electrically daisy-chained sensors on each side with a 40-mrad stereo angle between the strips on the two sides, and the hybrid read-out circuit is mounted around that. These sensors were supplied by Hamamatsu Photonics and the baseboards were constructed by ATLAS at CERN. The sensor-to-baseboard, and hybrid-to-module, assembly steps were delicate operations requiring high placement precision and reproducibility, and precise metrology at the μm level. A further delicate step was to make the (more than 3000) electrical connections between the sensors and from sensor to hybrid, using ultrasonic wire bonding. Barrel module assembly was carried out in four geographical "clusters": one in Japan, one in the United Kingdom, one in the United States and the remaining one in Norway/Sweden, each with an assembly site providing ABCD-to-hybrid assembly and testing, and module testing sites. Almost all the assembly and testing were carried out in collaborating institutes, with one cluster making use of an industrial partner. Extensive work was required to establish detailed quality assurance procedures, ensuring consistent production standards and performance. An overall barrel module yield of just above 90% was obtained.

The forward module design shares the main features of a TPG baseboard with sensors mounted on each side with a 40-mrad stereo angle, plus a hybrid

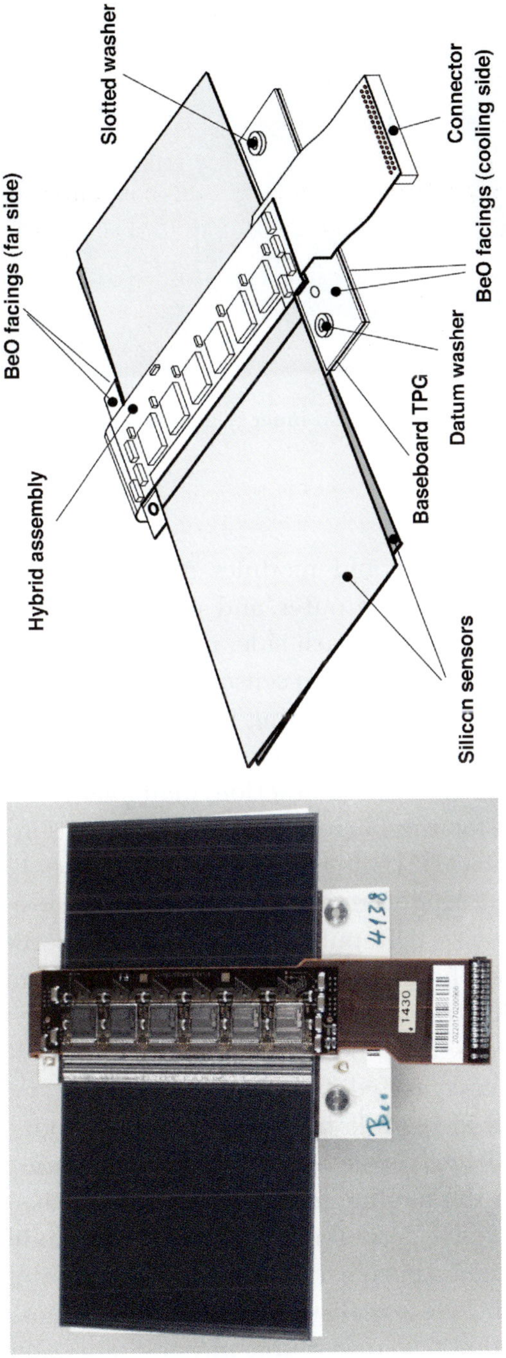

Fig. 5.7. A photograph (left) and schematic drawing (right) of the SCT barrel module.

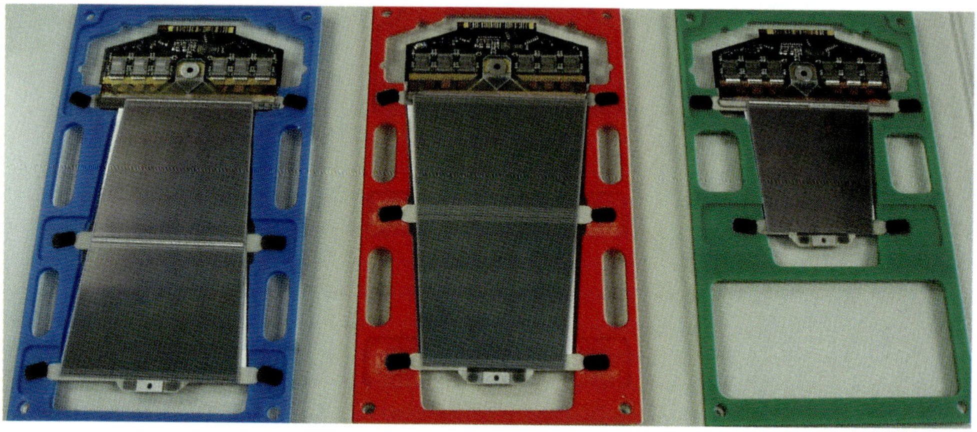

Fig. 5.8. Outer (left), middle (centre) and inner (right) forward modules, shown mounted on handling frames.

read-out circuit. The sensors and modules are wedge-shaped for efficient tiling of an annular disk. In the outer and middle forward modules, there are two daisy-chained sensors on each side, making four sensors per module, whereas for the inner module only two sensors are mounted. Figure 5.8 shows the three forward module types requiring five sensor variants. The hybrid is placed at one end of the module. The forward sensors were mainly fabricated by Hamamatsu Photonics, with a fraction manufactured by CiS Institut fur Mikrosensorik. As for the barrel modules, construction was distributed across geographically-based clusters of production sites, and hybrid and module assembly and testing were done in collaborating institutes, spanning Australia, CERN, the Czech Republic, Germany, the Netherlands, Poland, Spain, Switzerland and the United Kingdom. The significant time duration of the production is illustrated in Fig. 5.9, which shows the forward module completion count versus time between 2003 and 2005. The slow ramp-up at the start indicates the difficulties sites had in meeting stringent qualification standards in module production before production parts were released in quantity to any given site; the required $5\,\mu$m precision front-back alignment within the module having proved to be the most difficult to achieve reliably. The "pass" category corresponds to modules just outside the initial and very tight specifications, where with better understanding of design robustness and more realistic alignment strategies using tracks, we realised that there would be no discernible physics impact by accepting these modules.

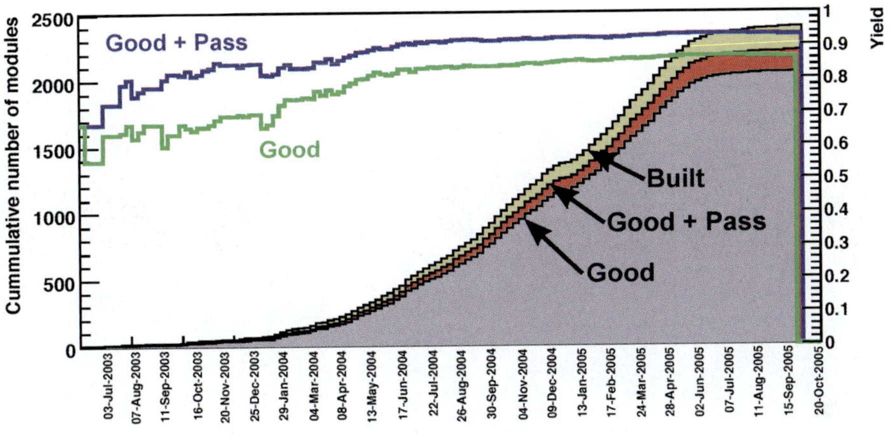

Fig. 5.9. Forward module production over time (solid), together with the cumulative overall yields of accepted modules (lines). The "Pass" category indicates a lower quality module than a "Good" one, but which is still acceptable for use in the detector.

The 60,000 ABCD ASICs for the SCT were manufactured in the DMILL 0.8 μm feature-size mixed analogue-digital radiation-hard technology. Other technologies were kept as common as possible between the barrel and forward modules, supports and read-out, subject to the differences in geometries. For example, both communities used copper-polyimide flexible circuits for the hybrid circuits carrying the module read-out ASICs, laminated to carbon-carbon substrates to provide thermal and mechanical stability.

Accepted modules were shipped to specific sites, one in the Netherlands and two in the United Kingdom, where they were mounted onto the barrel or endcap high-precision carbon-fibre support structures. Another major task was to equip these support structures with the required cooling and electrical services for all the modules. One assembled barrel and one endcap are shown in Figs. 5.10 and 5.11, respectively. The individual barrels and the two endcaps were shipped to CERN for final assembly and insertion within the TRT.

Overall, during design, construction, integration and commissioning phases, a whole range of challenging problems were found and fixed, although to describe them in detail is beyond the scope of this text. Nonetheless, the SCT has proved to be remarkably robust in operation at the LHC. In particular, the mechanical rigidity is superb as there have been no losses through many temperature cycles, with the fully working strip count at the time of writing still at 98.7%, and read-out performance remarkably similar to that at the start of LHC data taking.

Fig. 5.10. One of the four SCT barrels: the central portion of the structure shown is tiled with barrel modules, the silicon sensors can be seen (grey).

Fig. 5.11. Forward discs assembled into one of the SCT endcap structures.

5.2.3. *The Transition Radiation Tracker (TRT)*

The TRT concept follows extensive R&D carried out between 1990 and 1993 by the RD6 project. The TRT detector subsystem provides continuous tracking with about 35 precise ($\sim$150 μm) measurement points along the tracks of charged particles within its geometrical acceptance, complemented by information about the amount of transition radiation associated with the particle. Such a device provides a significant contribution to the particle momentum measurement, to fake track suppression, and to electron identification. The TRT detector is composed of a barrel and two endcap assemblies. The barrel and endcaps have a different design and can be considered as different detectors. They use common components such as straws, wire and electronic chips.

The basic detector element of the TRT is a cylindrical thin-wall proportional chamber (straw) of 4 mm diameter wound from a specially designed multilayer plastic film containing a composite layer of carbon and polyimide, which protects a thin aluminium layer deposited on a 25 μm thick Kapton film. This design retains the good electrical properties of the chamber cathode while protecting it from electrochemical erosion caused by positive ions during LHC data taking. The straws are filled with a $Xe/CO_2/O_2$ (70/27/3%) gas mixture. A crucial innovation for the entire TRT was to strengthen the straws by gluing four carbon-fibre filaments along the outer straw wall. This allowed the straws to be used as a part of the mechanical structure, providing an important reduction in the amount of material in the active detector volume. The initial development of the straws was carried out in Russia, as were most steps of their manufacturing process.

In order to operate at a charged particle rate per straw of a few tens of MHz with a sufficiently low threshold, a dedicated analogue chip was needed. It is followed by another chip that digitises the signals and sends the data to off-detector electronics. These two chips are housed on front-end boards, as shown in Fig. 5.12. The chips were the responsibility of groups in the United States, Sweden/Denmark and CERN. Digital parts of the front-end boards are shown on the left and right side of this figure as installed on the two ends of the barrel. The middle set of rectangular boards is from the TRT endcaps: analogue at the top and digital at the bottom. A careful grounding-shielding concept developed for the TRT and implemented at each step of the design, construction and installation, allowed us to reach an operating threshold very close to the thermal noise limit in real operating conditions. Institutes from Denmark, Poland, Sweden, the United States and CERN took part in the read-out and control systems production.

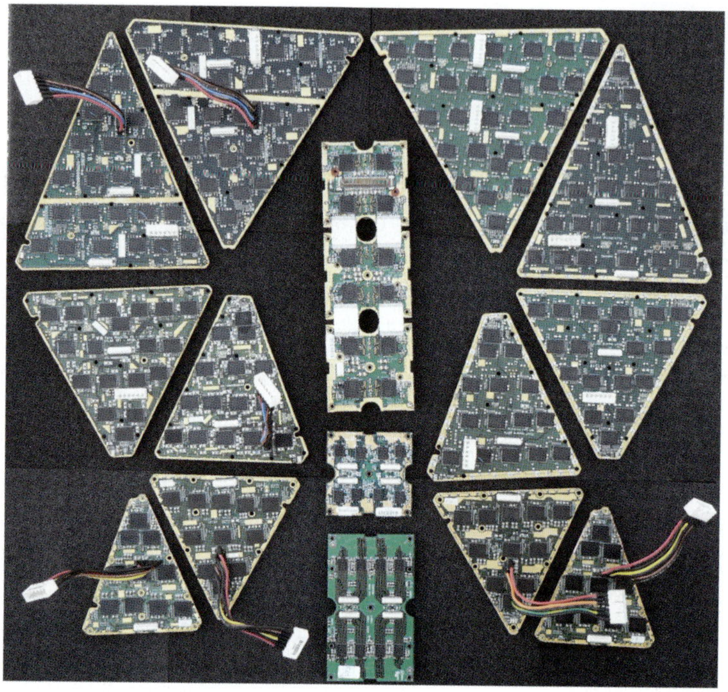

Fig. 5.12. TRT front-end electronics boards.

The barrel TRT contains 52,544 axial straws of about 144 cm length. It consists of three cylindrical rings, each containing 32 identical and independent modules. All the straws are embedded in stacks of polypropylene/polyethylene fibres that produce the transition-radiation X-rays used for electron identification (Fig. 5.13, right). The distance between the straws in barrel modules is on average 6.8 mm in both the radial and azimuthal directions. The straw layers are shifted relative to each other to ensure the best uniformity of the number of straws crossed by a particle. In order to reduce the occupancy, the wire in each straw was split into two parts using a glass wire joint and both sides of the barrel have independent read-out electronics, as shown in Fig. 5.14. The geometry and mechanical stability of the barrel modules are defined by a 400 μm thick carbon-fibre composite shell. The three layers of TRT barrel modules are supported at each end by the end-frames of the barrel support structure (Fig. 5.13, left), which also plays a role in the SCT barrel support. The process of assembling the barrel modules was done at three institutes in the United States, using straws shipped from Russia.

Fig. 5.13. (Left) Installation of the TRT barrel modules into the barrel support structure. (Right) Components at the end of the barrel modules (top) and the barrel module ready for installation (bottom).

The TRT endcap has two parts, with 122,880 straws each, one on each side of the barrel TRT. Each consists of two sets of identical and independent wheels (Fig. 5.14, right) where straws are positioned in a radial direction. The distance between straws in the R-φ plane varies from about 5.2 mm at the innermost radius of 640 mm to 8.4 mm at the outer radius of 1030 mm. Each wheel is assembled from identical sub-wheels of four straw layers (Fig. 5.14, left). The endcap module assembly was carried out in Russia, including JINR.

Successive straw layers in the endcap wheels are rotated from one layer to the next by 3/8 of the azimuthal straw spacing in a given layer. This provides optimal uniformity in the number of crossed straws for high-p_T tracks. The free space between layers is filled with 15 μm thick polypropylene foils separated by a polypropylene net that acts as a transition-radiation radiator (Fig. 5.14, lower left). Two sets of wheels have a different straw density in order to maintain the number of crossed straws at the required

Fig. 5.14. (Left) TRT endcap four-plane wheel during assembly. (Right) Two sets of identical, independent wheels for one part of the endcap, with connected front-end electronics boards.

Fig. 5.15. (Left) System test of the Barrel TRT and SCT after integration. (Right) Map of the noise level of all channels on one side of the barrel TRT during commissioning in ATLAS. Overall, the noise level is below 2%.

level but to keep the amount of material of the active detector approximately constant as a function of pseudorapidity η.

After assembly and integration with the SCT at CERN, extensive system tests were performed to demonstrate mutual noise immunity and that the performance reached expectations (Fig. 5.15, left). At the same time, the control and cooling systems were commissioned. Extensive tests helped to catch and fix problems in parallel with the assembly. A noise map for all

(52,500) channels on one side of the barrel TRT is shown on the right hand side of Fig. 5.15. Channels shown in green correspond to a noise occupancy level of 1.8% with a threshold of 250 eV in energy, satisfying the initial design goal (below 2%).

5.3. The LAr accordion electromagnetic calorimetry

As described in Chap. 2, the ATLAS electromagnetic (EM) calorimetry is based on an innovative lead/liquid argon technology, where the lead absorbers have an "accordion" shape, illustrated in Fig. 5.16. The central solenoid is integrated in the liquid argon barrel cryostat inter-wall space, with a separate vacuum. The design was strongly influenced by results of beam tests of a "two-metre prototype", which had most of the features of an EM barrel module over a 2 m length and of the "Spanish Fan" endcap prototype, as well as by extensive simulations. Features include a very fine granularity in pseudo-rapidity, η, in the first compartment of the accordion to help identify photons and reconstruct their direction, while the recovery of energy lost in the cryostat makes use of a separate presampler ("PS") detector with a coarser division.

Fig. 5.16. A stack of lead absorbers, honeycomb spacers and Kapton electrodes in an endcap LAr Calorimeter module.

The electronics chain employs an innovative solution with only warm electronics for the EM. The digitisation architecture is based on an analogue switched capacitor array ("SCA"). Inputs to the first-level trigger are based on on-detector analogue "tower-builders", whose output is led outside the cavern by cables (some of us wanted to be able at least to see trigger signals on an oscilloscope, in case of a large unexpected noise) which turned out to be extremely useful.

The mechanical support of the barrel EM has light outer and inner support rings. Bars along the edge of the absorbers use an innovative material with a tailored contraction coefficient different in the transverse and longitudinal directions. The EM endcap structure employs novel moulded bars that are precisely indexed with carbon fibre pins on epoxy rings on the front face to minimise material. Special care is taken regarding the mechanical behaviour during cool-down. The cryostat and feedthrough production are described in Sec. 5.5. The cryogenic system design emphasises temperature homogeneity and liquid argon purity.

A lot of effort was devoted to the design of the tooling, assembly procedures and quality control. The tooling included the absorber bending machines (barrel, end-cap), with their beautiful kinematics, lead cleaning, absorber gluing machines, module assembly benches, barrel and endcap assembly and rotation devices, etc. From the start, the full assembly procedure was considered and the integrated quality control was defined, constantly checking the impact on physics performance.

The liquid argon calorimetry system involved a relatively small number of institutes, as compared to most other systems, but included several large ones with strong technical teams, which eased the construction process and logistics, and allowed fast reaction to problems. Everyday communication between engineers, technicians and physicists allowed quality control to be implemented from the very start of the design and to be well accepted during construction. Measurements and tests were thoroughly performed, also by technical teams and physicists, during all construction phases, providing a wealth of knowledge and data that could be later used for operation and analysis.

5.3.1. *Common procurements from industry*

The lead absorber plates were ordered from a medium-sized company in Germany, which was more accustomed to rolling lead plates for buildings or radiology shieldings. ATLAS physicists came with a custom-adapted

X-ray system for the measurement of the lead thickness, to be adapted on the rolling mill. The operator of the mill was incredibly proud of adjusting the thickness to as constant a value as he could with this new, sophisticated tool!

However, the Kapton electrodes had a bumpy ride: the procurement of the polyimide base material proved difficult, as did the fabrication of the electrodes and the ink resistors, and the bending process. Other components also had various issues: the epoxy bars mentioned above could not be machined to the correct accuracy (20 μm) by the first company, the paper/polyimide honeycomb spacers, derived from aircraft technology but with difficult shapes and tolerance, were shown to be out of tolerance at first, and contained metallic dust.

5.3.2. *Construction of the modules*

The absorbers were processed at a French institute for the barrel and in a Spanish company for the endcaps. For the barrel, each of the 4096 lead plates had its thickness further measured on a $7.5 \times 7.5\,\text{cm}^2$ grid, a daunting job even with a specifically designed automatic machine. Using these measurements, a computer program optimised the arrangement of the plates in the calorimeter, so that an electromagnetic shower would see both "thin" and "thick" plates, and the overall response dispersion was minimised over the whole barrel. Then, each absorber was bent, glued, cured and measured. The Kapton electrodes from the factory were bent into the accordion shape and then tested in particular for the integrity of the delicate ink-resistors. The "cold electronics", i.e., cables, signal routing boards, etc., were also ordered from industry and measured in the laboratories. With all the components available, the barrel modules were assembled in three sites (two in France and the third at CERN) and endcap modules, in two sites in Spain and France. All 32 barrel modules were functionally tested in local cryostats in the assembly laboratories while the 16 endcap ones were tested at CERN. Many measurements were performed during the assembly: mechanical, electrical, high voltage and so on. The start was not very smooth: single planes had high voltage problems (solved by a better cleaning of the spacers) and the first barrel modules had a bad tendency for obesity (cured by a better electrode bending). After the first few modules, the rate increased to nominal, even stressing the logistics flow and the modules were sent to CERN for later assembly. Stacking of the barrel and endcap modules are seen in Figs. 5.17 and 5.18, respectively.

Fig. 5.17. Stacking a barrel module.

Fig. 5.18. Stacking an endcap module.

In parallel, teams from France, Morocco and Sweden built the PS electrodes (another stressful beginning), then assembled and cold-tested the PS modules, with an intense but smoother path. Similarly, a Russian institute produced in a very efficient way the endcap PS that was then the first to be inserted on the endcap cryostat front cold wall. Four (of 32) barrel modules and three (of 16) endcap modules were tested in a high-energy beam shortly after their construction, giving confidence that the construction process delivered the required performance.

5.3.3. *Assembly at CERN*

Each of the half-barrels was vertically assembled from 16 modules in an assembly hall at CERN. There had been some concern that the wheel would not "close". Because of this, a virtual wheel had been built in advance, accurately based on the dimensions of each module, and as each module was added to the previous ones, the overall dimension of the assembly was measured with the kind help of the geometers and checked against the virtual version. In the end both wheels did close smoothly (Fig. 5.19). The next step was the spectacular rotation of a full half-barrel with its support frame; in total a 70 tonne object. Then the PS modules were added, and the first half-barrel was inserted into the cryostat and connected to the feedthroughs.

Fig. 5.19. (Left) Closing a half-barrel. (Right) Assembly of one of the two endcap wheels.

Finally, after all modules were completed and assembled, the second half-barrel was in turn inserted in the cryostat and connected. Each endcap was assembled horizontally as a disk and then vertically lifted with a rotator. Next, for each endcap, the PS was inserted into the cryostat and connected to the feedthroughs, followed by the insertion of the endcap EM calorimeter disk and its tedious connection to the feedthroughs (Fig. 5.20). Finally the huge hadronic wheels were inserted.

During the whole assembly time at CERN, and even more with the EM calorimeter sitting in the cryostat, numerous measurements were made at room temperature as well as at cold temperature close to the nominal one, recording the response of every channel to electrical pulses, analysing reflections, etc. Cooling procedures of these unique objects were also tuned at that time. Quality control efforts, which had been performed right from the start and at all steps of the construction, assembly and early commissioning, paid off dramatically in our deep understanding of the calorimeter performance.

The cryostats were then closed and individually transported to the experimental site at LHC Point-1 on special trucks. The sight of the barrel EM calorimeter (Fig. 5.21) being lowered into the ATLAS cavern and detector was a very intense moment for the group and for the

Fig. 5.20. Connecting an EM endcap to the feedthroughs in the cryostat.

Fig. 5.21. The barrel LAr calorimeter, solenoid and cryostat assembly being lowered into the underground cavern.

Collaboration. The endcap cryostats were then lowered and inserted into the ATLAS structure. These two processes provided the perfect opportunity to capture some of the most beautiful and emblematic photos of the ATLAS construction concluding a major step in its building.

5.3.4. *Construction and installation of the electronics*

In parallel to the construction of the calorimeters, electronics that were needed to read out the >182,000 channels were produced and installed. The design and production of these electronics was part of a very wide collaboration, with the boards built in institutes in Canada, CERN, France,

Germany, Italy, Slovakia, Sweden, Switzerland and the United States. The electronics construction started with the "cold passive electronic" boards and kilometres of Kapton-covered shielded signal cables (explaining the golden hues in the photographs of the calorimeter), all located in the cryostats on or near the detectors. Then followed the feedthroughs, specific delicate objects connecting the inside to the outside for the signal, high voltage and monitoring cables. A large number of per mille temperature probes were carefully prepared and measured by an institute in Russia, and were spread throughout the detectors to provide precise monitoring of the liquid argon temperature.

After many prototype tests for radiation resistance, all the electronic components — preamplifier shapers, switched capacitor array analogue memories ("SCA"), optical transmitters (optical devices based on a then-emerging VCSEL (Vertical Cavity Surface Emitting Laser) technology now widespread in our field and in industry) — of the front-end boards ("FEB") were produced. The on-detector electronics for the liquid argon calorimeters also include calibration boards, which inject a very precise charge directly into the calorimeter, and the tower-builder boards, providing analogue energy sums to the first-level calorimeter trigger. All the ~1600 FEBs and other boards were assembled, tested and qualified in the institutes and at CERN. The front-end crates and base-planes that sit on the detector next to the signal feedthroughs were produced as well, as were the low voltage power supplies, which turned out to be the source of a long saga of difficulties that were finally solved later on in the operation. All these equipment items were installed on each cryostat in the cavern, and then commissioned. Extreme care was paid to noise control and its reduction to the expected limit. This was a long process that regularly involved removing and reconnecting the cables between the feedthrough to the base-plane, inspecting connector shielding, installing additional shielding, and so on. For example, special filter boxes were produced during the commissioning phase to improve the noise reduction on the high voltage cables.

The back-end electronics was also a masterpiece. These digital boards process the FEB raw data to calculate energies and times for each triggered event, and send them to the downstream trigger and acquisition stages. This complex system as well as its data acquisition online software counterpart was first installed and then commissioned in parallel with the FE crates. This constant interplay between the front-end and back-end installation was a key aspect in commissioning the LAr calorimeter read-out system, paving the road for future excellent physics results.

5.4. The Hadronic Endcap and Forward Calorimeters

The overall design and performance of the Hadronic Endcap Calorimeters ("HEC"), and of the Forward Calorimeter ("FCal"), are described in Chap. 2. Both calorimeters are realised using liquid argon as the active material with metal absorbers. The following subsections discuss the construction of the two devices in turn.

5.4.1. *Hadronic Endcap Calorimeter (HEC)*

The Hadronic Endcap Calorimeter consists of two separate assemblies, one at each end of ATLAS, with each composed of two separate wheels. The HEC was built in a joint effort mainly by Canadian, German and Russian institutes. Liquid argon is used as active material because of its intrinsic linearity, time stability and, most importantly, its intrinsic radiation tolerance. Copper plates are used as the absorber material for the HEC for space and financial reasons — 25 mm thick in the front part of the HEC, and 50 mm behind. An extensive test beam programme with electron and pion beams allowed the performance, including software compensation techniques, to be developed and optimised. To maintain a fast response time as well as efficiency for minimum-ionising particles, the read-out cells have a limited size, implemented by splitting a large LAr gap into four small sub-gaps. Figure 5.22 shows a schematic view of these sub-gaps between two copper plates: the central electrode contains the read-out pad structure and the outer electrodes serve as high voltage carriers. Individual cell signals are amplified and longitudinally summed to form read-out channels with active electronics placed as close as possible. This means that the front-end electronics need to be within the cold volume. This "electrostatic transformer" ("EST") arrangement has multiple advantages. For safety reasons, each sub-gap has an independent high voltage (HV) connection. The EST concept has the following benefits:

- The HV can be lowered by a factor of two due to the reduced gap size.
- The pad capacities are reduced due to the effective larger liquid argon gap. In consequence, the noise is substantially smaller.
- As the signals measured in the four sub-gaps are essentially identical, the loss of any HV line allows for a rather straightforward correction.

The calibration lines are brought directly to the strip-line entry of each pad, thus yielding an optimal signal calibration. All electrodes are covered with highly resistive layers, made of carbon-loaded Kapton with a

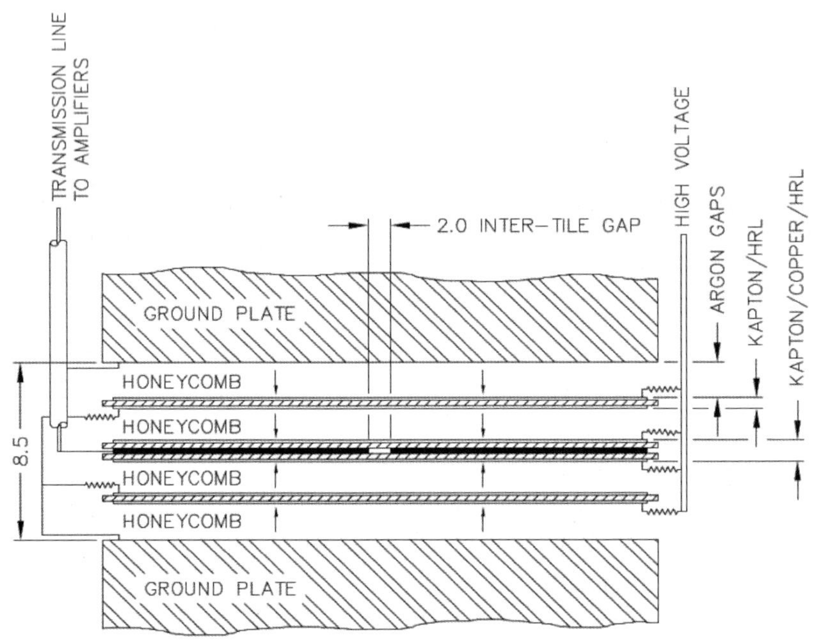

Fig. 5.22. Arrangement of the HEC read-out structure in the 8.5 mm inter-plate gap. The central board carries the pad structure for the read-out.

typical surface resistivity of about 1 MΩ/□. Thus, any HV-induced noise is substantially reduced and with the reduction of any large input current due to a potential HV discharge, the GaAs preamplifiers of the cold electronics are well protected.

Precise machining of the copper plates for the HEC was carried out by Canadian and Russian institutes. The electrode production was shared between Canadian, German and Russian institutes.

The signals are transmitted directly to the read-out motherboards with the amplifier and summing GaAs chips at the outer radius of each module. The front-end chip consists of eight preamplifiers and two line drivers. In contrast to a passive summation of the signals the active summation of the individual signals has the advantage of a superior signal-to-noise ratio. Even with the very low sampling fraction of 2.8% or 5.5%, a read-out cell calibration is possible to give a minimum-ionising particle signal a factor two above the noise level. A dynamic range of more than 3 TeV per individual read-out channel is also maintained. The output signals are fed to the cryostat feedthrough. With no access to the cold electronics during the lifetime of the detector, very detailed quality control was absolutely essential during the calorimeter construction. The signal read-out cabling and the cold

Fig. 5.23. One of the HEC wheels on the turning table prior to insertion into the cryostat.

electronics were the responsibility of a German institute. The high and low voltage cabling and the signal calibration were shared between German and Slovakian institutes.

All groups were involved in module construction and testing. With the very stringent requirements on quality control this last issue was in particular an outstanding task shared by all 14 institutes. Indeed quality control was by far the most challenging issue in the whole HEC construction project. Each module was tested cold after arrival at CERN. In the process of the wheel construction at CERN, each module was tested again for HV performance, correct read-out electronics response and calibration. Finally, all tests were repeated for the full wheel on the assembly table, after rotation to the vertical position (Fig. 5.23) in front of the cryostat, and once more after insertion into the cryostat.

5.4.2. *Forward Calorimeter (FCal)*

Two Forward Calorimeters, one at each end of the detector, extend the ATLAS solid-angle coverage to about 1° from the incident beams. Figure 2.10 in Chap. 2 shows the location of each FCal surrounding the LHC beam pipe about 5 m from the interaction point. This small-angle region is filled with the highest energy and density of particles, primarily from multiple soft collisions at LHC luminosities. Liquid argon ionisation calorimeters are well suited for detecting the particles of interest while tolerating huge backgrounds, but must deal with the slowly-drifting positively-charged argon ions, which build up in the liquid-argon gap and distort the drift field, and if unabated, will create a plasma. This plasma would neutralise much of the ionisation that the calorimeter is meant to detect. To avoid this extreme condition, the electrode gaps in the FCal modules are much smaller than in the other liquid argon calorimeters. To maintain high precision in the width of the gap, a different geometry is employed, as shown in Fig. 5.24 — a solid metal rod is suspended within a copper tube by an insulating polyetheretherketone (PEEK) fibre helically wound around the rod, thus making a snug fit. Liquid argon fills the gap between the rod and the tube. The ionisation electrons drift towards the rod which is held at a positive potential and the positive argon ions drift toward the inner surface of the tube, which is held at ground potential.

Each rod-tube electrode is deployed in a metal matrix in a hexagonal array, one unit cell of which appears in Fig. 5.24. Each forward calorimeter is made up of three modules, one behind the other, as shown in Fig. 5.25. The first module (built by a US institute), the one closest to the interaction point,

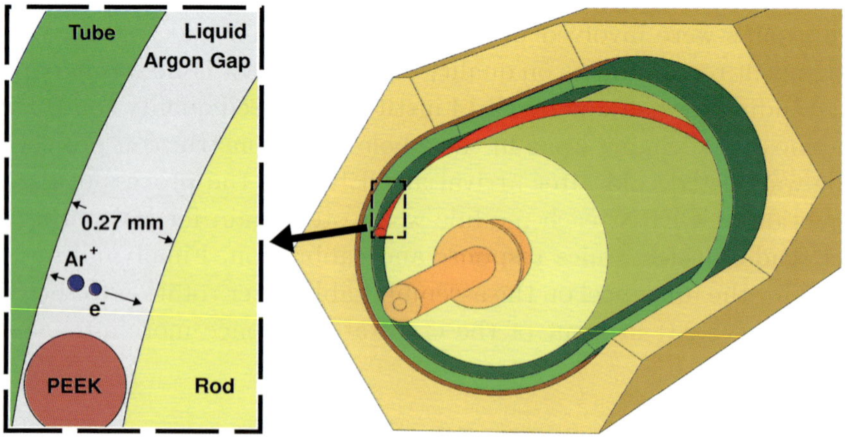

Fig. 5.24. Cut-away drawing of one unit cell of a Forward Calorimeter module. A solid metal rod (yellow) with read-out pin (orange) is centred within a copper tube (green), held in position by a PEEK fibre (red), all surrounded by a metal matrix (gold). LAr fills the gap between the rod and the tube.

Fig. 5.25. The three modules of one of the two Forward Calorimeters. The EM module closest to the interaction point is on the left, followed by the two hadronic modules. At the far right is a solid bronze shielding plug to reduce the rates in the muon chambers, which will be directly behind. The workers are routing the cables that will carry the signals from the face of each module to the summing circuits at the rear (not visible).

sees electromagnetic showers. Hadronic showers deposit some ionisation in the first module and the rest in the second and third modules. Copper is the metal used in the rods and matrix of the first module, in order to conduct heat efficiently away from the innermost region of the module, which is the hottest due to the energy deposited there from the shower particles.

Tungsten is the choice of metal in the second and third modules, as it minimises the spread of the showers. Figure 5.26 shows how the tungsten matrix was formed. Tungsten rods for these modules were supplied by institutes in Russia and China, while the middle and rear modules were separately constructed at two Canadian institutes. The completed modules for each end of the detector were shipped to a clean room at CERN where they were cabled up (Fig. 5.25), tested, calibrated in a test beam, and assembled into the cryostat support tube, a structural member of the endcap cryostat.

The coaxial read-out cables, the signal summing circuits and the high-voltage distribution system (all inside the respective endcap cryostat) were

Fig. 5.26. Laying down a layer of tungsten slugs, part of the matrix of the rear hadronic FCal module constructed in Canada. A layer of tubes can also be seen. Tungsten rods are yet to be inserted.

supplied by a US institute. The FCal uses the same electronics outside the cryostat as the LAr accordion electromagnetic calorimeter. The analogue inputs to the trigger are also the same as for the rest of the LAr calorimeters, except that the gains are adjusted to be proportional to the transverse energy before summing.

5.5. Cryostats and feedthroughs

The liquid argon calorimeters are located in three cryostats, as shown in Fig. 2.10 of Chap. 2. The barrel cryostat houses the barrel electromagnetic calorimeter and the two endcap cryostats each house an electromagnetic endcap calorimeter, a Hadronic Endcap Calorimeter and a Forward Calorimeter. Figures 5.27 and 5.28 show the barrel, and one endcap, cryostat during assembly. Each cryostat is composed of two concentric aluminium vessels, an inner cold vessel and an outer warm vessel. Each vessel forms a cylindrical torus centred on the beam axis and the space in between the vessels is under vacuum. To reduce material, the solenoid (providing the magnetic field for the Inner Detector) and the barrel cryostat share the same inner

Fig. 5.27. Two cylinders of the barrel liquid argon calorimeter cryostat.

Fig. 5.28. A cold vessel in preparation for one of the endcap LAr Calorimeter cryostats.

warm cylinder, which led to challenging logistics during testing because the two components were produced at two different factories. The services of the solenoid are therefore routed through the vacuum space. The barrel cryostat inner cylinders are tapered down towards both ends to further reduce the material traversed by particles. The different vessels are closed at both ends by bulkheads. All bulkheads are circular-shaped plates designed to minimise the amount of material and are welded in such a way so as to allow only the openings needed.

The signal feedthroughs bring all the signal, monitoring, calibration and spare lines through the insulating vacuum, from the liquid-argon cold volume to the front-end crates located at room temperature around and on the outside of the barrel and endcap cryostats. They maintain the integrity of the electrical signals, while minimising the heat loss of the cryostats and allowing for small relative motions between the cold and warm vessels. The endcap signal feedthroughs also carry the low voltage lines required for the HEC preamplifiers. A total of 64 identical feedthroughs serve the 122,800 lines of the barrel electromagnetic calorimeter, whereas a total of 2 × 25 feedthroughs serve the 2 × 48,000 lines of the two endcaps. All endcap feedthroughs are mechanically identical, but come in four different electrical configurations to service the various calorimeters in the endcap cryostats. A signal feedthrough consists primarily of a warm flange and a cold flange (see Fig. 5.29), with a flexible bellows welded between them. The volume between the two flanges is under vacuum. Each flange houses four gold-plated pin carriers providing a total of 1920 signal connections per feedthrough. The cold and warm flange pin carriers are interconnected with flat polyimide strip-line flexible cables. To minimise the lengths of the connection lines, the signal feedthroughs are distributed radially around each end of the barrel cryostat and at the back of each of the endcap cryostats, and hence must operate in all azimuthal directions.

Two high-voltage (HV) feedthrough ports are mounted on each cryostat. They are located close to the highest point at each end of the barrel cryostat and at the back of each endcap cryostat. A port consists of a single warm bulkhead mounted on the warm vessel, from which the wire bundle leads to the calorimeter via a tube connected to the cold vessel by stainless-steel bellows. The liquid argon rises up to the bellows and is followed, instead of vacuum, by a 70 cm high column of argon gas at 1.25 bar. With such an arrangement, the HV wires are never under vacuum and thus are not subject to corona discharge effects. The high-voltage wire consists of a solid

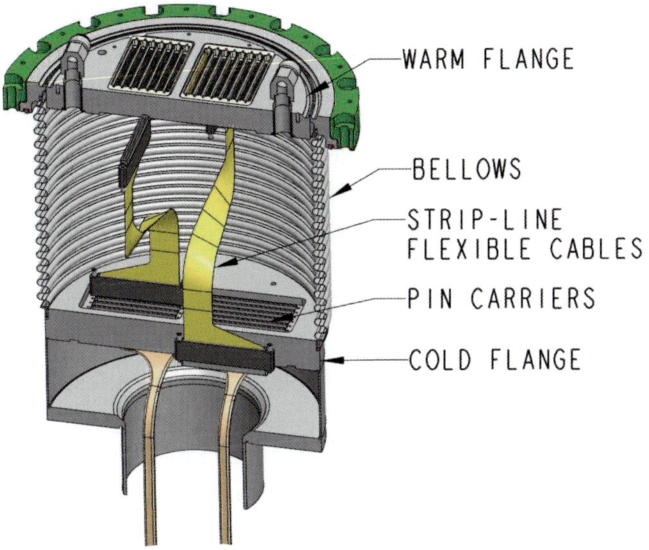

Fig. 5.29. A cut-away view of a signal feedthrough. The temperature at the top of feedthrough as shown is the ambient in the cavern and that at the bottom is of liquid argon. The outer diameter of the bellows is 30 cm.

0.41 mm diameter constantan (Cu/Ni alloy) conductor with a 0.30 mm thick radiation-hard plastic (polyether-ether-kctone) cladding. Each port holds approximately 840 HV lines, including spare lines that may be used to supply parts of the calorimeter that do not hold the nominal high voltage.

5.6. The Tile Calorimeter

The key elements of the Tile Calorimeter concept are shown in Fig. 2.11 of Chap. 2 — an iron absorber with slots for the active medium (scintillators) and grooves in which the wavelength-shifting read-out fibres run, bringing their light signals to read-out electronics located in the structural beam used to form the principal mechanical support. This design aims to maximise the azimuthal acceptance of the calorimeter by minimising the un-instrumented volume of the detector, and to maximise the instrumented depth.

5.6.1. *Submodule and module absorber structure*

The steel absorber structure is constructed in three longitudinal sections: a central barrel and two extensions in the endcap region ("extended barrels"), each comprising 64 azimuthal segments (modules), built up from smaller segments (submodules), each of which is constructed as a series of glued and pinned laminations. In total, the absorber structure of the Tile Calorimeter comprises approximately 80,000, 5 mm thick, long laminations (master-plates) and 450,000, 4 mm thick, small laminations (spacer-plates), each of which are trapezoidal in shape. Some 550,000 wavelength-shifting optical fibres capture the light from the scintillator tiles and propagate it to the ~10,000 photomultipliers ("PMTs") located in the module structural beam (girder).

The basic unit around which the overall construction is made is the submodule. The construction concept, though straightforward, required great accuracy for its realisation due to the high precision demanded, with typical dimensional specifications of 50 μm. Moreover, the master-plates and spacer-plates are not simple trapezoids. The master-plates contain 22 precision holes to match pairs of corresponding holes in the 11 different sizes of spacer-plates. These highly complex high-precision shapes were fabricated by die stamping, such that a single master-plate could be fabricated in ~60 seconds from the time the blank plate was loaded into the die to the time the formed plate was removed. In the case of the spacer-plates, die stamping was also used and multiple spacer-plate types were fabricated from the same plate blank in a single pass. At the submodule construction sites the plates were loaded onto the assembly fixture in sequence, spacer-plates were pinned to master-plates using spring-like C-shape pins inserted into the holes in the plates and glue was applied between the plates to control the spacing. A submodule nearing the final phase of construction is shown in Fig. 5.30.

Submodules were constructed at assembly sites in Spain, Russia, Belarus, Italy and the United States, in a total of ten locations. Module assembly occurred at sites in the United States, Spain, CERN and Russia using specially designed assembly and alignment tooling. The intrinsic uniformity of the plates and the use of a common design for the submodule and module assembly fixtures allowed the construction of identical submodules and modules across all construction sites.

5.6.2. *Module instrumentation*

To provide an estimate of particle energy, the Tile Calorimeter measures the amount of scintillating light produced by the passage of charged particles through plastic scintillators. Therefore, considerable attention was paid to the design of the scintillator tiles and their coupling to wavelength-shifting

Fig. 5.30. A submodule under construction on the assembly fixture.

fibres, and the transport of the light signal to PMTs. The scintillators and fibres and other optical components were designed and produced by institutes in Russia, Portugal, Italy, France and the United States.

The calorimeter design required production of more than 400,000 scintillator tiles of high optical quality. The tiles are trapezoidal and of 11 different sizes, with two holes to allow a path for a cesium calibration source. Injection moulding the tiles resulted in enormous savings in construction cost and time. The injection-moulding machine, shown in Fig. 5.31, mixes commercially available optically transparent granulated polystyrene with wavelength-shifting dyes, melts the mixture and injects it into the mould at very high pressure. This production takes less than two minutes per tile. The mould design keeps tolerances on the tile dimensions to within 0.1 mm. Light yield uniformity and attenuation length of the produced tiles is measured with X-rays and a ^{90}Sr radioactive source.

Wavelength-shifting fibres are the part of the optical chain that transports the light produced by charged particles transiting the scintillator to the PMTs. The fibres were manufactured commercially. The production was staged in batches and the optical quality of each batch is controlled by measuring the light attenuation length of a sample of fibres — a substantial effort in measuring and bookkeeping. To increase the light collected, a mirror is applied to the end of each fibre. This mirroring is performed by bundling

Fig. 5.31. A scintillator tile and the injection mould used for its fabrication.

~100 fibres together and simultaneously depositing aluminium on all of them using vacuum deposition.

Three fibres must be located in each channel formed by the absorber structure and must be installed such that each fibre only sees light from specific scintillator tiles. To accomplish this, the fibres are located in a plastic sheath termed a profile, similar to a straw. Loading these profiles with over half a million fibres was realised with a custom-design automated machine using synchronised rotation of loading barrels to control the fibre to be loaded at a given location, and compressed air to blow the fibres into the profile. This loading machine is shown in Fig. 5.32. In less than a minute, a profile would be loaded with its three fibres. As was the case for the moulding of the scintillator tiles, this technology led to an enormous reduction of cost and time, both in the loading and in the module instrumentation.

The installation of the scintillator tiles and wavelength-shifting fibres into the absorber structure took place in sites in the United States, Spain and CERN. The insertion of scintillator tiles in the absorber structure used to its advantage the fact that calibration tubes run the length of a module. Therefore, once inserted and located relative to the edge, the tiles are mechanically held in place by the calibration tubes. The two opposite sides of each scintillator tile are coupled to wavelength-shifting fibres, which absorb the light from tiles, re-emit it at longer wavelength and transport

Fig. 5.32. The machine that loaded the half-million wavelength-shifting fibres into profiles.

a fraction of it to PMTs on each side of the module, where the light is converted into an electrical signal.

The cell structure of the calorimeter is obtained by an appropriate grouping of the fibres for a given depth and position along a module to the same PMT. Again an inventive approach was used to realise the necessary alignment between the fibres and the PMT. Rather than machining the holes in the support beam to very high precision, which would have been costly, a fixture was designed to match the pairs of transition pieces inside and outside of the girder, the girder-rings, to the locations of the PMTs. Fibres were then grouped together to form read-out cells and glued into a small plastic tube that could be fastened by screws into the girder-rings. Fibres on one side of a module that have been bundled together to form read-out cells in this fashion are shown in Fig. 5.33.

5.6.3. *Electronics integration*

All front-end and digitising electronics had to fit in a limited space inside the girder of each calorimeter module. The electronics boards and PMTs are located on a removable element, the drawer, containing up to 24 pairs of PMTs and their associated read-out electronics. Each front-end electronics channel is housed in a so-called PMT block that is positioned in front of its corresponding fibre bundle. Its external part is an iron cylinder that

Fig. 5.33. Grouping of the wavelength-shifting fibres to form calorimeter cells and routing these to the insertion pieces located in the module support beam.

plays several roles: the housing of its optical and electronic components, the magnetic shielding of the tube, and heat transfer to the cooling system. The internal components are a light mixer, the PMT surrounded by a further μ-metal magnetic shield, its high voltage divider, and the very front-end card. Approximately 10,000 of these PMT blocks were assembled, tested and installed in drawers, by collaborating institutes in France, Greece, the United States, Sweden and Brazil. Figure 5.34 shows the insertion of one such drawer into the girder.

In the drawer electronics, the scintillation light signal produced by charged particles is converted into an analogue electric signal by the PMTs. Using dedicated boards, the signal is shaped, digitised and stored in buffer memory. At the same time, the sum of analogue signals from a few neighbouring cells is sent via a separate path to the first-level trigger system, which identifies potentially interesting proton-proton collisions. The calibration of the response of the full path from scintillation light to digitised electronics signal is provided by three complementary calibration systems. A radioactive Cs calibration system provides a full tomography of individual scintillator tiles, a laser system calibrates the PMT and front-end electronics, and charge injection systems calibrate the digitising electronics.

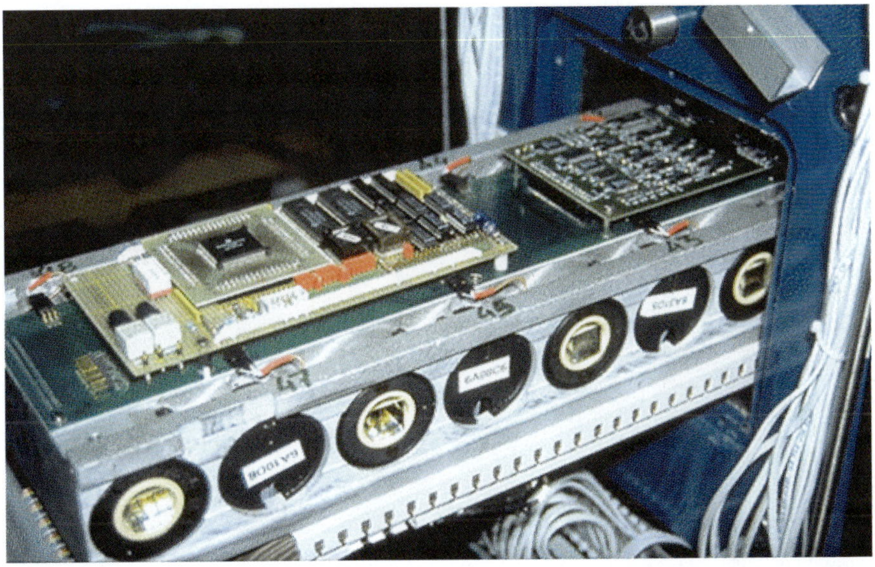

Fig. 5.34. The electronics components in a drawer, slightly extracted from the inner volume of the module support beam.

5.6.4. *Pre-assembly and installation into ATLAS*

The Tile Calorimeter is a self-supporting structure mounted on saddle supports that transfer the load to the ATLAS main rails through hydraulic jacks. Individual modules in each calorimeter section bear on each other only at the support beam and at the inner radius. Extensive engineering analysis was carried out to determine the different load conditions, to determine the tension pre-loads in the bolts connecting submodules to the support beam and also to determine the forces in the region below the support saddles where the connecting plates between modules require to be pinned to take the shear load of the entire calorimeter section. The insertion of these pins and bolts is shown in Fig. 5.35. The holes in which the pins are to be inserted are large and must fit perfectly. The holes for the pins were match-drilled through the connecting plates into the girder and the pins inserted after first reducing them slightly in size by immersing the pins briefly in liquid nitrogen. Thereby a perfect pinned connection was formed bearing the entire load of the calorimeter, that once formed would never be taken apart until the ultimate conclusion of the ATLAS experiment.

To ensure that the calorimeter would meet its required radial envelope inside the ATLAS toroids of +5 mm and –10 mm, a full assembly was carried out on the surface for one of the endcap calorimeters as well as

Fig. 5.35. The crucial task of inserting bolts and pins in the region below the support saddles.

the barrel calorimeter. These surface assemblies were used to verify the shim thicknesses to be used at the interfaces between the modules, and to test geometrical control techniques and the support of the endcap electromagnetic calorimeter. Although time-consuming, these full surface assemblies proved to be invaluable. They established the basic assembly plan, provided initial thicknesses for shims at the module interfaces, and provided essential training for the assembly and survey teams. Without these preliminary assemblies, it would have been much more of a challenge to assemble the entire Tile Calorimeter in the ATLAS cavern in the short time taken.

Surface assembly of the Tile Calorimeter sections was carried out in a location some 3 km from the experimental area. Figure 5.36 shows the movement of one of the pre-assembled extended-barrel sections on its assembly frame, modules and saddles on its way to the experiment, where the assembly frame would be connected to the main crane and lowered into the cavern.

Testing occurred at many stages of the construction. For example, the thickness of the raw steel sheet was checked to meet a tight thickness specification, gauges were used to check the geometry of the master and spacer-plates, raw fibre boules were tested for transparency and the final drawn fibres tested for attenuation length and light output, and the reflectivity of the mirrors applied to the fibre ends. Electronics components

Fig. 5.36. Transport to the ATLAS experimental site.

were similarly tested, for example, PMTs had their gain and other properties recorded, electronics boards were measured on test benches, and components were tested for their sensitivity to ionising radiation and neutrons.

Tooling was designed to minimise variations in constructed components. This included stamping dies, scintillator moulds, assembly tooling and survey schemes. Survey data taken during the surface pre-assembly were incorporated into models used in controlling the final assembly in the cavern where careful shimming resulted in a calorimeter structure over 8 m in diameter whose gaps between elements was of order 250 μm. In addition, extensive testing of production modules was carried out in high-energy particle beams to measure the response, build a calibration reference relative to radioactive sources and provide a library and response benchmark for use in simulating the response of the calorimeter to physics processes.

The ~15 years from the first ideas to prototypes and then the final design and construction of the Tile Calorimeter would not have been possible without the continuous dedicated effort of physicists, engineers, technicians and students from 16 countries spanning over half of the globe — Sweden, France, Italy, Slovak Republic, Czech Republic, Italy, Romania, Greece, Spain, Portugal, Armenia, Georgia, Belarus, Russia, the United States,

Brazil and CERN. It has truly been a universe of human relations full of interactions of people from different professions and was filled with many unforgettable moments.

5.7. Muon Spectrometer chambers

An advantage of the ATLAS experiment's air-core toroid magnet system is that multiple scattering is reduced and therefore a precise transverse momentum (p_T) measurement is possible in the external muon system. The design goal was to reach a relative p_T resolution of 10% for $p_T = 1\,\text{TeV}$. As discussed in Chap. 2, Monitored Drift Tube chambers ("MDTs") are used over most of the acceptance, completed by Cathode Strip Chambers ("CSCs") around the beam directions, for the precise measurement of muon tracks in the bending plane of the toroid magnets. Triggering chambers that use Resistive Plate Chamber technology ("RPCs") in the barrel region and Thin-Gap Chambers ("TGCs") in the endcaps complement these.

5.7.1. *Muon momentum measurement: MDTs and CSCs*

After an intensive period of design studies, test measurements and detailed evaluations, it was decided to use MDTs to provide precision measurements of muons at large scattering angles, and CSCs at small angles. As shown in Fig. 2.9 of Chap. 2, the MDT chambers consist of two multi-layers of three or four individual drift tube layers. Each drift tube consists of thin-walled aluminium tubes with outer diameter of about 30 mm, filled with a 3 bar Ar-CO_2 (93:7%) gas mixture. Each tube provides a position resolution perpendicular to the wire of 80 μm. The second coordinate, along the wire, is provided by RPCs in the barrel or TGCs in the endcaps. With all 6 (or 8) drift tube layers combined, a position resolution of about 50 μm can be achieved for one chamber. The spacer support structure between the multi-layers provides mechanical stability and contains internal alignment rays (12,000 in total), which allows the monitoring of chamber deformations. A further advantage of the spacer structure is that the distance between the measurements in the two multi-layers is increased, leading to an improvement in angular resolution. MDT chambers are very robust since the failure of one tube leads to only a small deterioration of the combined measurement; the remaining tubes provide enough redundancy to achieve a sufficiently good measurement.

Critical R&D was performed during the late 1990s to achieve the required precision in tooling and procedures for tube and chamber construction. Work moved to the mass production phase in 2000. About 1200 MDT chambers were built, with 18 main types, for a total area of 5500 m^2. Production sites were institutes in China, Greece, Germany, Italy, the Netherlands, Russia and the United States. Specially designed and constructed wiring machines were built, which provided a semi-automatic scheme for the assembly of about 370,000 drift tubes using aluminium tubes, end-plugs with a precision wire locator and a 50 μm diameter gold-plated tungsten-rhenium anode wire. After assembly, each tube underwent a detailed quality control to check, for example, the wire location within the tube, the gas tightness, the stability under high voltage and the noise rate.

In order to reach the required mechanical precision, each chamber production site used a "clean room" with a large, massive and precise granite table. Very precise jigs and optical alignment devices located on the granite table guaranteed that all tubes inside the chamber were positioned and glued with a precision of better than 20 μm, with respect to each other. After chamber gluing, the services, gas, high voltage, low voltage, read-out and alignment electronics were mounted. The complicated electronics of the MDT read-out were developed and constructed by a common effort of Japanese, Dutch and US institutes. The electronic scheme uses fast on-chamber electronics to convert the drift tube signals to digital data, which is then sent onto the next stage, which collects the data of the whole particle collision event.

Each chamber production site had to undergo a qualification process for the first chamber and several chambers during the regular production, with the goal of checking the mechanical precision of all drift tube wires with respect to each other. For this purpose, a special and very large X-ray tomograph was constructed at CERN. This tomograph could take X-ray images, which allowed wire positions inside the drift tubes to be reconstructed with a precision of 10 μm. The measurements showed that all production sites reached the goal that nominal wire positions were on average within 20 μm of their nominal positions. Thanks to the many years of great effort from the teams across 13 production sites worldwide, the muon MDT chambers were built with high precision, unmatched by any muon systems built in the past for high-energy physics experiments.

After the chamber production the MDT chambers were transported to CERN. The rectangular barrel chambers were assembled together with RPC chambers into packages with gas connections, cables and electronics, ready for individual installation. The trapezoidal endcap chambers instead were

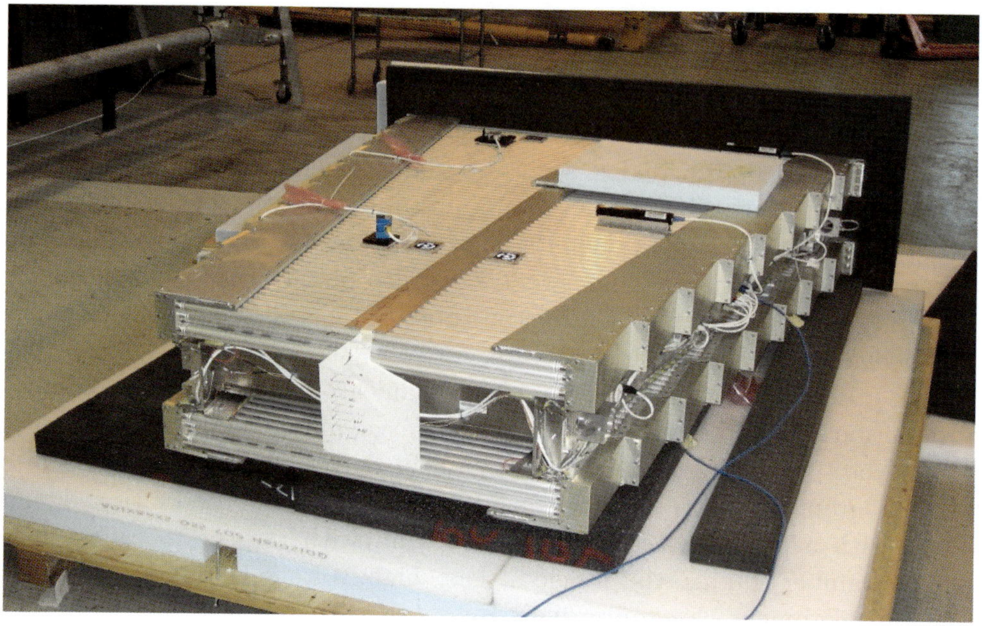

Fig. 5.37. A typical small wheel MDT endcap chamber.

mounted on pre-fabricated sectors of the endcap wheels and connected to their services. Figure 5.37 shows a fully assembled endcap MDT chamber for the small wheel.

The limit for safe operation of the MDT is at counting rates of about 150 Hz/cm^2, which will be exceeded at small scattering angles in the forward and backward direction close to the beam pipe. A different technology with higher rate capability was needed. The Collaboration therefore decided to use Cathode Strip Chambers, which consist of many closely spaced anode wires that can operate safely at rates up to 1000 Hz/cm^2. The signal is extracted from cathode strips placed near the wires, with two planes rotated by 90°, providing two-dimensional position information. A precise position resolution can be achieved by using the relative signal strength of adjacent strips. A signal-to-noise ratio of 100:1 was achieved with custom integrated front-end electronics providing a resolution of order 85 μm from a 5.3 mm strip pitch. The CSC system was designed and built by Russian and US groups. The chambers were installed in the inner part of the small wheel (Fig. 5.38). The chambers are inclined by 11° with respect to the plane of the small wheel, such that particles emerging from the interaction point penetrate the chambers at normal incidence.

Fig. 5.38. View of the CSC chambers in the inner part of the small wheel.

5.7.2. *Muon trigger chambers: RPCs in the barrel*

An essential feature of the muon system is its capability to trigger on muon tracks. In the barrel region of the muon spectrometer, the Resistive Plate Chamber technology is used. The RPC is a gaseous parallel-plate detector that combines a spatial resolution of about 5–10 mm, determined by digitally read-out strip panels, with a time resolution of about 1.5 ns that ensures a reliable beam-crossing identification well inside the 25 ns LHC bunch spacing. Moreover, it provides the measurement of the second coordinate in a direction orthogonal to the one measured with the precision MDT chambers. The trigger chambers cover a total area of about 3650 m^2 in the barrel and the total number of read-out channels is about 370,000. The RPC detector has a simple mechanical structure, it uses no wires and is therefore simple to manufacture. It was invented in Italy in 1979–80.

The basic RPC unit has a narrow gas gap, 2 mm thick, formed by two parallel melamine-phenolic resistive plates, separated by insulating spacers. The spacers are glued on both plates at 10 cm intervals. A 7 mm wide frame of the same material and thickness as the spacers is used to seal the gas gap. The primary ionisation electrons are multiplied into avalanches by a high, uniform electric field of typically 4.5 kV/mm. Amplification in avalanche mode produces fast pulses of typically 0.5 pC. The gas mixture is based on tetrafluoroethane ($C_2H_2F_4$), a non-flammable gas that allows for a relatively low operating voltage. The signal is read out via capacitive coupling by metal strips on both sides of the detector. A trigger chamber is made of two rectangular detector layers, each one read out by two orthogonal series of pick-up strips (Fig. 5.39).

The first-level trigger scheme is based on coincidence logic between signals coming from strip panels with a time resolution shorter than the bunch-crossing time interval. For the barrel it is organised in two stations. The first RPC trigger station is placed in the middle of the toroid, at a radius of about 7 m and consists of two RPC chambers sandwiched together with an MDT chamber. The low-p_T trigger requires a coincidence window between signals from the two RPC chambers placed in the middle station in both projections with a 3-out-of-4 majority.

The second RPC trigger station is placed at a radius of about 10 m and consists of only one chamber, mounted on the MDT of the outer station. The high-p_T trigger requires a time coincidence between the RPC strip-panel

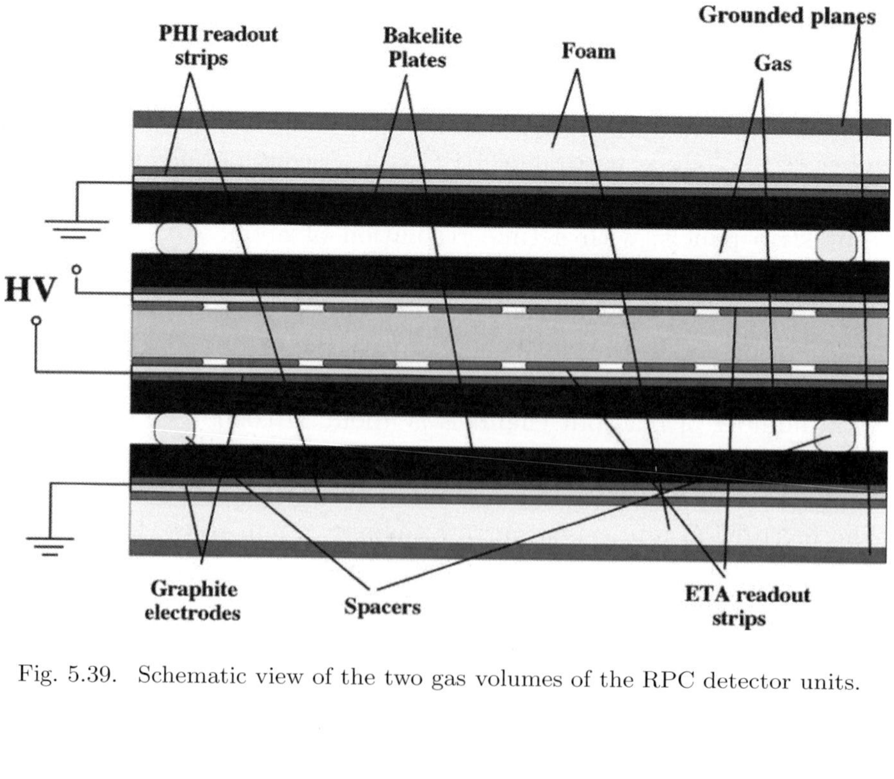

Fig. 5.39. Schematic view of the two gas volumes of the RPC detector units.

signals of the middle and outer stations. Again the coincidence condition must be satisfied in both projections with global majorities of 3-out-of-4 for the middle station and 1-out-of-2 for the outer station.

The RPC chambers that have been built for ATLAS, although they look very similar to earlier designs, are the result of more than ten years of R&D work done mostly in the framework of the RD5 project at CERN, making use of a test-beam. The most important result of this R&D work was the change from the streamer working mode to the avalanche working mode in order to improve the detector rate capability and cope with the high expected counting rates due mainly to soft neutron and gamma background. The new mode of operation had a strong implication on the front-end electronics and required the development of a powerful fast amplifier, developed in 0.7 μm GaAs technology, to compensate for the lower charge deposited in the gas. Moreover a new gas mixture, composed of $C_2H_2F_4$ (96.5%), Iso-C_2H_{10} (5%) and SF_6 (0.5%) was chosen to suppress streamers.

The RPC construction and testing work was shared between four Italian and one Russian institute. Italian companies produced most of the components of the detector while the support panels were produced and assembled in Russia. The front-end electronics boards were tested and mounted on the strip panels in an Italian institute while the RPC chambers were mechanically assembled in another. All the detector components had to pass strict quality assurance requirements.

Dedicated cosmic-ray telescopes, with tracking capability for cosmic muons, were used by three groups in Italy to certify the performance of the 1100-odd chambers built. Then they were shipped to CERN to the "BB5" construction facility for assembly together with the correspondent MDT chambers and services. In BB5, the muon trigger stations were cabled to the electronics and trigger boxes for a final cosmic ray test before their installation in the experiment. A view of RPC chambers mounted with MDTs, as installed in the experiment, is shown in Fig. 5.40.

5.7.3. *Muon trigger chambers: TGCs in the endcap*

While the RPC technology is used for the muon trigger in the barrel region, a different solution was adopted for the endcap regions. The primary reason is the much higher radiation level in the location where the endcap trigger chambers are installed. Common requirements for the trigger chambers include good timing resolution for bunch identification and

Fig. 5.40. An assembly containing MDT and RPC chambers installed into the experiment.

adequate position resolution for the measurement of muon momentum. The Thin-Gap Chamber technology was chosen for the endcap trigger detector. Figure 5.41 (right) schematically shows an endcap trigger detector (TGC-1 in Fig. 2.7 of Chap. 2). A TGC chamber has a trapezoidal shape of 1–2 m size, and contains two or three detection layers depending on the chamber type.

The basic structure of a TGC chamber is a multi-wire proportional chamber. The detail has been optimised for ATLAS. The structure of a single TGC detection volume is schematically shown in Fig. 5.42. It consists of an array of 50 μm gold-plated tungsten wires separated by 1.8 mm, and resistive cathode planes with a thin graphite layer at both sides of the wire plane at a distance of 1.4 mm. The volume between the cathodes is filled with a gas mixture of CO_2 and n-pentane (55:45%). A high voltage of 2.8 kV is applied to the wires, while the graphite layers are grounded. There are a total of 1488 chambers in the big wheels and 90 chambers in the endcap inner station. The total number of gas gaps are 3408 and 180, respectively. When a muon passes through the gas volume, it ionises the gas along its trajectory. The electrons are pulled toward the nearest anode wire, where the strong field produces electron multiplication, overall by a factor about 10^5. Signals are

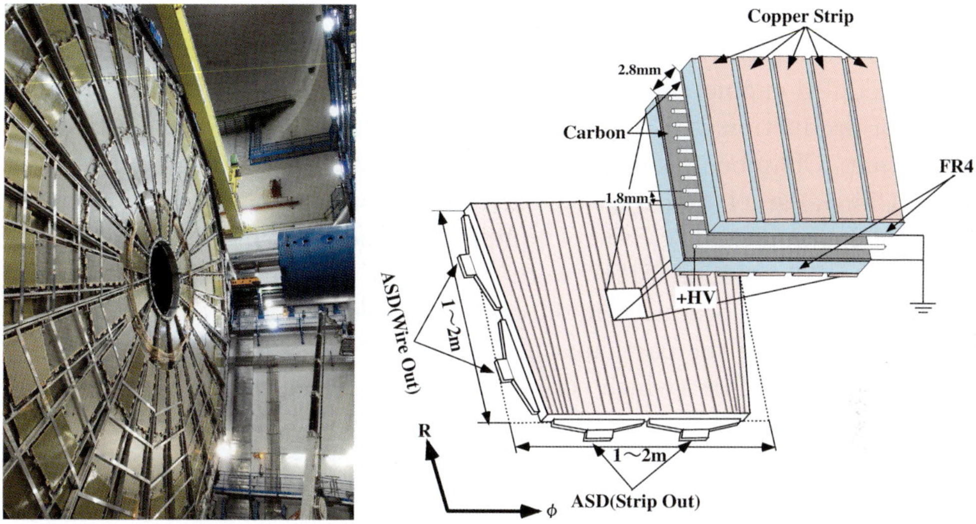

Fig. 5.41. (Left) A TGC "big wheel", as seen from the side facing the pp collision point. (Right) Schematic of a single TGC chamber.

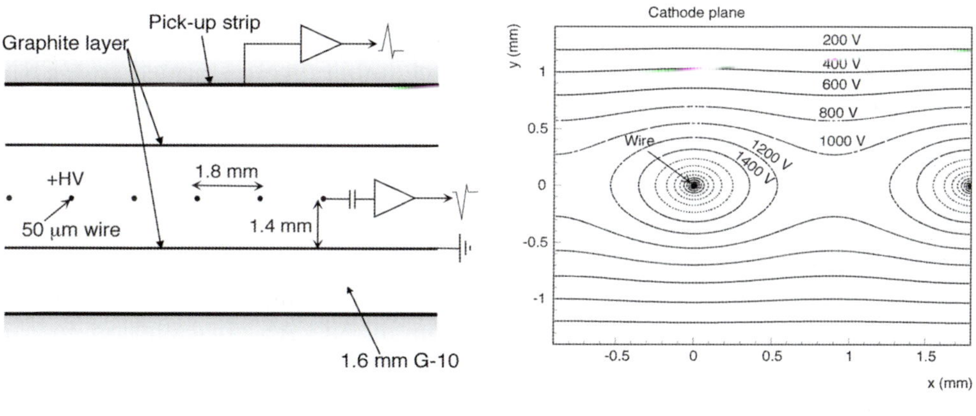

Fig. 5.42. (Left) TGC structure showing the anode wires, graphite cathodes and readout strips orthogonal to the wires. (Right) Equipotential lines in the TGC chamber.

measured on the wires and pick-up strips located behind one of the cathode planes, providing muon position information in two orthogonal coordinates. The signals are amplified and sent to electronics circuits for further signal processing.

Groups from Israel and Japan were engaged in the R&D leading to the TGC detector from the early 1990s. Small, then large, test chambers were produced to test and optimise the design. Extensive beam tests were performed using the test beam facility at the KEK laboratory in Japan.

A detailed engineering design was done by 1997, when detailed preparations for production started. Quality assurance and control were particularly important. Chamber production and testing was shared among institutes from China, Israel and Japan. Each group prepared its production site with tooling and quality control equipment. The main mechanical chamber components (made of FR4) were produced in Israel. The gold-plated tungsten wire was produced in Japan. Production of other components was also shared among the participating groups. It took 3–4 years to establish a production procedure (which contains over 100 assembly steps) and quality control process to be done after each relevant step. During this period several generations of prototype chambers were produced, refining the production method. Chamber series production started in 1998 and took about six years to complete.

All chambers were tested using cosmic rays. Dedicated test stands were prepared in Israel and Japan. Performances were carefully checked for efficiency and uniformity. A further certification step, performed on most of the detectors, included irradiation with a gamma source to check the chamber stability in a high rate environment. The pass-rate yield after all the tests was high (>98%). The certified chambers were shipped to CERN for detector integration. During the detector integration phase, the chambers were first assembled in a sector structure on the surface before installation in the ATLAS underground cavern. There are a total of 72 TGC sectors. The electronics were also mounted and connected to the chambers. The chambers and electronics were further tested at this stage. This provided an opportunity to gain experience of operating the large scale, complex detector system well in advance of the installation in ATLAS. Installation in the ATLAS cavern was completed in 2007.

5.7.4. *Alignment system*

In the muon spectrometer, the MDT and CSC chambers and their auxiliary alignment objects are installed with a precision of about 5 mm and 2 mrad with respect to their nominal positions. In contrast, the actual chamber locations must be known to a precision of 30 μm in order to achieve the required momentum resolution — this is more than two orders of magnitude tighter than the positioning accuracy. To reach this precision goal, an alignment system was built that relates the position of each chamber to that of its neighbours. With the internal accuracy of the MDTs guaranteed by construction, and their planarity monitored by optical in-plane alignment sensors, this alignment system forms a dense, stable grid for monitoring (in the barrel) or reconstructing (in the endcap) the position of each MDT wire in the muon system. A schematic representation of the alignment system is shown in Fig. 5.43.

The alignment system consists of various types of optical three-point straightness monitors, which continuously track the relative position of chambers within the same layer and adjacent layers. In the barrel, the chambers inside an MDT row are referenced to each other by chamber-to-chamber alignment sensors ("praxial", see Fig. 5.43, and axial systems), while the projective system connects the inner, middle and outer MDT layers, simulating the trajectories of an infinite-momentum track originating from the interaction point. This optical network is able to reliably detect relative changes in chamber position at the 20 μm level. Additional optical lines link MDT chambers to the barrel toroid cryostat, or relate adjoining chambers in large and small sectors.

In the endcap muon system, alignment bars are used to provide a precise reference grid, which is used to measure the three-dimensional position of each chamber with proximity monitors. Alignment bars are large precision objects, built inside and around an (up to 9.6 m long) aluminium bar of

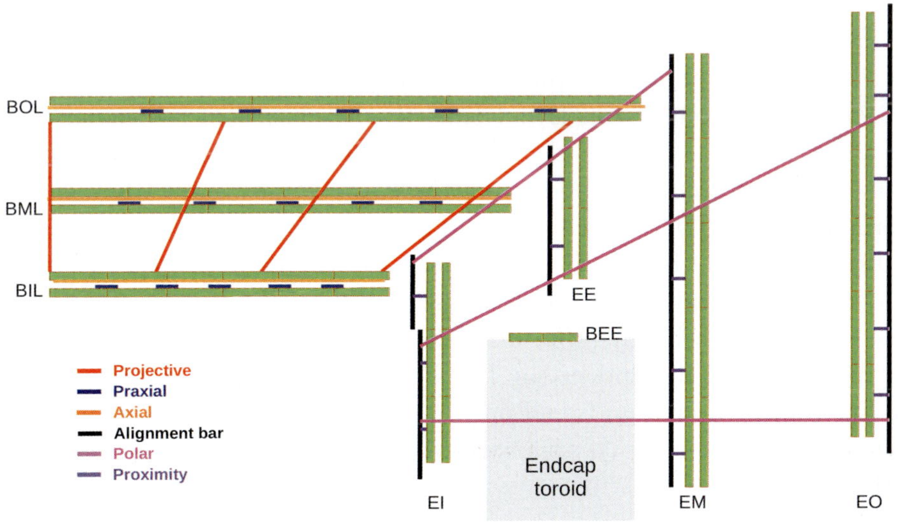

Fig. 5.43. Schematic representation of optical alignment paths that measure the relative position of the three layers of MDT chambers (large sector).

80 or 85 mm outer diameter. On the outside of the tube, platforms with sensor mounts are clamped to the bar, on which optical sensors are mounted. On the inside, overlapping straight-line monitors and temperature sensors are installed. In the endcap there are approximately 6000 sensors, which comprise about 11,000 logical devices.

5.8. Forward detectors

The forward detectors presently available in ATLAS (and earlier introduced in Chap. 2) are known as LUCID, ZDC, ALFA and AFP. The construction of the forward detectors started later than for the other detector systems, when the main construction was almost completed. The detectors are relatively small, allowing such a strategy.

The first forward detector to go into production was LUCID, begun in 2006 and completed in 2007, and installed in the forward shielding in time for the first beams. The LUCID detector measures the luminosity delivered by the LHC machine to ATLAS, that is, the total number of proton-proton collisions produced by the LHC at the ATLAS interaction point.

The first version of the detector consisted of two arms placed on the two opposite sides along the LHC beamline. Each arm was formed by 16 aluminium cylindrical tubes of 1.4 cm diameter and 150 cm length, arranged around the beam pipe and pointing to the geometrical centre of ATLAS. The tubes were filled with a C_4F_{10} radiator gas, and Cherenkov light produced by charged particles was collected at the end of the tube. The light signal was read out by a photomultiplier device and converted to an electric current fed into the read-out chain. Every time the amplitude of the signal current was above a pre-set threshold, a hit was counted. A precise (percent-level) luminosity measurement could be extracted by counting the number of hits recorded in LUCID at each colliding bunch. The radiation level at the location of LUCID is very high. For this reason the detector was provided with a back-up read-out system consisting of 3 m long quartz fibre bundles placed around the beam pipe, and also used as Cherenkov radiators: the light produced at the end of the fibres close to the beam-pipe was guided to a low radiation area and read out by photomultiplier tubes. With rapidly rising LHC luminosity, pile-up exceeded the capacity of the original detector by the end of 2010.

The LUCID detector was the result of a joint effort of groups in Canada, Italy and Sweden, who also developed an updated version of the detector, dubbed LUCID2, which was installed during the first long LHC shutdown and provides a precise ATLAS on-line and off-line luminosity measurement in Run-2.

The two arms of the Zero-Degree Calorimeter ("ZDC") are situated between the incoming and outgoing beam pipes at a distance of 140 m from the centre of ATLAS. Each of its arms is embedded in the LHC neutral-particle absorber in that location. The primary purpose of the ZDC is the detection of forward neutrons and photons with $|\eta| > 8.3$ in heavy-ion collisions, but the detector was occasionally also operated in proton-proton collisions. The ZDC was available to join ATLAS data taking in 2010. Each arm is composed of four calorimeter samplings, the first (closest to the centre of the detector) electromagnetic (EM, about 29 radiation lengths thick) and the following three hadronic (HM, each about 1.14 interaction lengths thick). The modules are composed of a tungsten converter with an embedded matrix of quartz rods, whose Cherenkov light is collected by PMTs. The modules are 9 cm by 9 cm laterally to the beam, and 15 cm thick. The ZDC was designed and built as part of a joint effort between groups from the United States, Chile and Brazil.

Since its first year of operation the detector has suffered severe radiation damage due to the high doses that it receives, affecting the performance of the read-out quartz bar and photomultiplier. Nevertheless, thanks to regular detector refurbishment and careful monitoring of stability during running, it is able to trigger and collect valuable data during heavy-ion collision data taking. These data are useful to investigate the features of ultra-peripheral collisions ($\gamma\gamma\rightarrow\mu\mu$, di/multi-jet production) and to provide information about the centrality of heavy-ion collisions event-by-event. An upgrade is planned to provide more radiation hardness in Run-3.

The ALFA detector was initially designed for calibration measurement of the absolute luminosity from elastic scattering in the Coulomb-nuclear interference region, using the optical theorem in special runs with rather unfocussed beams ("high β*"). Because of the anticipated rather short operation duration (100 hours) and limited resources, a fairly simple and conventional technology based on scintillating fibres coupled to multi-anode photomultipliers read out by an already available chip was selected. Several years of R&D and prototyping focusing on fibre tracker optimisation and development of customised electronics followed, before the series production was launched in 2008; institutes in the Czech Republic, France, Germany, Poland, Portugal and CERN carried out the detector production. The production of eight Roman Pot detectors with about 12,000 fibres was completed in 2009 and an extensive final test-beam campaign for detector qualification was carried out in 2010. The installation of the detector took place in the end-of-year shutdown in early 2011 and was commissioned in the first months of data taking in 2011, alongside the development of the high-β^* beam optics. The first special run with $\beta^* = 90\,\mathrm{m}$ was recorded in autumn 2011. Since then, further dedicated runs were taken with ALFA, to measure not only elastic scattering but also diffractive processes, going well beyond the initial goal of the luminosity calibration.

The AFP detector is designed to enable measurements of exclusive central diffraction and other processes with forward protons during standard proton-proton data taking. AFP uses two horizontal Roman Pot stations on either side of ATLAS for the detector insertion. The tracking is done with 3D pixel sensors read out by the FE-I4 chip, both components also in use in the IBL detector (see also Sec. 10.2.1). For operation with a high pile-up, the downstream station of AFP is equipped with a time-of-flight detector using Cherenkov light emission in fused silica bars coupled to multi-channel plate photomultipliers. The production of the system was mainly done in 2015 by institutes in Canada, the Czech Republic, Norway, Poland, Portugal, Spain and the United States. Given the very tight schedule, a two-stage approach was used for the installation: the first arm was installed in 2016, followed by the second arm and the timing detectors in 2017. After commissioning, AFP took data in nominal high-luminosity runs and in special runs with low pile-up and high β^*.

5.9. The trigger and data acquisition systems

5.9.1. *The trigger system*

Much of the conceptual design of the ATLAS trigger was performed in R&D projects while the experimental collaborations were coming together in the early 1990s. A three-level trigger system was adopted, with each stage keeping only around one percent of the incoming events and thus giving the following stage a hundred times longer to work on the surviving data. The first-level trigger needs purpose-built electronics, while the remaining levels use highly tuned fast algorithms running on carefully selected commercial computers, but with custom link technology to import the data from the detector and first-level trigger. The three-level system is schematically shown in Fig. 5.44.

One of the most important design considerations for the first-level trigger was rapid signal propagation. An interval ("latency") of only $2\,\mu\mathrm{s}$ was allocated from the time of the event to make this initial trigger

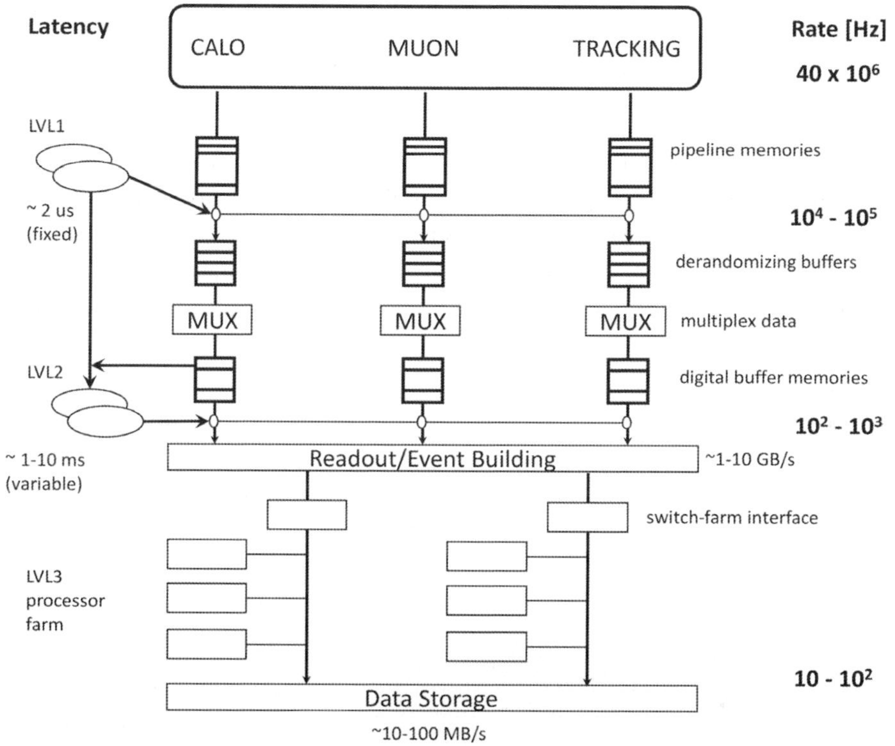

Fig. 5.44. Estimated rates and latencies at different stages of the ATLAS trigger and DAQ system. The event size was initially estimated to be 1 MB.

decision. Half of this time was used just to transmit data from the detector to the underground trigger area and then to return the decision back to the detector electronics. It was clear that, even within the first-level trigger, data would have to be transported serially. The technology chosen, including the Hewlett-Packard (later Agilent) "G-Link" chip, came after considerable R&D and careful component selection. Lessons learnt in this early R&D included the need for precise control of signal timing and printed circuit board manufacture, and remain essential also for the trigger upgrades (see Chap. 10), where the link speeds are a factor 10 to 100 higher.

Once the digital data have arrived, the first-level trigger processor applies selection algorithms to pick out wanted events. ASICs — custom silicon chips in which the algorithms were fixed into the silicon structures on the device — are used in some parts of the trigger. However, for flexibility algorithmic processing is typically done instead with field-programmable gate arrays ("FPGAs"), which have the required capacity and speed and affordable prices. FPGAs can be reprogrammed, so errors can be corrected and hardware can be developed before the full algorithm is completed. Equally importantly, algorithms can be updated later to match changing physics requirements. This was needed in many parts of the first-level trigger to provide the data for the topological processor added for Run-2. An example of a first-level trigger electronics board is shown in Fig. 5.45 — this module, the "CPM", is responsible for finding localised energy clusters in the calorimeter consistent with originating from a high-momentum electron or photon.

The first-level trigger decision is distributed to the detector's front-end electronics, and for accepted events, information on which areas of the muon spectrometer and calorimeters contain interesting data (the Regions of Interest, or "RoI") is forwarded to the second-level trigger by means of

Fig. 5.45. A "CPM" electronics board from the first-level calorimeter trigger. The eight large dark grey heatsinks hide the FPGA chips beneath.

the so-called "RoI-builder". This component was originally developed using FPGAs and was later converted into a software application running on a Linux-based server for Run-2.

One of the most important strategic choices for the second- and third-level triggers (the latter known as the "Event Filter", EF; the two together are known as the "High-Level Trigger", HLT) was the decision to use the same software framework as run offline to run (customised) data selection, once it was clear that these trigger levels could be software-based. This reduced the software development effort. Even more importantly, this allowed a much more uniform view of the data throughout the online and offline worlds. Initially, the second-level trigger and EF were built as independent computer farms exchanging only the second-level results to seed the EF. As of Run-2, the two trigger levels were merged into a single HLT system.

Development of custom trigger hardware was distributed widely round the globe. For example, the first-level calorimeter trigger was developed by groups from Germany, Sweden and the United Kingdom, while the muon trigger work was shared between institutes in Israel, Italy and Japan. The final trigger-decision logic and the precision timing infrastructure for the experiment were developed by CERN. Developments in the HLT were truly global, involving almost all countries participating in ATLAS.

Each grouping needed a mixture of physics, electronics and software experts. With such a distributed team, precise specification documents were essential and frequent (eventually weekly) videoconferencing was also an invaluable tool.

5.9.2. *The data acquisition (DAQ)*

When the design of the data acquisition (DAQ) for ATLAS was launched, the challenges were manifold and commercial off-the-shelf technologies were not yet able to deliver the required performance at an affordable cost. The system requirements were outlined in the ATLAS Technical Proposal and are shown in Fig. 5.44. In particular, critical aspects of the DAQ that required dedicated R&D were:

- Buffering elements capable of receiving data fragments at 100 kHz over point-to-point links, the read-out buffers ("ROB");
- Networking technologies;
- Storage technologies (estimated 100 Hz of 1 MB events);
- Scalability of control, configuration and monitoring.

Initial R&D on the ROBs relied on VME technology, which demonstrated that the system was in principle, feasible, but the cost and dependency on the evolution of the VMEbus market were not optimal. The establishment of Linux as a high-quality operating system for DAQ as well as the advent of higher performance PCs with high speed CPU clocks (GHz) and multiple PCI bus slots, capable of accommodating different network cards as well as custom designed PCI boards, provided a big step forward. The eventual read-out system design merges the ROB, second-level trigger and event-builder interface functionality into a single software process, with highly optimised scheduling and buffer management; the ROB functionality was implemented in custom PCI boards (known as "ROBINs"), each receiving data over up to four point-to-point links (three such cards were hosted in a single rack-mounted PC); and the local crate controller was transformed into a software API embedded into the read-out system process and communicating over network to the overall control, configuration and monitoring system.

ATLAS took data with this system throughout the experiment's commissioning and during the first collision run period from 2009 to 2013. During the first long shutdown, LS-1, in 2013–2014, a new generation of ROBIN cards was introduced, based on PCIe and supporting higher link density (12 links per card). The software design was upgraded to better exploit modern multicore CPU architectures and to support 2 ROBIN cards per rack-mounted host PC.

The second technological challenge for the DAQ was the network. Despite the ingenious introduction of RoI-based event selection for the second-level trigger, which greatly reduced the requirements on the network, it was not obvious which of the communication technologies would scale to the needs of the full experiment. Multiple independent networks with an aggregated bandwidth of up to 10 GB/s each were devised: one to transport RoI data (few % of the 1 MB event at 100 kHz) to the second-level trigger farm, another one to carry out event building at ~1 kHz, a third one to transport accepted events from the Event Filter to the storage system and a fourth one for configuration, monitoring and control of the overall trigger/DAQ system.

While it was always assumed that the control network would be based on Ethernet, a lively R&D programme took place for the data networks: technologies such as SCI, HIPPI, Fibre Channel, ATM and Ethernet were all studied and benchmarked heavily. Different types of network topologies were proposed and analysed, in order to be able to achieve the overall aggregated throughput avoiding large central switches, which were deemed prohibitively expensive.

The solution to this problem came from the formidable progress of networking devices and technologies. The introduction of Gigabit Ethernet wiped all concurrent switching technologies away, for the purpose of the ATLAS DAQ system. By the time that the DAQ system was being built, large telecom routers surrounded by smaller top-of-rack switches provided the required performance and functionality.

After the first period of data taking, the network was completely refurbished during LS-1, introducing 10 Gb/s at the ROS and a large degree of redundancy, thus making the network, i.e., the backbone of the DAQ system, much more reliable and fault tolerant. The change in architecture is illustrated in Fig. 5.46.

The third challenge for the DAQ was the storage system that was needed to record and preserve a copy of the physics data until it was safely transferred to permanent storage, both at the main CERN computer centre and outside CERN, on the worldwide LHC computing grid. Storage technologies advanced rapidly in the decade before data taking. As a result, the ATLAS requirements became easily achievable with a few high-performance storage servers. However, careful software development and optimisation was still required to make efficient use of the underlying technology.

Last but not least, the trigger/DAQ system needed to be coherently controlled, configured and monitored. A component-based, software architecture was designed from top-down, followed by a detailed design of each core component. At the heart of the back-end software was the choice of a common inter-process communication technology: CORBA. While components have been re-designed and re-implemented several times, profiting from new software techniques and adapting to hardware and operating system platforms, the overall structure of the back-end software is largely unchanged. This is an example of a great success of a large distributed software project, which has maintained its shape throughout its twenty years of technological evolution.

DAQ expertise in ATLAS, similar to that of the trigger community, is spread over many participating countries, from Russia to the United States,

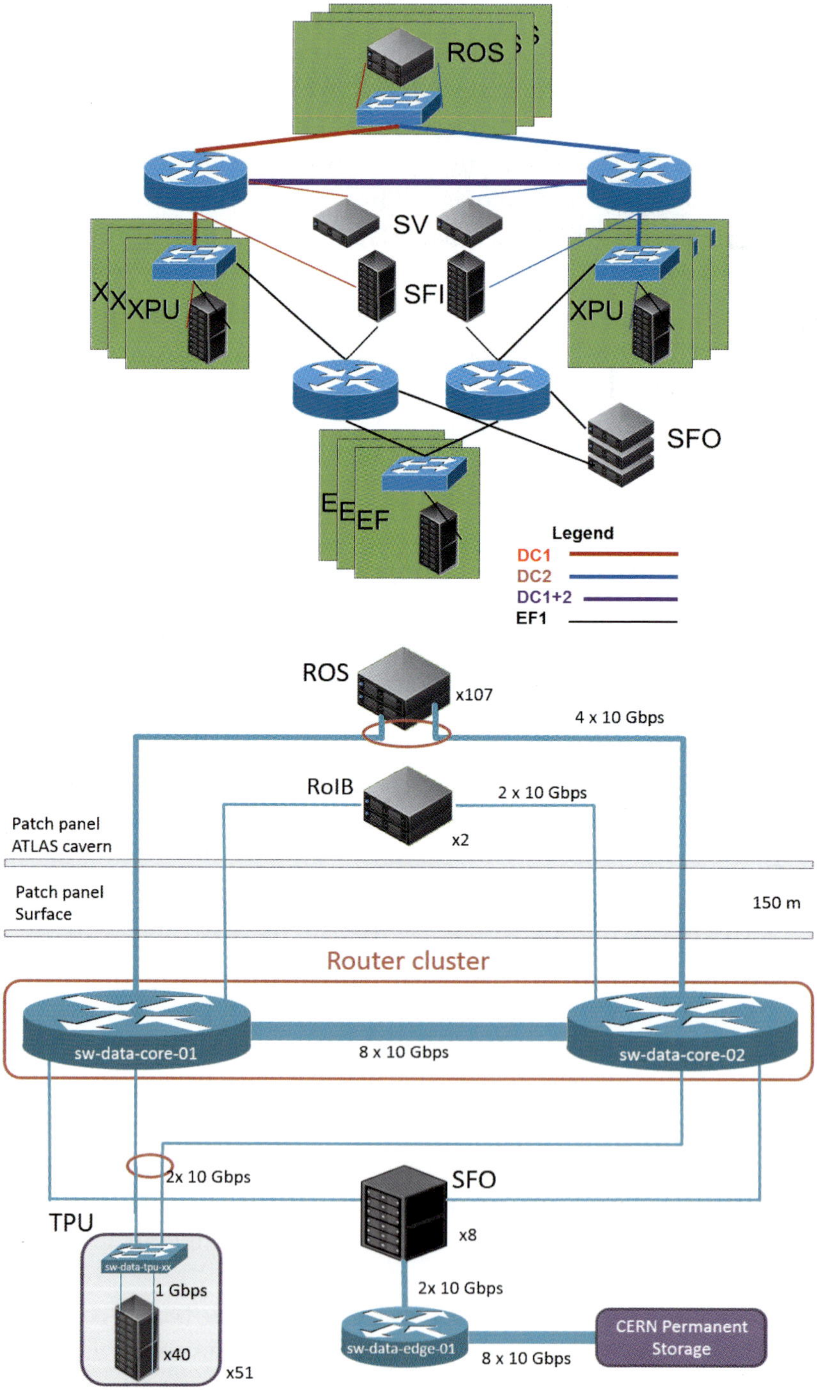

Fig. 5.46. Trigger/DAQ network architecture during Run-1 (top) and Run-2 (bottom).

Fig. 5.47. Experts monitoring the DAQ from the satellite control room during commissioning with the first LHC beam collisions.

from Romania to France, from the United Kingdom and the Netherlands to Italy, to mention a few. The expertise is strongly centred round CERN, especially for leadership of R&D, integration and testing, as well as for support to detectors for test-beams, installation and commissioning (see also Fig. 5.47).

Inner-Detector material radiography

The Inner Detector, despite being one of the smaller components of ATLAS (2 m high and 6 m long), still weighs over 4 tonnes! Having such a massive detector has consequences for its performance, as the particles arising from the LHC collisions will interact with this material in ways which may degrade the detector's ability to measure particle trajectory and energy. Since interactions are unavoidable, one must understand how and where they occur.

In order to understand particle interactions with matter, a computer model of the detector is needed. This model, first created from the engineering diagrams of the detector, requires detailed knowledge of both the material composition and location of that material. Unfortunately, this did not provide a perfectly accurate representation of the detector *as-built*. For the model to be accurate, all changes even as simple as the location of a cable had to be taken into account. As the knowledge of these changes resided in disparate databases or in the notes and minds of many scientists, engineers and technicians, the challenge was to find all the design modifications. This required discussions amongst numerous colleagues while examining photos and engineering diagrams.

To validate the model, the weight of the Inner Detector was checked to ensure that it was the same in the model and in real life. Once verified, the location of the material needed to be validated — a task made more difficult by the fact one couldn't see much of it after installation.

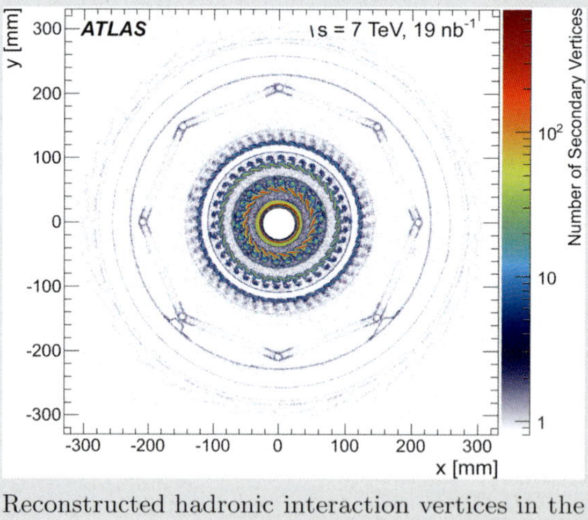

Reconstructed hadronic interaction vertices in the barrel Pixel Detector showing detection devices, cooling pipes and support structures.

Data from LHC collisions were used to "see" inside the detector. Over a dozen unique measurements sensitive to particles interacting with material were performed. Some of the most precise included reconstructing the position of photons converting into electron-positron pairs or hadronic particles colliding with the nuclei of matter in the detector giving rise to additional particles created at the impact point. These measurements provided detailed three-dimensional "radiograms" of the inner portions of the Inner Detector (see figure). Other measurements checked the total amount of material in the Inner Detector by selecting well understood collisions and measuring how energy was distributed in the calorimeters.

These measurements took three years to complete but have allowed for a better understanding of the material in the Inner Detector. This effort has been one of pieces of the puzzle needed to achieve the ATLAS physics objectives.

The road to the first minimum-bias publication

Finally! On 23 November 2009, ATLAS recorded the first LHC collisions. Within days, close to 500,000 events at an energy of 0.9 TeV had been collected. After years of simulation studies, ATLAS faced the reality check.

Within two months, a team of some 40, mainly early-career, physicists, finished analysing these events and ATLAS submitted its first collision-data publication, which the anonymous journal referee considered "a very thorough piece of work, which I read with great pleasure".

The goal of the paper was to measure the properties of "minimum-bias events" sensitive to some of the least understood aspects of strong-interaction physics. Furthermore, these events contribute noise to the high-energy processes which are the focus of much of the LHC program.

The consistency observed between simulation and data was remarkably good. The analysis provided a stringent test of the simulation model of the detector response and the material inside the tracking volume, which was derived from data using complementary methods. A small problem was identified in the very forward region, which intense scrutiny revealed was due to services such as power and data cables being shifted by a few cm in the simulation to lie outside the tracking volume. In reality, particles had to transverse more material in the forward regions reducing the tracking efficiency. Thus, the simulation needed to be modified and some 10 million simulated events generated again.

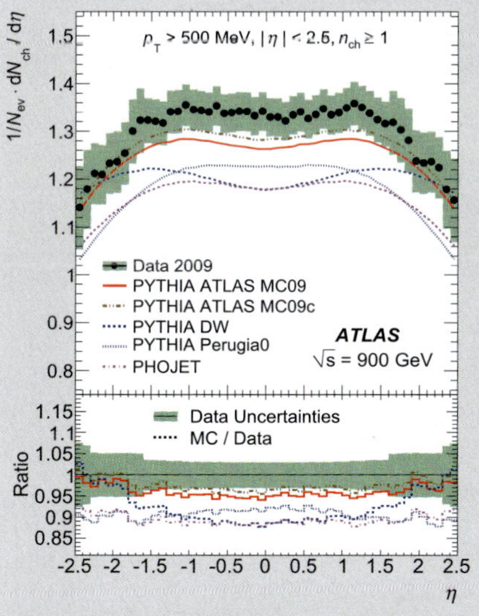

Charged-particle multiplicity as a function of pseudorapidity, confronting data to models.

This and many other issues had to be solved. "It was amazing to see how problems that normally would have taken weeks or months of study were resolved in a few days by the consistent effort of all the group," one analyser commented. But it took nights and weekends, and even the Christmas break was not completely spared, to finalise the analysis. A paper draft was submitted to the Collaboration on 5 February 2010 together with 250 pages of detailed background material. After receiving 71 comments by the Collaboration, the first ATLAS collision-data physics result was presented at the La Thuile conference in February and shortly afterwards published.

This was a memorable and extremely useful experience for all those involved in the effort, and for ATLAS in general. It was a major stress test for the data preparation chain, underlined the importance of the close connection of detector experts and analysers, and exercised the whole collaboration review process under heavy time pressure.

Chapter 6

Installation of the Detectors and Technical Coordination

6.1. Cavern and civil engineering

The civil engineering guidelines for the LHC accelerator and its experimental programme were to make optimal use of existing LEP and CERN infrastructures, and to minimise costs whenever possible. As the ATLAS detector with a radius of 12 m and a length of 44 m was much more voluminous than the CMS detector, the correspondingly much larger cavern had to be positioned into the favourable underground geology of Point-1, opposite the entrance of CERN.

However, no large surface building could be accommodated there to pre-assemble the ATLAS detector. Therefore, ATLAS used various existing CERN hall space, scattered over the site, for assembling its detector components.

Several actions were undertaken to speed up the civil engineering works for ATLAS, so as to allow for as much time as possible for its installation and integration:

- The cavern excavation was taken to −12 m from the cavern vault during the LEP operation; pre-stressed steel cables from auxiliary side excavations to the access shafts to the vault were provided to suspend the concreted vault ($\sim 10^4$ tonnes) firmly in its final hanging position. The stress was relieved years later when the supporting sidewalls of the cavern were built.
- Machine components of LEP in the tunnel below the excavation were regularly readjusted to compensate for the upward movements of the LEP tunnel (about 1 cm) caused by the elastic decompression of the diminishing rock above.

- A pathway and the necessary authorisations were established to dump the spoil ($\sim 3 \times 10^5$ tonnes or $\sim 10^5$ m^3) of the excavation some hundred metres away from Point-1, without having to pass through police and custom checkpoints and by using Swiss and French roads.

Underground excavation started in February 1998, during the LEP operation that lasted until November 2000. Except for the upper parts of the shafts all underground structures were excavated in molasse rock using rock breakers, without having to resort to explosives so as to avoid problems with LEP and neighbouring habitations. After excavation, they all received a temporary lining, thorough water proofing with proper drainage system, and a final lining made of in situ cast concrete. Where necessary pre-formed channels or pits were included for services and other purposes. The floor of the main cavern was cast horizontally, necessitating sloping supports for the experiment to follow the inclination of the LHC machine. By April and June 2003, the underground cavern and all underground works were, respectively, handed over to CERN for the installation of its infrastructure. Figure 6.1 shows the layout of Point-1, while Fig. 6.2 shows the inauguration of the cavern on 4 June 2003.

The underground work included the excavation of:

- The PX14 and PX16 shafts, respectively 18 m and 12.6 m in diameter, both 60 m deep.
- The PX15 shaft was an existing shaft that required concreting only.
- The main cavern UX15 (50 m long, 30 m wide and 35 m high), which houses the ATLAS detector.
- The electronics room and services cavern USA15, which are accessible during LHC machine operation.
- The seven linking galleries for personnel access and the distribution of services between the UX15 and USA15 caverns.

The surface buildings include the following:

- SX1, a wood clad steel frame building housing the travelling cranes with maximum capacity of 280 tonnes, on top of the main shafts. This building was the first to be structurally finished, thus allowing the infrastructure works to take place in dry and controlled conditions.
- SDX1, located on top of the personnel access shaft and used for personnel and material access control to the underground areas. It also contains the uninterruptible power supplies and the DAQ room.

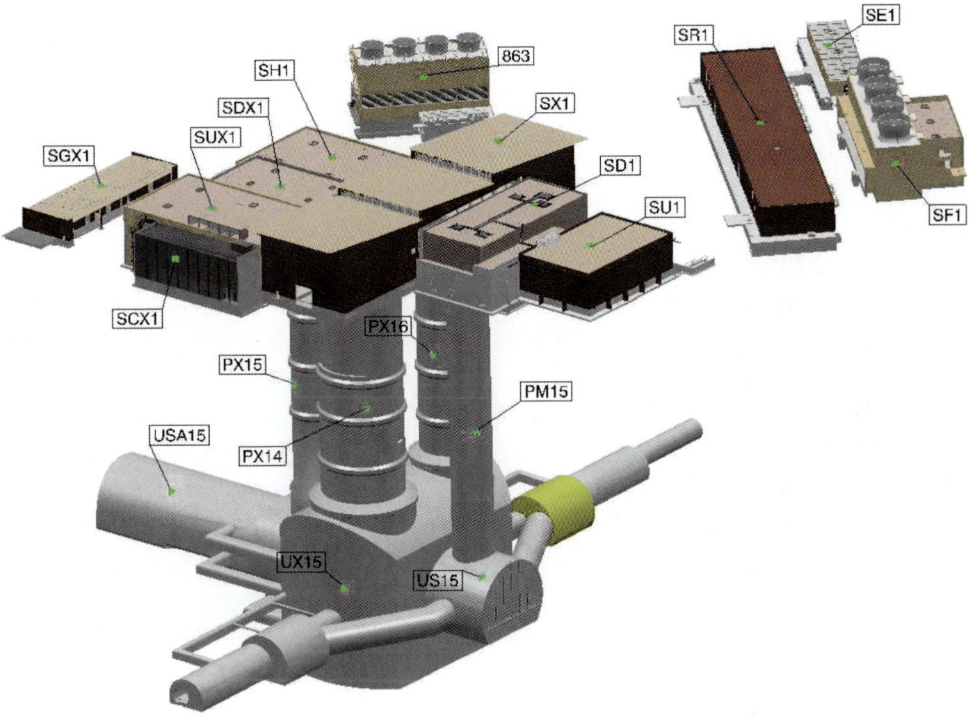

Fig. 6.1. Layout of the surface buildings and access shafts to the ATLAS cavern at Point-1. The main areas of underground activity are the main cavern (UX15) and the main electronics room and service cavern (USA15). The main control room is in building SCX1 on the surface.

- SUX1, the ventilation building containing water chillers and air-conditioning units.
- SCX1, the main ATLAS control building and visitor centre.
- SGX1, mixing and storage of gases for ATLAS detectors.
- SH1, a concrete building containing the noisy cryogenic compressors and other cryogenic equipment.
- SF1, the new cooling towers for the final dissipation of heat recovered from all the underground areas and surface buildings via the water cooling circuits.

Stability measurements performed in UX15 between August 2003 and August 2007 in various places inside the concrete floor and parts of ATLAS (>85% of its weight) showed upward movements of up to 1.4 mm and then remaining stable. The prediction by the civil engineering team had been up to 6 mm upward movements.

Fig. 6.2. Inauguration of the ATLAS cavern on 4 June 2003, in the presence of Mr. P. Couchepin, the President of the Swiss Confederation.

6.2. Feet and rails, access structures, shielding

One of the basic elements of the detector is a system of supports on which the most massive components — the central solenoid with the Inner Detector, the calorimeter system, the toroid magnet system, the barrel muon detectors and the small wheels — of ATLAS are mounted. The total weight of these elements is about 7000 tonnes. This support system, called "Feet and Rails" (F&R), must satisfy the following requirements:

- Minimum magnetic permeability, so as not to spoil the characteristics of the magnetic field;
- Sufficient strength;
- High machining accuracy;
- Ability to withstand emergency loads, for example, rapid cooling with cryogenic liquids;
- Convenience for mounting and maintenance of the muon barrel detectors;
- Acceptable cost.

The main components of the F&R are the bedplates, providing an inclination of ATLAS equal to that of the LHC plane, and nine pairs of feet with rails on top, for the longitudinal movement of detectors during the installation phase and the maintenance periods.

In 2000, the ATLAS Technical Coordination was nearing the completion of the definition drawings, and the Collaboration was about to decide on placing a manufacturing order in the industry. It should be noted that F&R manufacturing deadlines were very tight, since the system had to be first mounted before any other parts of the installation could follow. This project, along with other infrastructure components, was a Common Project (CP) item, thus neither a Funding Agency (FA) nor a responsible institution were initially chosen. The Russian participants of ATLAS were then invited to perform this work at one of the Russian enterprises, as an in-kind contribution to CPs. IHEP Protvino was appointed as the main Russian institute responsible for the project.

After considering possible manufacturers, "Izhorskiye Zavody" (IZ), a firm in St. Petersburg that could perform all the above work, plus the production of materials (Fig. 6.3, left), was favoured. Thus, this centralisation of all work aspects helped IHEP significantly simplify the following processes: coordination of technical documentation, work schedules and production control. IZ is one of the oldest industrial enterprises in Russia, founded in 1722 by a decree of Peter the Great. It is a large enterprise with its own metallurgical plant, rolling mills and advanced metal processing. In its portfolio are equipment for nuclear power plants (including vessels of nuclear reactors) and petrochemical equipment, metallurgical castings from steels with special properties, and bridge and road metal structures.

Fig. 6.3. (Left) Mechanical cleaning of a 80 mm steel plate at IZ. (Right) Assembled Feet and Rails in the ATLAS cavern.

For this plant, the F&R order was relatively small, "only US$4 million for 500 tonnes of stainless steel products", and non-standard, so one could expect a variety of hard-to-predict problems. It was not surprising that this proposal was not met with much enthusiasm. However, after studying technical documentation and further communications with colleagues from CERN and IHEP, it was found that this project was not only feasible, but also interesting for the development of the company's technological potential and prestige. One of the interesting tasks was the production of the necessary stainless steel. CERN's designers relied on European standards, which differed from typical Russian ones. It took a thorough analysis of the technical requirements and the manufacturing capabilities of the plant to achieve the parameters for the metal that would satisfy all the magnetic and strength characteristics needed.

To prepare the fabrication drawings, the IHEP made a separate agreement with the IZ design bureau. However, the main issue with the drawings done then was the uncertainty of the welding deformations, and at the same time, the attainability of the required accuracy in the subsequent machining of the parts. Another contract was concluded for the production of a trial batch of steel plates and for the production of welded joint samples from these plates. The contract for the manufacture of the F&R stipulated the supply of six batches of components for the ATLAS support structure during 2002 and 2003. This made it possible to control the schedule and quality of work to some extent.

The quality of welds was controlled by several methods, including ultrasonic inspection, and when any defects were found, the seams were repaired. For an independent control of welding works, IHEP attracted the St. Petersburg Maritime Register under a separate contract. The flatness of the bedplates was controlled by an independent GRADAN company specialising in optical measurements. The list of control operations was summarised in the Quality Control plan for the project. It is worth acknowledging the readiness of IZ to carry out and document all necessary checks and measurements, as well as their liberal attitude towards the changes requested by the customer in the drawings, including some significant ones. The cost of the project did not increase in comparison with the initial contract.

The transportation of the components to CERN was simple, since neither the dimensions nor the weight of the shipped components were too big. The ATLAS F&R were successfully assembled in the pit from October 2003 until February 2004 (Fig. 6.3, right).

6.3. Magnet system installation

The magnet system was installed on the stainless steel feet and rail system (see previous section) that carries the barrel toroid (BT) directly and the two endcap toroids (ECTs) indirectly via the central rails (see Fig. 6.3, right). Using rails enables one to retract the two ECTs, thereby creating space for access to the detector interior for maintenance and repairs. The central solenoid, integrated in the common barrel liquid argon calorimeter cryostat, rests on the rails as well. The BT installation started in October 2004 with the lowering of the first coil into the cavern (Fig. 6.4). The toroid layout after the installation of six coils is shown in Fig. 6.5, and the completed barrel toroid is pictured in Fig. 2.6 of Chap. 2.

A warm structure comprising eight inner and eight outer supporting rings mechanically interconnect the eight coils. Each ring consists of eight support beams ("struts"), and every strut is assembled between pairs of coils. In total, 64 struts, high-strength bolting and shims are required to complete the structure. The struts are 5 to 6 m long and the largest version has a connection flange section of $500 \times 800\,\text{mm}^2$. The warm structure is made of Al alloy and produced by forging and stamping (inner ring struts), and extrusion (outer rings struts). For the duration of the BT installation, large temporary platforms were erected in the cavern to facilitate the work; they

Fig. 6.4. Descent of a 25 m long barrel toroid coil through the shaft into the underground cavern.

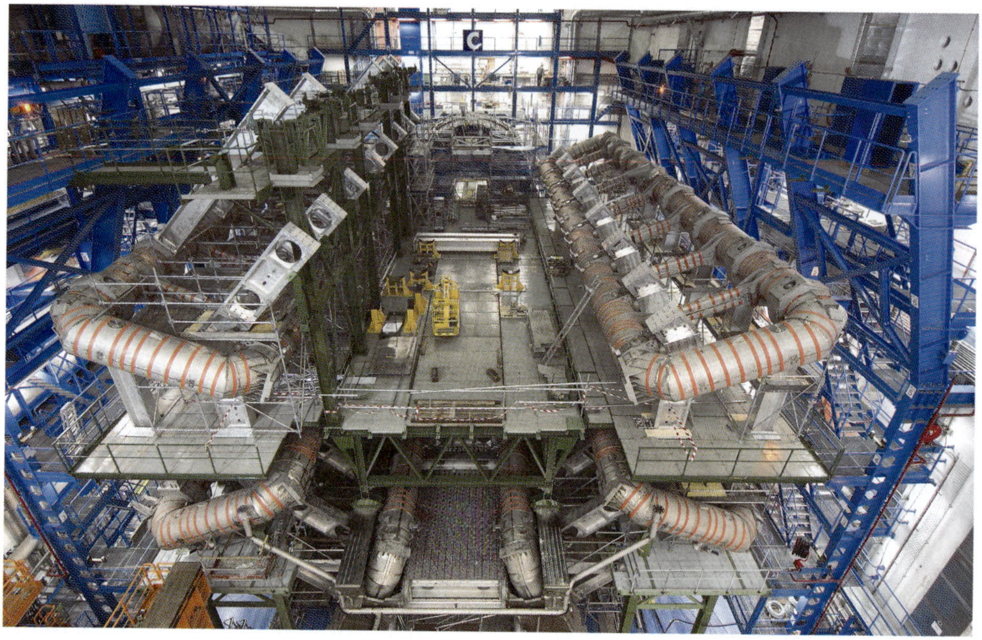

Fig. 6.5. Barrel toroid installation in progress (June 2005), status when six coils are in position.

are well visible in Fig. 6.5. The warm structure parts were produced mainly in industries in Russia and the Belarus Republic, under the leadership of the JINR Dubna workshop.

Quite exciting was the vertical deflection of the BT structure, measured at the top, during the load transfer. Since the toroid bore must be cylindrical at its end, all coils were installed in calculated positions on an oval with +30 mm in vertical direction. After release of the temporary support structure, and loading the toroid with its self-weight of 830 tonnes plus the 400 tonne weight of the muon chambers, the final shape must have a perfect circular cross section. After the release and removal of the installation supports the toroid did come down by 26 mm, demonstrating that the design was very well calculated and that the installation of the 25 m long open structure with a diameter of 22 m was precise to within a few millimetres.

For the two ECTs, the coil modules, thermal shields, multilayer insulations, coil suspension parts and vacuum vessels were produced in the industry, but the integrations of the cold mass, cryostat and connections to services were performed at CERN. Figure 6.6 shows the stage of integration of a cold mass in the vacuum vessel before applying the superinsulation and

Fig. 6.6. Cryostat integration of the endcap toroid, the cold mass (140 t) with its heat shield is inserted in the vessel (80 t).

closing the vacuum vessel with its end plate. Clearly recognised inside the castellated vacuum vessel are the eight coil modules interconnected by eight keystone boxes. The frame visible around the vessel keeps it in shape during the assembly. The completed ECTs were then transported from the assembly hall on the CERN site to ATLAS in Point-1.

Service connections for helium, current, vacuum pumping and controls are on the top, arranged in a so-called turret. The transport and craning of the 240 tonnes ECTs to their final position in the ATLAS cavern was completed by July 2007. Figure 6.7 shows an ECT after transport, just as it enters the surface hall on top of the shaft.

When energising the three toroids the ECTs are pulled towards the centre by a net Lorentz force of about 240 tonnes, thereby compressing the BT at either side. A precise alignment is necessary to avoid excessive asymmetric forces on the individual BT cold masses.

Following the mechanical installation, all interconnections between coils for the helium transfer lines and the current bus bars were put in place. After completion of the BT installation, the Central Solenoid was moved together with the barrel calorimeter on the rails and locked in the centre of the detector. In autumn 2006, the BT was gradually powered up to the

Fig. 6.7. The endcap toroid after transport, in the surface building giving access to the shaft and cavern 100 m underground. Special hydraulic tooling on rails is attached to lift-and-slide the toroid in the building.

operating current of 20.4 kA. By summer 2007 the two ECTs were lowered down into the cavern and positioned on the rails, thereby completing the magnet system's installation.

Parallel to the installation of the magnets, all the following services required to operate the system were installed and commissioned. The barrel toroid and both endcap toroids are connected in series to a 24 kA/16 V switch-mode power supply through 200 m of bus bars and switchgear. Across the BT and the ECTs, diode/resistor units guarantee safe slow dumps of the magnets, thereby taking up the 1.6 GJ magnet stored energy. The maximum voltage is 40 V in the case of a fast dump triggered by a quench in the superconducting coils. The toroids are charged-up in three hours, and a slow dump takes three hours as well. A similar system is present to serve the Solenoid with its maximum current of 7.73 kA.

The magnets are conduction-cooled and two-phase helium is circulated through cooling tubes attached to the various cold masses. The proximity cryogenics system, feeding the coils with helium, can be seen in Fig. 2.4 of Chap. 2. A 6 kW He liquefier fills an 11,000 litre buffer dewar, from which the helium is taken and circulated by a cold helium pump of capacity 1.3 kg/s. A distribution valve box in between pump and coils enables one to control the flow in each of the 10 magnet units (8 BT coils and 2 ECTs). A second 30 kW refrigerator serves for the cooling down and keeps the thermal shields at about 40 K. The Magnet Control and Safety System provides the process controls for vacuum, cryogenics, electrical systems and the quench protection system triggering fast dump, respectively.

The magnet system connected to all its services was thoroughly tested and commissioned in summer 2008, just on time for the first beam operation of the LHC in September 2008.

6.4. Installation of detector components

Starting the description from the centre, the ***Inner Detector*** (ID) installation, with its four main preassembled components (SCT and TRT) for the Barrel and the two endcaps (ECs) plus the Pixel Detector, took place between August 2006 and May 2008. After years of design, construction, integration and commissioning, the four individual ID parts were individually moved from the SR1 cleanroom to the ATLAS cavern. The detectors were relayed across the surface area by an assembly line of cranes, from the doorway of the clean room to the entrance of Point-1 and then 100 metres down into the cavern itself. Special measures were put in place to minimise shocks and vibrations during all handling operations. Accelerometers were fitted to each subdetector to provide real-time monitoring. No values above 0.2 g were recorded, in line with the transport specification for these extremely precise and fragile detectors.

The final insertion and commissioning of the barrel started on 23 August 2006. The barrel detector was lowered down, placed on the alignment stage sitting on the so-called minivans, moved and rotated towards the barrel calorimeter (Fig. 6.8, left). The rails of the transport trolley were then linked and precisely aligned with those of the LAr cryostat at the level of a few tens of microns. Precise templates and a robotic measurement arm were employed to verify the tight installation clearances. The detector was made to slide on pads via a controlled movement mechanism, whilst being constantly monitored for full integrity against any possible skewing, twisting or unbalanced reactions of supports. With only a few millimetres of clearance, the detector was successfully inserted into the heart of the ATLAS detector, surveyed and adjusted to its final position with respect to the nominal beam line. Once the barrel services were deployed, connected

Fig. 6.8. (Left) The insertion of the ID barrel into the LAr Cryostat before it will be moved towards its final position around the interaction point. (Right) Barrel commissioned, PST-installed, and ready for the insertion of the ID-EC and the Pixel Detector.

and fully commissioned, the pixel support tube (PST) was installed, and the LAr cryostat bore was ready to receive the two ID endcaps and the Pixel Detector (Fig. 6.8, right).

The EC's installation process, following a very similar procedure as done for the barrel, started with the EC side-A on 24 May 2007, followed by the EC side-C on 18 June. The lowering and insertion of the 7 m long Pixel package took place immediately after that on 26 June, before partially closing the detector, which was required for the lowering of the toroid endcap magnet. Time was then allocated to connect and fully commission both ECs, before making those inaccessible with the deployment and connection of the dense Pixel services. The entire commissioning of all ID detectors and services was completed in about eight months, followed in May 2008 by the final closing of the ID by the end-plate on both sides, providing sealing of the dry gas environment for its full operation.

The ***LAr and Tile Calorimeters*** were the first detector components installed in the ATLAS cavern, beginning in April 2004 with the pre-assembled eight bottom modules of the Tile central barrel resting on their temporary support cradle. These were positioned on the extension of the ATLAS rails directly below the ATLAS cavern shaft, as seen on the left hand side of Fig. 6.9. In the following months, additional Tile modules of 20 tonnes each (modules 9 to 32), were installed, each bearing on the one below at the inner and outer radius to form a self-supporting structure. The calorimeter "saddles" that transfer its load to the main ATLAS rails were installed when 18 Tile modules were assembled. As a result of this support condition, the module interfaces at the outer radius are in tension/compression below or

Fig. 6.9. (Left) First eight modules of the Tile Calorimeter central barrel lowered in the ATLAS cavern. (Right) The LAr barrel cryostat being placed onto the Tile Calorimeter half barrel.

above the saddles. The overall system was designed so that no load is carried across the two Tile modules at the top of the cylinder.

In October 2004, after the Tile bottom half cylinder assembly, the LAr barrel cryostat, containing the barrel electromagnetic calorimeter, was lowered onto the half Tile barrel and supported by feet resting on the Tile cradle, as seen on the right hand side of Fig. 6.9. The mounting of modules for the upper half-barrel of the Tile Calorimeter (modules 33 to 64) was continued and the full barrel calorimeter was completed two months later. The installation of each Tile module and of the LAr cryostat was carefully monitored with a set of surveying tools, all along the assembly process, to form an 8.5 m diameter and 6 m long, almost perfect, cylinder. The global radial envelope was set at +5 mm outward and −10 mm inward. The prior pre-assembly of the Tile cylinder on the surface allowed the adjustment of the thickness of the shims placed in between the Tile modules at the inner and outer radius of each module, and was a crucial step for the successful final installation, respecting the overall dimensional envelope.

After the final assembly and while still positioned on the ATLAS rails extension platform, an important part of the mechanics to hold the detector electronics and services was put in place. The LAr "boxes" (Front-end Crates) to contain electronics read-out boards were installed on the "pedestals" where the signal cables exit the cryostat through cold-warm feedthroughs. Also cable trays to accommodate cables as well as cooling pipes for the Inner Detector, and for both LAr and Tile Calorimeters were placed on the Tile Calorimeter surface.

In November 2005, now as one object, using a combination of hydraulic cylinders and an air pad lifting system creating an almost frictionless surface between the calorimeter supports and the rails of ATLAS, the 1600 tonnes of the barrel calorimeter cylinder were pushed-pulled onto the rails into its final position at the centre of the detector. After this displacement, the location directly below the shaft of the ATLAS cavern was freed to receive the endcap part of the calorimeters. Following a similar assembly procedure as done for the barrel cylinder, the two endcap calorimeter systems were completed, with the lowering of components being done using the shafts on either sides of the cavern. In the case of the endcaps the movement system remains in place today, providing access to the Inner Detector system for maintenance at each longer technical stop. The calorimeter assemblies were completed in May 2006. At this point all services and cables for the calorimeters were installed and routed to the electronics cavern USA15 (see Sec. 6.5).

One of the critical elements for the calorimeter installation was the manufacturing and commissioning of the "cable chains", which connect all services and cables from the endcap calorimeter assemblies which need to move (up to 14 metres in the large access opening scenario) during shutdowns. Three chains carry these connections on each side and were filled with cables and services. They are visible on the left hand side of Fig. 6.10. In parallel to the services, the installation and partial commissioning of electronics on the detector was performed (see Fig. 6.10, right), typically with interim connections and temporary test systems. The final closing of the endcap calorimeters took place in May 2008, when all the electronics had been installed and the first commissioning carried out.

The ***Muon Spectrometer*** is composed of two sections, a barrel and the two forward region spectrometers, respectively. They together form the outermost part of the ATLAS detector. While the components in the forward regions rely on mechanical assemblies that are different for all layers, the supports for the barrel stations (MDT and RPC chambers linked together for the middle and outer layers) make use of the BT warm structure.

At regular intervals, along the BT length, brackets are used to fasten extruded aluminum rails. These rails of various forms were designed to minimise deformations generated by the different shapes and loads of the stations. On surface these were slid into frame rails by means of their kinematic supports and positioned inside a tooling reproducing the given

Fig. 6.10. (Left) The endcap assembly before being moved towards its final position. On the cylinder surface note the services and cables. (Right) The barrel and endcap calorimeters ready to be closed; the electronics installed and powered, can be seen on the barrel calorimeter.

final configuration. By using the cavern cranes, the assembly was lowered into the trench, then rotated to the correct installation angle, and finally lifted and positioned facing the BT rails (Fig. 6.11, left). The assembly rails permitted the transfer of the chambers to the inside of the BT with two winches. Additional barrel chambers, installed after 2008 for completing the Muon Spectrometer coverage, closely followed the described technique.

No such general installation procedures, as for the barrel, could be applied for the placement of the forward chambers. The various forward layers, namely the Small Wheel (SW), the transition region chambers (called EIL4, EEL and EES), the Big Wheel (BW) and the End Wall (EO), had to be placed considering different approaches and procedures (Fig. 6.11, right). To permit the housing of the SW chambers (CSC, MDT and TGC), installed with special rotator tools, dedicated iron and aluminum support structures had to be designed. Completed in a surface hall, the two SW were transported fully assembled and lowered into the cavern. They were displaced via air pads on the main detector support rail system, and positioned. The installation of the special EIL4 section was inspired by the barrel techniques. It was slid inside the BT and vertically placed by rotating it with tools temporarily fixed to the BT.

For the transition region installation, the chambers were inserted through an opening, located at the top of the BT that was sufficiently large but not

Fig. 6.11. (Left) Installation of a barrel muon chamber station. (Right) Installation of a forward region MDT sector, with a TGC Big Wheel in the background.

high enough, so that the assembly had to follow an atypical pathway to be introduced and vertically placed. The chambers (EES type) were lowered and locked in position by rolling on two circular rings linked to supports fixed onto the BT. In some other cases (EEL type), the method was similar to that used for the SW, also employing one of its rotator tools.

The BWs are composed of sectors. These were assembled, stored on surface and transported into the cavern to be attached to supports temporarily fixed to the end wall (HO) structure. When all the sectors were connected together forming a BW, the latter was detached from the HO and moved away using two chariots located on rails along the top of the cavern on both sides.

Like the BW, the EO detectors are also divided into sectors. Their mechanics was assembled on surface, transported into the cavern, installed and fixed onto the HO structure. The installation of the chamber was made in the cavern using the rotator tool used for the SW.

6.5. Cables and pipes

Cables are like the blood vessels and nervous system of ATLAS, reaching each of its elements. With the help of all these cables, the ATLAS detector is supplied with power, controlled and its data read out.

Though physicists started to think about services at the very earlier stages of the ATLAS detectors development, the cabling project turned into its active phase in 2002. The following five years were very intense, passing through the projection and computer-aided design (CAD) phases, then the installation of cable trays and finally, the cables. Close collaboration between Technical Coordination and the detector experts at all stages was essential for the success. The complex environment, especially inside the detector volume, required a thorough CAD workout, a true challenge.

Installation of cable trays in caverns and galleries started in 2003 and then continued inside the detector volume, reaching its peak in 2005–6. The cable installation campaign itself started in May 2005, when Technical Coordination created a cabling team, involving technicians from several institutions from different countries, with the majority coming from Russia. On average, the installation team included 21 technicians over the next 2.5 years, reaching a maximum of 36 people in the middle of 2006.

Here are some numbers to illustrate the scale of the ATLAS cabling task. More than 25,000 optical fibre channels are used for reading the information from the subdetectors and delivering the timing signals. About 3000 cable trays are present in the cabling database. The total number of installed cables and flexible pipes, not including the local cables on the sub-detectors, is about 78,600; their total length is approaching 3500 km. The development and realisation of the cabling project took more than 60 people-years of intensive work.

Interestingly enough, the smallest size of all subdetectors, the Inner Detector (ID), possesses a third of all cables, approximately the same quantity as the largest one, the muon system. Their cable installation complexity is opposite: the ID cables are concentrated in a very small volume, requiring detailed design and precise installation, whereas the muon cables are distributed over the whole detector volume, sometimes making installation tricky. The other ATLAS subsystems share the last third of the cables.

The cables and tubes for some subdetectors had to be in place before the components were lowered into the pit and moved to their final position. This was definitely necessary for the ID. Cables would need to start at the closest point possible to the detector and reach racks of electronics in the pit or in the underground service areas. In some cases, they would need to be segmented, at breaking points called patch panels. Patch panel 1 is the closest point to the detector inside the cryostat in the case of the connections for the integrated SCT+TRT barrel. The connection inside the cryostat was very delicate because of the work being confined inside a cylinder, and handling the very tiny connections.

In parallel to the actual installation of the cables, the cabling campaign included significant work on the commissioning of the cables, with tests performed at every step along the way, namely after connecting to patch panels, to power supplies or to read-out boards. The final test was the actual commissioning of the whole chain, once the detector cables were connected to the services already in place. Physicists (ranging from students to professors), engineers and technical staff from all the institutes who were part of the construction of the detector participated in connecting and commissioning the cables. Some were even involved with the actual installation too, either by providing mappings or preparing material for the installation.

In the case of the ID, one estimates that 200 people from different institutes participated in the cabling campaign. One can say that the cabling campaign brought many ATLAS communities together for the first time. It was a very exciting time.

The cabling project is still alive these days, serving for the upgrade of ATLAS. As an example, about 2000 new cables and flexible pipes were installed for most of the ATLAS subdetectors and systems during the first long LHC shutdown.

6.6. Detector closing and opening

A detector movement system has been developed with two goals. The first one is the installation of large and heavy parts of the ATLAS detector, such as the barrel and the two extended barrel calorimeters, the two small wheels, and the two endcap toroid magnets. The second is to perform the opening and closing of the detector during the yearly maintenance period, so that these systems can be accessed.

Because the ATLAS rails support all these systems, it has been possible to use the same concept for their movement. They are permanently equipped with air pads that take the load (up to 1000 tonnes for the extended barrel calorimeters) and provide a low friction factor of 1%. Horizontal hydraulic cylinders push and pull the system supported by the air pads (Fig. 6.12). In addition to the large size and mass of moved objects, each operation is very challenging because of the small clearance with the central beam vacuum pipe, the accuracy required for system positioning, the fragile environment and the slope of the ATLAS rails (see Sec. 6.2). This last point is of particular importance as the slope (0.708 degrees corresponding to 1.24%) is very close to the 1% friction factor of the air pads, consequently there is a big difference in the behaviour of the system depending on whether the movement goes upstream or downstream.

The Muon Big Wheels need to be moved as well. Because their technical specifications were different, their movement devices are different from the ones in the other moveable systems. The air pads are replaced

Fig. 6.12. First movement of the extended barrel calorimeter during installation, air pads (in orange colour) support the nearly 1000 tonne calorimeter. The traction cylinder and its anchorage point to the rail are shown in the foreground.

with heavy load rollers and electrical engines produce the translation. However, all these movements are performed thanks to the same system that integrates hydraulics, pneumatics, electricity and a PLC monitoring of all the parameters.

Moreover, all these moveable systems are permanently connected to a large number of cables, gas and cryogenic pipes. Disconnecting them was not a realistic option, hence they have been equipped with flexible chains housing all the services. Whilst some of these chains are passive and follow the movement of the components, other more sensitive chains have their own movement devices synchronised with the main movement.

Since 2005, more than 150 movements have been performed successfully. Although the basic functions remain, the movement system was upgraded to increase its reliability and the accuracy of positioning. During the first long shutdown (2013–2014) a set of 50 CCD cameras has been installed, able to measure positions in two dimensions with an accuracy of 50 μm. As an example, the extended barrel calorimeters, which require the best positioning, have been moved back into run position with an accuracy of 0.3 mm over the last three years.

6.7. Control room

When it comes to operating ATLAS, the control room takes centre stage. The detector components are switched on and off, and data recording is started and stopped from it. But there is much more. The controls of the detector safety systems, fire alarms, access control and communication units with the LHC accelerator are all located in the control room and constantly supervised. In addition, the various detector components have their place in the control room to observe their systems and intervene in case of problems. When designing the control room, all these various functions and the number of people that might need to be present at the same time had to be taken into account. Of course, communication among those present is essential, but the layout of the room should also enable experts to concentrate on their work without too much disturbance.

The solution chosen for ATLAS is an arrangement of six semi-circular islands of two or three workstation desks, each dedicated to a particular set of tasks or detector components (Fig. 6.13). All desks face the same way towards a wall where projectors display information that is relevant

Fig. 6.13. The interior of the ATLAS Control Room. Here, two central desk workstations face the projector wall. Two similar workstations (not visible) are located to the left and right of this picture.

for everyone and tailored to the current situation, for example, during data taking or maintenance work during a shutdown period.

The control room is almost always occupied. During data taking, a full crew will be present around the clock, separated into three shifts of eight hours per day. Hence, care was taken to create a working environment as comfortable as possible, despite the large amount of computing and electronics equipment needed to control the experiment. For example, the computers powering the desktop workstations are housed in a different room nearby, with only computer screens and keyboards present in the control room, thereby removing the noise and heat generated by these machines from the room.

This control room has been in operation throughout the final commissioning and all running periods, and has served the experiment very well.

6.8. Technical Coordination and schedules

The Technical Coordination (TCn) has played a central role through all the phases of the ATLAS Experiment. Without it, neither the detector's conception, construction, installation and operation, nor the planning of future upgrades would have been possible. Its past activities included (most visibly) all aspects of coordination and follow-ups on the distributed detector component construction, as well as the monumental task of successfully installing the detector in the underground cavern. All Common Projects have been constructed under the direct and ultimate responsibility of TCn, as are all the operations and maintenance of the detector at the LHC Point-1 site. Therefore, as described before, the Technical Coordinator is part of the highest-level ATLAS management team.

This central role of TCn started very early on in the 1990s during the progressing R&D and overall detector optimisation and design phases. The various detector communities were highly competent in developing their own individual detector system components, but it needed strong and active guidance from the TCn to ensure that these systems could be integrated into an optimal overall detector concept. Of the many activities only a few can be mentioned here.

Even though the Collaboration reported its design progress regularly to CERN's LHC Experiment Committee (LHCC), these reviews were not sufficient enough to ensure engineering soundness, schedule compliance, required quality control and anticipated physics performance of the component constructions. So the responsibility fell onto TCn to conduct countless reviews during all the prototyping and construction years, both on the mechanics as well as the electronics aspects of the detector. Such reviews were typically made at the conceptional design stage and then just before the "green light" for construction with the so-called Production Readiness Reviews, based on full-size final prototypes, often called Module-0.

During the component construction TCn followed-up very closely on the series construction progress, both in terms of quality and importantly of the agreed schedule. Scheduling the distributed construction work was a very complex and difficult task, and TCn needed to exercise much active guidance and flexibility with inventing solutions in order to "concert" the timely arrival of all parts to build the whole ATLAS detector. Over many years, TCn monitored the detailed construction progress in monthly meetings (which preceded the Executive Board (EB) meetings) with a special Technical Management Board. Special tools were developed. For examples, Project Progress Tracking planned and kept track of the actual construction progress, and a so-called "Top Watch List" for critical items with schedule (or technical) difficulties that were discussed at every EB meeting. There was also a large set of LHCC milestones that was followed, as shown in Fig. 6.14. But contrary to the LHCC, which would only observe and give warnings, TCn was actually committed towards finding solutions to overcome difficulties! Another of its responsibility was to maintain a database with the necessary technical documentation, and the configuration change management, of the detector.

The work and responsibilities of the TCn grew considerably in the early 2000s, with the activities at Point-1 moving into the focus of the ATLAS experiment. The entire ATLAS infrastructure on the surface and in the underground cavern, from design to realisation, including its safety aspects, formally fell under the sole purview of the TCn after the cavern was handed over from CERN to ATLAS in June 2003 (see Sec. 6.1). It was a great challenge to organise hundreds of activities over the five years of installation starting from summer 2003, and that the detector was successfully installed in time for the start-up of the LHC in September 2008 is a fantastic achievement of TCn and all those who contributed to the installation. It also demanded a great deal of ingenuity from the TCn team and its institute partners to finding solutions when components accumulated delays, sometimes unavoidable for such a complex undertaking like the distributed construction of the ATLAS detector. The last years of the installation were particularly impressive with many activities ongoing in the cavern in parallel, in order to meet the LHC schedule. A glimpse of the planned activities in

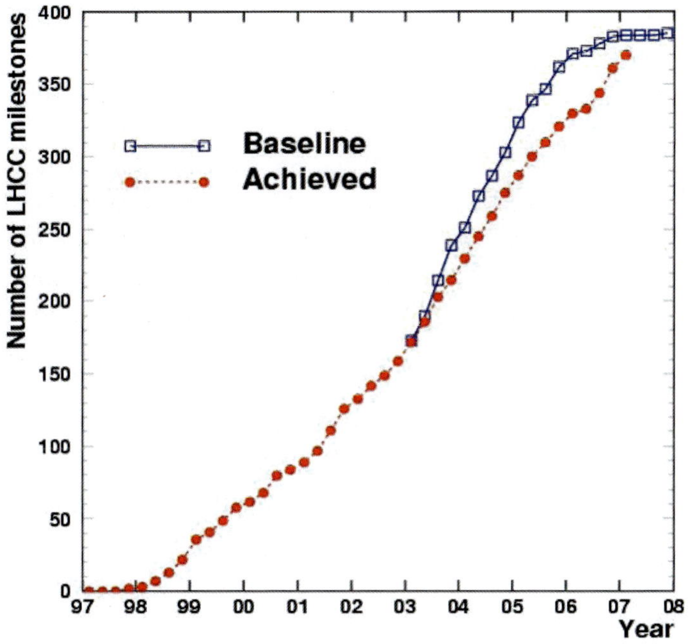

Fig. 6.14. An example of the ATLAS's construction milestones, as reported to the LHCC, covering the years from 1997 to 2008.

the cavern is given in Fig. 6.15 as an example, valid at a given time, which illustrates the parallel works going on in the barrel as well as at both end regions of the cavern. Of course, highly professional tools for the scheduling activity supported this useful visualization for the Collaboration.

Amongst the other important activities of the TCn, two more examples must be mentioned here, both relating to the interfacing with the LHC machine. A major issue of the LHC experiments is the harsh radiation environment at a hadron collider, coming from particles arriving in the cavern due to various backgrounds from the machine as well as from the high (secondary) interaction rates in the detector originating from the high luminosity at the interaction region in the ATLAS detector. A whole suite of shielding thus had to be designed to minimise radiation backgrounds in the detector and the cavern, based on very detailed simulations of the various radiation sources. These shielding elements ranged from the very massive, in their material composition specially optimised, absorbers at the LHC tunnel exits to other elements that were integrated into the detector components. All this work was placed under the responsibility of the TCn.

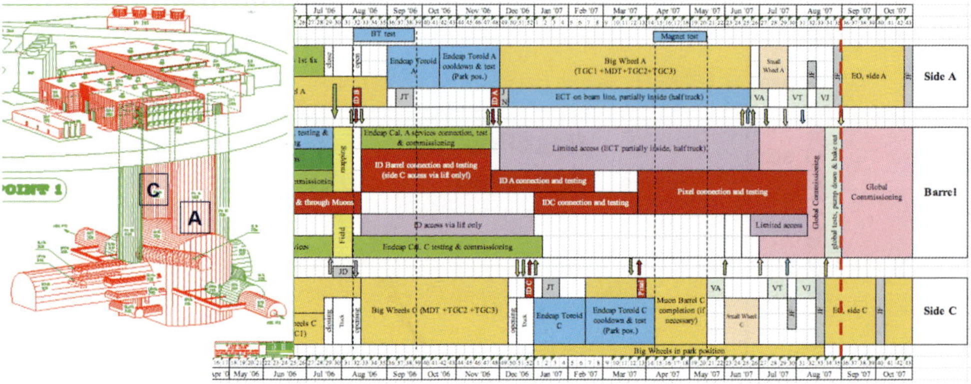

Fig. 6.15. An example of the visualisation for the planning of the parallel activities in the underground cavern, in the barrel and the two end regions A and C, during the installation phase of the detector, from June 2006 to October 2007.

The other example is the TCn's crucial role in liaising with the LHC machine and more generally with CERN's machine and technical departments and services. There are many direct common interfaces with the LHC collider, for examples, on the technical side, the vacuum beam pipe and the ATLAS forward detectors installed in the LHC tunnel, and then on the organisational side, coordination work on scheduling and operational conditions. Furthermore, the operation and maintenance of the Point-1 infrastructure demand coordination with all the technical services of CERN, in particular also with CERN as the Host Laboratory of the LHC experiments. All these tasks were — and still are today — assumed largely by the TCn.

A very different sort of ATLAS Upgrade Week

In general, upgrade people like to plan their projects in detail and very much in advance. For example, Phase-II of the ATLAS upgrade for the high-luminosity phase of the LHC will commence only in 2026 yet all of the Technical Design Reports were due in 2017. In this business, risk management is always an important topic. One has to think of all possibilities, including the most unlikely events, which could spoil the upgrade projects causing serious financial damage or significant delays. As anyone who has ever worked on a detector project knows, in the end, something unexpected always goes wrong. For example, one would never have thought to plan for a volcanic eruption in a CERN non-member state when scheduling a meeting, like the Eyjafjallajökull in Iceland that upset the 2010 ATLAS Upgrade Week in Hamburg quite a bit.

On the positive side, ATLAS people are creative and adventurous so that even though no planes were flying in the northern hemisphere, still a lot of people made it to Hamburg in Germany. The flight corridors were closed on the Thursday before the meeting so that people had one or two days to reschedule travel — and some found creative alternate journeys to Hamburg.

One CERN colleague realised Friday that no plane would bring him to Hamburg. Thus, he jumped into his car and drove the 1100 km to Hamburg with only one stop-over in Cologne to look at the cathedral and to drink one or two beers in a very famous brewery beside it. How many beers was it? The facts are not well documented... He was the first to arrive in Hamburg Sunday afternoon.

A team from Liverpool saw it as a "Beatles moment" and decided that going to Hamburg was extremely important. They took a train and so, besides seeing a lot of England, they also got to see a lot of France and much of Germany, as they travelled from Liverpool through London, Calais, Paris, Strasbourg, Karlsruhe, Cologne to finally get to Hamburg: a 17-hour train trip instead of a 1-hour flight. The Beatles probably had an easier trip to Hamburg in 1960...

Rumours say that one participant was brought in by secret military transporter, probably flying below the radar. Checking with him later revealed a slightly less exciting story: he claims to "just fly a few days later with a regular flight". It will forever remain a secret just how he really got there.

In the end, 70 of the 240 registered colleagues made it to Hamburg.

ATLAS as seen from the upgrade perspective is as diverse as it is interesting. Upgrades are an essential part of the overall endeavour!

2010 ATLAS Upgrade Week in Hamburg, Germany.

Test beam adventures

Test beams are small-scale experiments as they include beams, detectors, data-acquisition systems, reconstruction and analysis software and, more importantly, people! Many test beams were done in the preparation phase of ATLAS and many are still ongoing for upgrade projects. But as in large experiments, people have to work hard and in closely-knit teams to get things done. Indeed, a piece of hardware that is not working, a missing tool, or a detector module that simply refuses to send data out can give birth to strong bonds of friendship.

For example, many students have joined the ATLAS ITk test beam groups over the years, ranging from undergraduate to PhD students. Thanks to their youthful presence, the atmosphere in the counting room is often light and happy, and sometimes even music is playing on an otherwise boring night shift!

Over the years, many funny stories have happened around these test beam campaigns. In May 2016, the ATLAS ITk strips team was trying to test the first module irradiated to HL-LHC levels. As with most test beams, one can make the most elaborate plans beforehand. However, planning does not always equate to success. For five days, a reference sample was being tested, but at the same time a cooling system for the irradiated samples was being set up. However, nothing tried could achieve the desired low temperatures. Many different chillers, Peltiers, boxes, and other known tricks were tried to get the cooling system to cooperate. Nothing worked.

It was then pointed out that most likely a lot of cooling power was lost due to the piping used. The team then worked through the night to prepare and fit large, heavy, but well insulated pipes, to the samples box. Finally, on completion they managed to run some tests. They waited patiently to see the temperature decrease, and then, for the first time, saw the temperature approaching the desired goal! Everybody cheered after such a hard-won fight to reach the goal. However, it soon dawned on them that in their efforts to install the heavy, bulky cables, they had used a ladder to prop them up between the box and the chiller.

But there was just one tiny problem...

New uses for a ladder during an ITk strips test beam.

Chapter 7

Trigger, Software and Computing

Developing the trigger selection and offline software is one of the major activities in building a modern high-energy physics experiment like ATLAS. As for any experiment at a hadron collider, the online event selection in the trigger farm requires sophisticated event reconstruction and selection algorithms, in order to reduce the immense amount of collision data produced every second to a rate that the offline computing infrastructure can process and deliver to physicists for analysis. Data selected by the trigger system are subject to careful calibration and a prompt reconstruction at computing facilities at CERN. The data in different formats are distributed worldwide for analysis on the LHC Computing Grid, a federation of more than 130 computing centres across the globe at national laboratories and universities. The real data are complemented by simulated data, with generators modelling collision events of all types of physics processes at the LHC. Those generated events are fed into a highly detailed detector simulation that models the experiment response and produces the same data format as that coming from the real detector. The simulated data are used in physics analysis to understand and correct for detector effects. Figure 7.1 illustrates the ATLAS data processing chain.

7.1. The ATLAS software for reconstruction and simulation

The ATLAS Collaboration developed a complex software suite to carry out all required data processing steps. The same software suite is used for the online trigger event selection and for the offline processing, deploying customised reconstruction strategies for both applications. The ATLAS software suite also provides the tools required by physicists to analyse the data and to extract physics results. Numerous technical, performance and

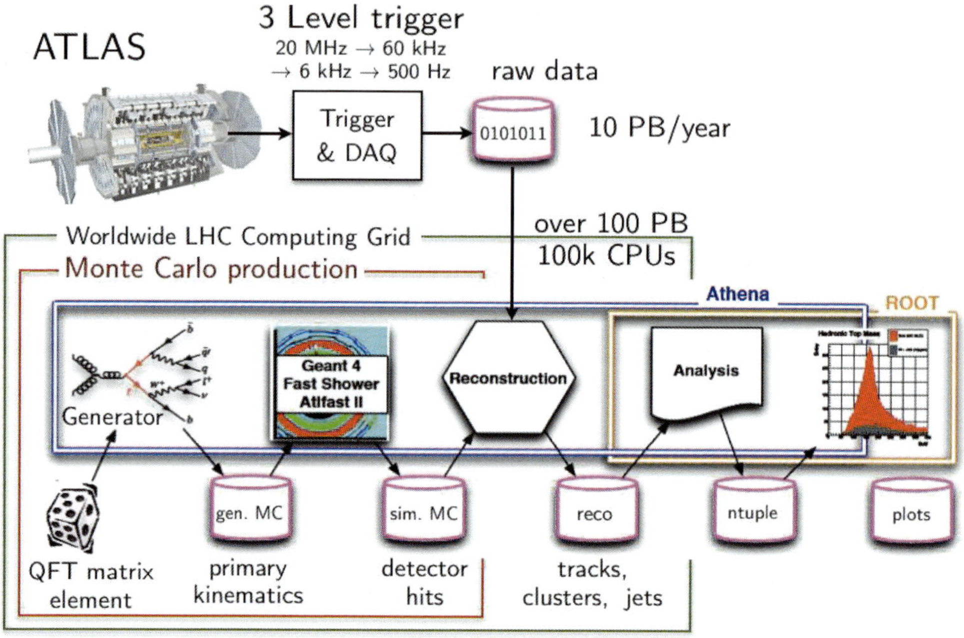

Fig. 7.1. ATLAS data processing model in Run-1 from the detector and simulation (Monte Carlo production), to the physics results.

physics requirements have to be met, including the processing time and memory consumption per event necessary to meet financial constraints, and the physics performance to reproduce the details of the underlying physics processes. Real data are compared to simulated data to understand and correct for detector effects. The detector simulation is also a tool to test the physics performance of the software against our knowledge of the underlying physics processes. The simulation was also used to optimise the design of the detector that we have today. The complexity and scale of the ATLAS detector requires that the software be highly modular and robust. Furthermore, it needs to be flexible enough to meet the needs of the experiment throughout its operational lifetime.

The design and implementation of this software requires the establishment of a software development environment capable of supporting a distributed development model appropriate for the widely geographically dispersed ATLAS developer community, and the provision of testing and quality assurance tools to ensure that the software meets both the technical and physics performance requirements. ATLAS adopted an object-oriented development methodology, based primarily on the C++ programming language, with a component-based model whereby applications are built up from collections of plug-compatible components based on a variety of configuration files. This capability is supported by a common framework that provides common data-processing support. This approach results in great flexibility in meeting the basic processing needs of the experiment and in good response to changing requirements throughout its lifetime.

A rather lightweight development process has been adopted, with emphasis on guidance and use of automated tools, and extensive testing and validation rather than a rigid enforcement of strict protocols; experience has shown that this approach is better matched to our community. However, there are clearly potential disadvantages to such an approach in the areas of robustness and reliability, and more emphasis has been placed on the formal development process in core areas of the software.

7.1.1. *The Athena software framework*

The Athena software framework is an enhanced version of the Gaudi framework that was originally developed by the LHCb experiment. Athena and Gaudi are concrete realisations of a component-based architecture, which was designed for a wide range of physics data-processing applications. The fact that it is component-based has allowed flexibility in developing both a range of shared components, and where appropriate, components that are specific to the particular experiment and better meet its particular requirements.

Major design principles that influenced the development of Athena were:

- Abstract interfaces, which allow the provision of different implementations providing similar functionality but optimised for particular environments (e.g., high-level trigger and offline, different persistency technologies).
- A clear separation between data and algorithms. At first sight, this separation appears to violate one of the main principles of object-oriented software, encapsulation, but the opposite is actually true. By separating the complex algorithmic code that is responsible for the construction of the event objects from the much simpler and more stable data representation of an event, the dependencies between the producer of a data object (the algorithmic event reconstruction code) and its consumer (the analysis code) are drastically reduced.
- The recognition that there are many types of data having different lifetimes within the software. Event data are directly associated with a single ATLAS event, but detector information such as the geometry,

alignment and electronics calibration is in general stable across many events. Similarly, statistical data accumulated in histograms are long-lived compared to the data of an individual event.
- A clear separation between persistent and transient data. In general, the algorithmic code operating on the data should be independent of the technology used to store it, which might vary over the lifetime of the experiment, or depend upon the local environment (e.g., within the trigger data acquisition and offline environments).

7.1.2. *The ATLAS simulation*

To simulate the ATLAS detector response correctly, the Athena framework together with the Geant4 simulation toolkit, which models the detector geometry and the passage and interaction of particles with the detector material, is used. The software accesses the geometry and conditions database for an accurate description of the as-built and as-operated detector, including the magnetic fields. It is done in rather compact software consisting of about a quarter of a million lines of code. The majority is written in C++. For run-time configuration, Python is used and libraries are dynamically loaded.

The simulation is broken into three steps: the generation of the primary particles from the proton collisions ("event generation"), the simulation of the detector response due to the passage of particles ("detector simulation"), and the simulation of the read-out electronics ("digitisation"). Many different event generators are used to create the primary particles of interest using state of the art calculations. Information of what has been simulated ("truth information") is recorded at various stages. The primary particles from the generation are fed into the detector simulation step, which produces "Hit" objects for the energy depositions in sensitive portions of the detector. Many possible interactions between the passing particles and the various detector materials (including passive, i.e., non-sensitive materials) must be modelled and the charged particles must be propagated correctly within the magnetic fields. The ATLAS detector is large and complicated, requiring five million geometry volumes and more than three hundred different materials to fully simulate.

The Hits produced from the detector simulation of the signal of interest are then overlaid with pre-simulated pile-up Hit events. These are simulated background events coming from other proton-proton collisions occurring within the same, previous or even subsequent bunch crossings. Each signal event requires thousands of pile-up background events, which are digitised altogether to give the correct read-out response. During this digitisation stage the charge collection specifics of each subdetector including electronic noise and channel dependencies are modelled. These and the other simulation algorithms have been heavily tuned using laboratory, test-beam, cosmic ray and LHC data tuning. Finally, Raw Data Object (RDO) files are outputted, which can be read by the same reconstruction algorithms as for real data processing.

Each event takes up to ten minutes to simulate and we produce of order ten billion events per year. To minimise the CPU time required there is a hybrid approach to the detector simulation stage allowing mixed simulations, where different simulators may be selected to match different spatial regions and particle types. So far most of ATLAS simulation has been done using full Geant4 simulation, but with pre-simulated shower libraries for the forward calorimeter; or a mixed mode where a fast, parameterised simulation, FastCaloSim, is used for the showers of particles in the entire calorimeter, and full simulation is used for the rest of the detector.

The accuracy of the simulation has been checked through numerous tests, e.g., as the large flux of particles traverses the detector they can produce rare hadronic interactions with the detector material. Figure 7.2 shows physics interactions with the material in the ATLAS Inner Detector for both real data and simulation. The structure and geometry of the LHC beam-pipe, pixel modules, cooling pipes and support structures can be clearly seen from the position of these interaction vertices (also see the Inner Detector material radiography story).

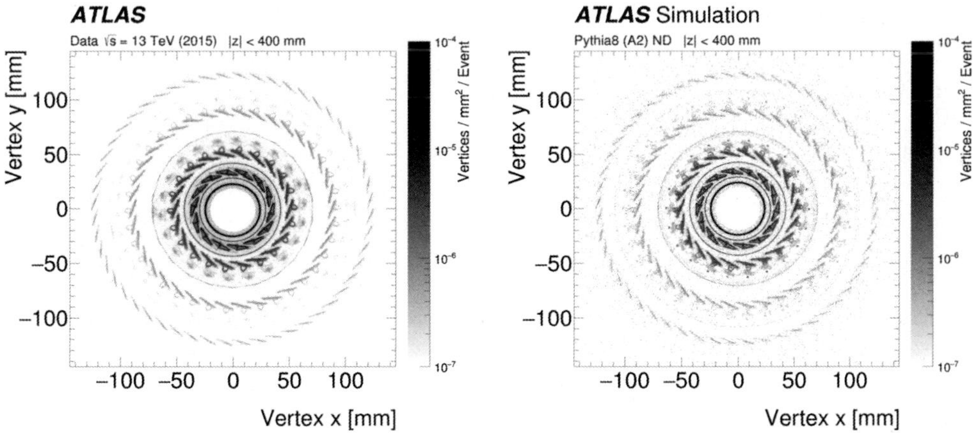

Fig. 7.2. Hadronic-interaction vertex candidates in the Inner Detector for real data (left) and simulation (right).

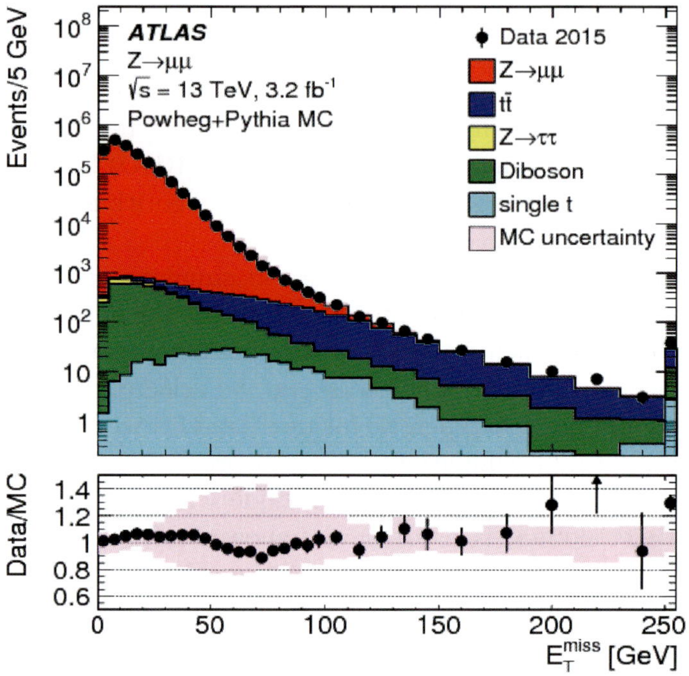

Fig. 7.3. The missing transverse momentum (E_T^{miss}) for a sample of Z boson events decaying to two muons and compared to the simulation including all the relevant backgrounds.

One of the most difficult variables to model correctly is called the missing transverse momentum, which is a measure of the undetected particles in the detector. It is determined by considering the signals of all the detected particles and looking for an imbalance in the energy of the event. It is challenging since it involves all detector subsystems and requires the most complete and unambiguous representation of the signal interaction in the tracking and calorimeter detectors. Figure 7.3 shows the excellent simulation to data agreement over many orders of magnitude in precision. Missing transverse momentum requires a very accurate modelling of both the generated primary particles and their interactions in the detector.

7.1.3. *Event reconstruction*

The role of reconstruction is to derive from the stored raw data the information about the produced physics objects in the events. For much of the ATLAS proton-proton physics programme those objects are leptons (electrons, muons and taus), jets, missing transverse energy, as well as information about the production vertex and the quark/gluon content of produced jets. Sophisticated pattern recognition algorithms are used to combine the information of all detector subsystems in an optimal way to maximise the reconstruction performance.

The development of the C++ offline reconstruction software was based on the experience from the Fortran based code used for the original Technical Design Reports. The reconstruction is structured along the detector specific software tasks to reconstruct calorimeter clusters and charged particle tracks. The calorimeter software uses a common approach for both, the Liquid Argon and the Tile Calorimeter. Similarly, the Inner Detector and Muon Spectrometer software is based on a common suite of tracking and vertexing tools that implement the event data model, the mathematical tools for track propagation, track fitting and vertex finding.

The output of the detector specific reconstruction is fed into the combined reconstruction to identify all physics objects in the events. For each set of physics objects, advanced identification codes have been developed. Electrons and photons are reconstructed, combining electromagnetic clusters found in the calorimeters with tracking information from the Inner Detector. Muons are reconstructed combining the Muon Spectrometer tracks with the Inner Detector information in an optimal way. Tau candidates are identified using narrow jets, combined with tracking information. Jets are primarily reconstructed from the calorimeter clusters and bottom (charm) jets are identified using the precision impact parameter information from tracks associated to them. Combining the information about all physics objects in the event, the missing transverse energy is reconstructed, which is sensitive to high-energy particles produced in the collisions that leave no signal in the detector.

The first large-scale test of the offline software was the so-called "combined testbeam" in 2004 (see also Chap. 8), where the new C++ based reconstruction was used to analyse the data from a full detector slice that was tested with beam data. Figure 7.4 shows a display of a reconstructed event taken at the time. Starting in 2007 the offline reconstruction was used for the first real detector tests with cosmic muons, taken first at the surface building with different components of the detector prior to installation at the pit, and later with the installed detector down in the cavern.

Like the commissioning of the detector, the readiness of the offline and trigger reconstruction software for collisions was a major concern at the time, and in particular the robustness of the software. The continuous trigger and offline operation during data taking requires running the reconstruction code

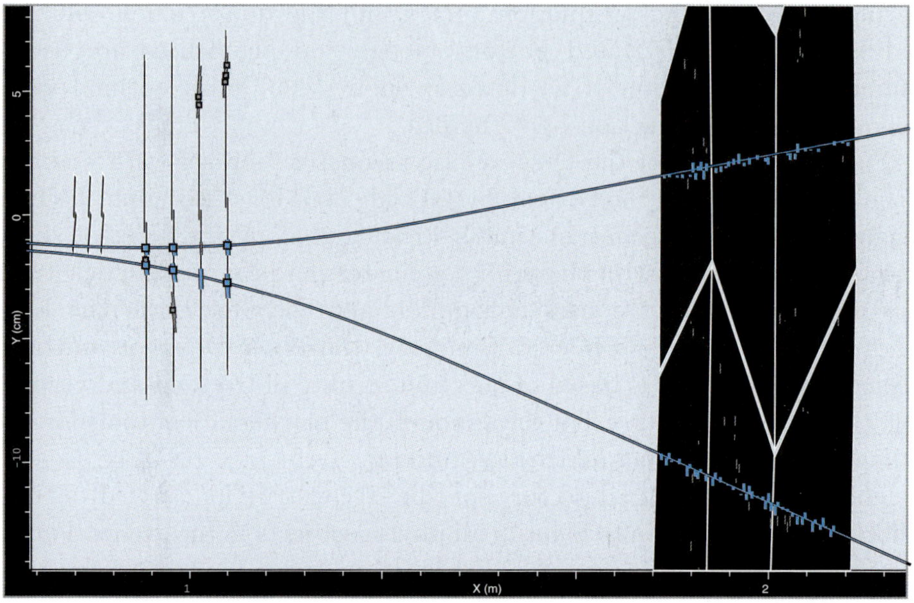

Fig. 7.4. Display of a reconstructed combined testbeam event taken in 2004.

on billions of events, with only a handful of software crashes. From the large-scale reconstruction tests done at the time it became clear that reaching this goal was not a trivial task. A scheme of "descoping" the reconstruction was invented, which would enable switching off parts of the reconstruction in case of crashes. Fortunately, ATLAS never had to actually make use of this feature! From day one of proton-proton collisions the software achieved stable running in both the trigger and offline context. Figure 7.5 shows an event display of one of the first collision events recorded in ATLAS in 2009.

Already during the first years of data taking the raising instantaneous luminosity provided by the LHC machine was posing a new challenge to the detector and to the reconstruction software. More luminosity meant more pile-up of multiple proton-proton collisions per bunch crossing. This does not only increase the raw event size, but it results in a rapid increase of the combinatorial complexity in reconstructing the event. The reconstruction has been step-by-step optimised for higher and higher levels of pile-up, while preserving and in fact enhancing the physics performance, based on experience gained from analysing real data. An excellent example of such enhancements is two updates that were introduced in the end-of-year shutdown between the 2011 and 2012 data taking periods. At the time, the

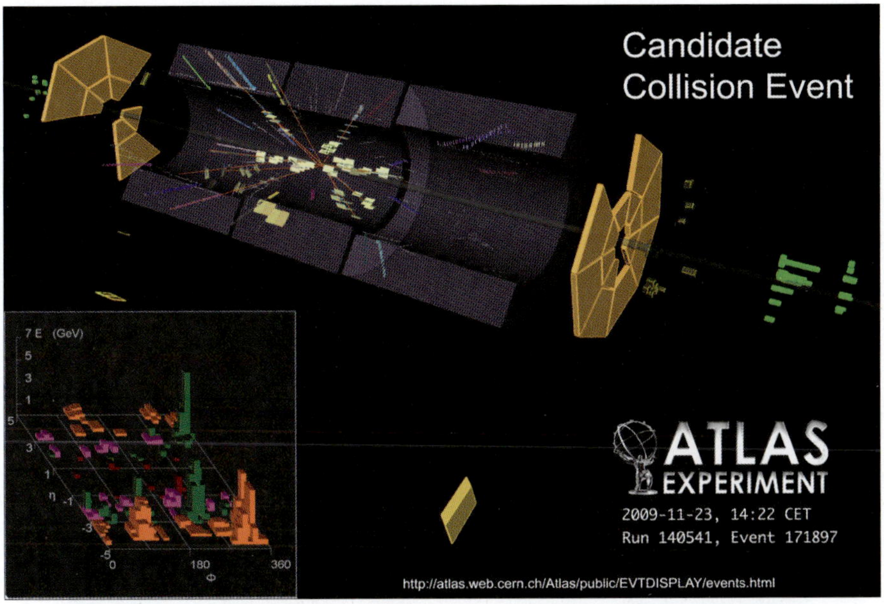

Fig. 7.5. Display of one of the first collision events recorded during the pilot run in 2009.

first indications for a light Higgs boson at 125 GeV had been found. Novel photon conversion reconstruction algorithms and a track fit that is able to allow for the Bremsstrahlung of electrons in the detector material were deployed for the 2012 data taking. Those new algorithms helped improve the experimental sensitivity for the Higgs to two photons and Higgs to four leptons in the final states, in which the Higgs boson was discovered in 2012.

7.2. Event selection at the trigger level

The goal of the ATLAS trigger systems is to select the events that are needed for further analysis, with minimal bias to the physics studies that will be performed on the selected events, while being as economical as possible with hardware and computing resources (see also Chap. 2). The trigger is designed so that it can efficiently reconstruct energetic electrons, photons and muons as well as taus, jets and missing transverse energy. Electrons, muons and photons are among the decay products of W, Z and Higgs bosons. These bosons range in mass from 80 to 125 GeV and often decay to final states with leptons with transverse momentum above 25 GeV. Events with multiple objects can produce multiple affirmative trigger decisions, any one of which is sufficient to store the event for further analysis. The trigger system must

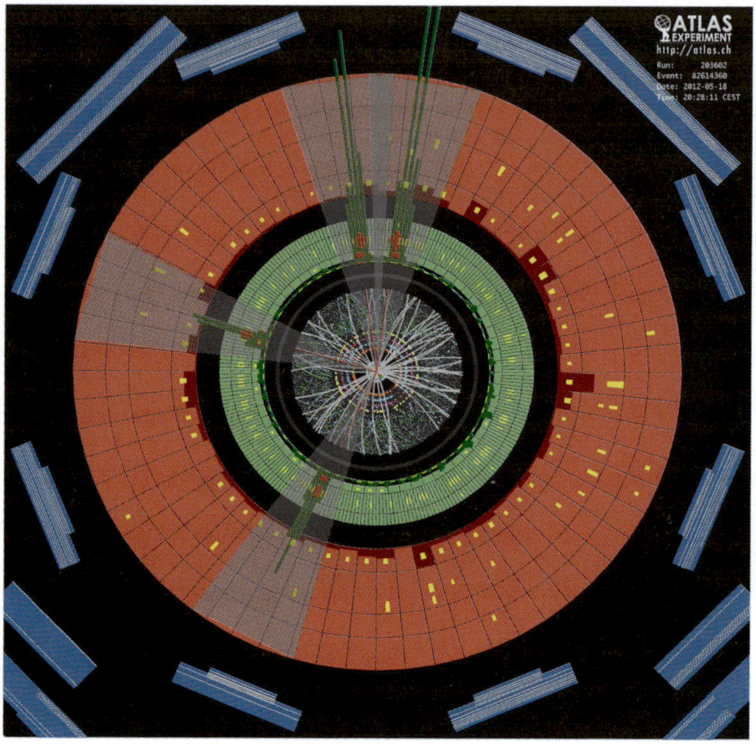

Fig. 7.6. Event display where the regions of interest are highlighted in grey.

also decide to accept an event quickly; for example, the Level-1 system has a latency of 2.5 μs.

The trigger selection in ATLAS starts with the objects that are identified by the Level-1 hardware trigger, which is based on energy deposits from electrons, photons, taus and jets in the ATLAS calorimeters and on the pattern of ionisation (hits) left by penetrating muons. It is then followed by a software selection that implements fast and refined algorithms reconstructing and identifying the various trigger signatures upon which the final decision to store the event is based.

Three important concepts underlie the trigger strategy. The first one is regional reconstruction. While in the standard offline processing, whole events are reconstructed in as much detail as possible, in the trigger reconstruction, where possible, the hardware trigger objects are used as a guide to search for objects in very narrow regions of the detector, known as a "region of interest (RoI)". The regional trigger strategy is illustrated in Fig. 7.6 where an ATLAS proton-proton event taken in 2012 is displayed.

Shown in green and red circles are the electromagnetic and hadronic calorimeters while the Muon Spectrometer is shown in blue. Four electron candidates are visible in the electromagnetic calorimeter with the deposited energy shown by dark green lines. The electron candidates are geometrically matched to tracks, shown as red lines, in the central tracking detector. Narrow regions around the electromagnetic clusters identified by the first level trigger are shown in grey. For the selection to be fast, it is only the data corresponding to these regions that is used in the first step of the electron selection. The second strategy, applied here, is early rejection; for only those events with candidates fulfilling the trigger requirements is the full detector data reconstruction performed to further refine the selection. After each step of reconstruction, the candidate trigger object is required to fulfil at least one requirement. The trigger reconstruction is abandoned if the requirement is not fulfilled. The third strategy is to employ efficient, but fast algorithms that can be later refined once the object has fulfilled the initial selection. The effect of these three trigger strategies is to reduce the time needed to process an event by a factor of 50 from approximately 25 seconds for the standard offline reconstruction to roughly 0.5 seconds per event in the trigger.

The trigger selection of electrons is shown in Fig. 7.7 where the relative fraction of events fulfilling a given step is displayed with a proportionally thick arrow, illustrating the early rejection strategy described above. The input to the selection process is the Level-1 object, in this example case an isolated electromagnetic cluster with transverse momentum (p_T) of at least 22 GeV (noted as EM22i). The first step is a selection based on the shape of the deposited energy in the calorimeter, which has been tuned using machine learning techniques. In the second step, for the remaining candidates, fast tracking is done in a small area of the Inner Detector. The resulting reconstructed tracks are then extrapolated to the calorimeter and a tight matching between the position of the electromagnetic cluster and the extrapolated track is required. This is followed by a detailed reconstruction of the calorimeter and tracker data. The resulting information is used in steps four and five to reject fake electron candidates according to the transverse momentum of the reconstructed electron ($p_T > 26$ GeV) and a likelihood function based on all of the available information. Finally, in step six precision tracking that includes low transverse momentum tracks is applied around the electron candidate. The tracks that are consistent with coming from the same primary vertex as the electron, can then be used to veto electrons that are accompanied by jets of particles as opposed to the isolated electrons expected from W, Z or Higgs bosons.

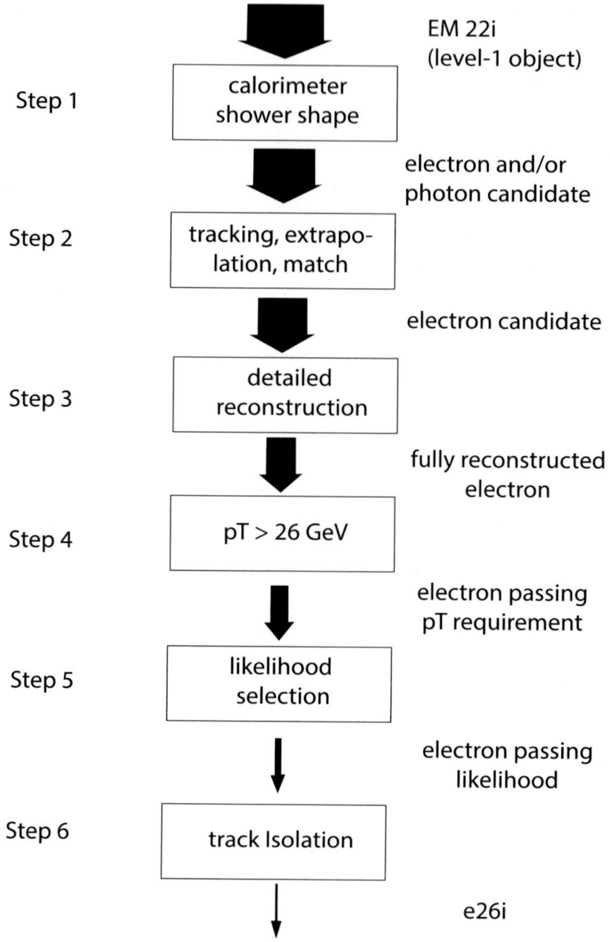

Fig. 7.7. Successive steps in the identification of electrons in the ATLAS software-based trigger.

The muon trigger starts from the trigger objects identified in the Muon Spectrometer. When a high-p_T muon segment is reconstructed in the Muon Spectrometer, it is used to define a small area in the tracker where tracks are reconstructed. If a reconstructed track is found and matched to the muon segment, a more precise determination of the p_T can be made and a requirement of $p_T > 25$ GeV can be imposed. A final isolation step is applied in the same way as for electrons.

Triggers which involve the products of strongly interacting quarks and gluons, as well as the hadronic decay of the tau lepton, are based on objects much less well separated from the background than the triggers from electrons, photons or muons. Jets of particles that result from the production of quarks and gluons in proton-proton interactions are abundant in LHC collisions; only the very most energetic objects or objects with unique topologies can be selected by the trigger. The trigger uses a sequential recombination algorithm to reconstruct jets, similar to the one used offline.

Finally, one of the most important triggers for searching for new physics is the missing transverse energy trigger. This trigger takes advantage of the fact that the vast majority of background events leave energy deposits that are spatially well balanced around the beam axis. Events with a significant imbalance in the deposited energy are evidence for one or more invisible particles. In the Standard Model these invisible particles are neutrinos, but dark matter would also appear to be invisible in the detector. Therefore, many searches for dark matter particles rely on missing transverse energy triggers. The missing transverse energy algorithm used in the trigger is based on the sum of the energies of clusters in the calorimeter. It includes a sophisticated method to correct this measurement for effects from pile-up.

In addition to providing the single-object triggers that are used for the majority of analyses, the trigger menu has grown in complexity and is able to provide stable operation for ATLAS in a wide variety of luminosities and configuration of the LHC machine. The number of triggers in the menu has grown from about 400 in Run-1 to almost 2000 in Run-2 and ATLAS ran very efficiently. In the main data taking years of Run-1, more than 3.6×10^9 events were recorded with a mean rate of between 200 and 400 Hz. By the end of 2017 more than 6.8×10^{15} collisions were produced by the LHC and 12.6×10^9 events were recorded with a mean rate of between 1.0 and 2.0 kHz. To date, more than 12.6×10^9 events were recorded in Run-2, with a mean rate of between 1.0 and 2.0 kHz. The expected future increase in luminosity during the High Luminosity LHC operation is challenging the trigger to keep the objects momentum thresholds as low as possible.

7.3. Data preparation

The data preparation activity area of ATLAS was introduced in 2005 with a mandate covering all aspects of preparing the triggered data collected by the experiment for physics analysis. The main areas involved are calibrating the data, checking the data quality, and the reconstruction of the data to produce reconstructed physics objects that are ready for use in physics analysis. Many of these tasks are also applicable to the simulated event samples. Based on these tasks the data preparation area by definition has very strong links with the other ATLAS activity areas, in particular with the software and computing, and physics analysis.

The data preparation activities for collision data follow the following set of steps:

- The detector-level data quality is assessed during data taking by shifters in the ATLAS control room, allowing for a fast turnaround for problems observed at this time.
- The express stream (comprising of a small fraction ~5% of the collected events but containing a mixture of triggers useful for data quality and calibration purposes) is processed promptly at the CERN "Tier-0" computing centre. This provides reconstructed data about 12 hours after the data are collected.
- The reconstructed express stream data are used for more detailed data quality assessment, and for deriving various offline calibrations. An important calibration that is determined from these data is the beam spot position, and size, which is used as a constraint in the track reconstruction in the later processing of the data. The tracking detector alignment (see also Sec. 8.2.1) can also be corrected using the express stream data. In addition, noisy channels in the different detectors identified from the reconstructed express stream data, can be masked for the next processing.
- After the above data have been analysed in terms of data quality and calibrations, the full set of triggered data are reconstructed, taking into account the updated calibrations derived from the express stream processing. This usually happens 48 hours after the data were collected.
- A final data quality check is carried out on the full processed data, to make sure these data can be used for high quality physics results.

All of the processing of the data in the above steps is done at the CERN Tier-0 and allows the data to be available for physics analysis approximately one week after the data were collected. Figure 7.8 shows a schematic diagram of the prompt processing and data quality assessment of the data. The processing described above has been adapted during the operation of the experiment.

In addition to the prompt processing of the data described above, the data are typically reprocessed with improved reconstruction software, and improved conditions (especially related to the tracking and muon detector alignment) about once per year. The reprocessing is carried out on the Grid, and uses a large amount of computing resources. It is therefore essential that

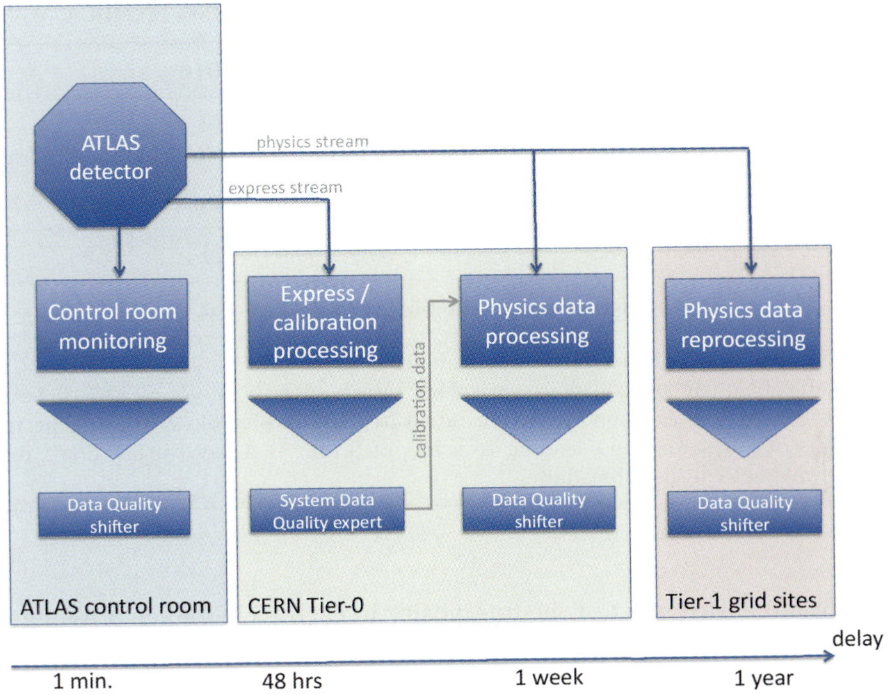

Fig. 7.8. Data processing, calibration and data quality assessment of ATLAS data.

the software and conditions used are thoroughly validated. The reprocessed data are also subject to a thorough data quality assessment.

The data quality assessment is carried out at the granularity of the luminosity block (usually corresponding to one minute of data); the system is set up such that only luminosity blocks corresponding to good data quality are used for physics analysis. Typically, the fraction of good data is high (>90%), and an example of the data-quality efficiency as a function of the different systems for the 2016 data taking can be seen in Table 7.1.

In addition to the main activities described above, the data preparation activity area is also responsible for measuring the luminosity delivered to the experiment and providing this luminosity to the physics groups. ATLAS has a dedicated luminosity detector, LUCID (see Chap. 2). Initially, the LUCID detector was filled with gas to increase the Cerenkov radiation produced by relativistic particles, but as the luminosity increased in Run-1, the gas produced too much signal and was removed. For Run-2, bare photomultiplier tubes are used and the signal comes from Cerenkov radiation produced in the glass windows of the photomultipliers themselves. The LUCID electronics

Table 7.1. Data quality assessment for pp collision data in 2016.

ATLAS pp 25ns run: April–October 2016											
Inner Tracker			Calorimeters		Muon Spectrometer				Magnets		Trigger
Pixel	SCT	TRT	LAr	Tile	MDT	RPC	CSC	TGC	Solenoid	Toroid	LI
98.9	99.9	99.7	99.3	98.9	99.8	99.8	99.9	99.9	99.1	97.2	98.3
Good for physics: 93–95% (33.3–33.9 fb^{-1})											

Luminosity weighted relative detector uptime and good data quality efficiencies (in %) during stable beam in pp collisions with 25ns bunch spacing at $\sqrt{s} = 13\,\text{TeV}$ between April-October 2016, corresponding to an integrated luminosity of 35.9 fb^{-1}. The toroid magnet was off for some runs, leading to a loss of 0.7 fb^{-1}. Analyses that don't require the toroid magnet can use that data.

allows for a measurement of the luminosity in each 25 ns bunch crossing, and the luminosity observed in each pair of colliding bunches is averaged over one luminosity block and stored for offline use. Similarly, per-crossing data are also measured by the Beam Conditions Monitor (BCM), a set of small diamond sensors that were initially designed to monitor radiation levels near the Pixel Detector.

The response of LUCID and BCM to inelastic proton-proton collisions is calibrated using beam-beam scans following the method of Simon van der Meer, which had been originally developed for the ISR in the 1970s. These beam separation scans (otherwise known as "vdM" scans) measure the convolved beam widths at the interaction point directly and can thus be used to provide an absolute calibration of the luminosity at a fixed moment in time when combined with a measurement of the number of protons in each bunch. These vdM scans take several hours for each interaction point and are typically performed only once or twice in any given year. An LHC-wide effort has gone into improving the accuracy of this technique, including special beam optics and injection settings, and systematic uncertainties on the vdM technique approaching 1% have been achieved.

As the vdM scans are only performed occasionally, and the luminosity calibration is performed at a relatively low collision rate (usually less that one pp interaction per crossing), a considerable amount of effort goes into tracking the stability of this calibration at higher interaction rates and over many months of LHC operations. A wide range of complimentary methods have been developed to cross-check the primary luminosity detectors. These methods include properties of the physics data, such as counting the number of tracks in randomly triggered crossings, or counting the number of Z bosons produced, but also meta-information from the detector. The average current drawn in the Tile Calorimeter, for instance, turns out to be a very linear and sensitive monitor of the average luminosity, due to the high flux of particles produced in inelastic collisions every 25 ns that hit this detector. Similar average current measurements made in the forward LAr calorimeter power supplies also have been shown to track the luminosity very well. Being able to compare many uncorrelated measurements allows drifts in any one measurement to be identified and corrected and provides a high degree of confidence in the linearity and stability of the final luminosity result. Relative uncertainties in the luminosity determination for physics data of 2–3% have been achieved for each year, which was widely believed to be unachievable when the LHC programme began.

7.4. The worldwide computing grid and ATLAS computing

The ATLAS Collaboration began conceiving its computing model shortly before the millennium, many years before the first detector data. The expected running conditions of the LHC and the early estimates on data volumes indicated from the outset that computing would present a challenge without precedent in high-energy physics, and indeed in scientific computing in general. In previous large-scale CERN experiments at LEP, computing resources were provided primarily at CERN in the form of rather homogeneous clusters. Computing outside the host laboratory was sporadically utilised, typically for very special needs and with a very low level of automation. For LHC, the computing resources at CERN, even scaled up, would be insufficient to address the needs. Rather than making very large investments at CERN to achieve adequate capacity, it was decided to adopt a completely different and at the time revolutionary approach. Computing resources would be provided at many distributed administrative domains, located in many of the countries participating to the experiments. Such resources would be by nature heterogeneous and loosely coupled, in the sense that they would retain a high level of autonomy. Resources must however operate in concert with a high level of automation, so they would be connected through the network as a "grid" and made accessible by open, well-defined protocols available to workload and data management systems designed to orchestrate the processing. Funding Agencies saw this as an opportunity to evolve the information technology of their respective countries and provide a shared data processing and storage infrastructure for all sciences. The LHC experiments agreed to rely on the grid model, making computing a challenge shared across the institutes of the Collaboration. Grid initiatives were funded in various regions and in 2002 the Worldwide LHC Computing Grid (WLCG) project was created with the mission to provide global computing resources to distribute, store, process and analyse the data generated by the LHC.

The WLCG technical design report was published in 2005, incorporating inputs from the four LHC experiments and projecting the computing needs for the first LHC run. The ATLAS computing technical design report describing the experiment's initial computing model was published a few months earlier. The model was based on a hierarchical structure of the computing centres, also called tiers. Different tier levels would play different roles and were expected to provide different kinds of resources and quality of service. CERN, the Tier-0, was responsible for storing the master copy of the RAW data into archival media (tape libraries) and running the first pass reconstruction. The Tier-1 centres, large national facilities in the major countries participating in computing, would be responsible for re-processing the data and storing a second copy of the RAW data. Tier-2 centres, mostly located at universities, would host a copy of the processed data for the analysers and contribute to simulation production. Calibration and alignment were also done at CERN, with the conditions data stored in a central Oracle database and distributed from there to Tier-1s and Tier-2s. Because of their critical responsibilities, Tier-0 and Tier-1 centres are expected to provide continuous (24/7) support for most services, while the expectations are lower for Tier-2s, as specified in the WLCG Memorandum of Understanding signed by the respective Funding Agencies. The initial computing model was based on the MONARC model developed in the late 1990s on the expectation that the network would remain a scarce resource; a dedicated LHC Optical Private Network (LHCOPN) would offer large capacity between the Tier-0 and Tier-1s, covering the most important and demanding use cases (data delivery from CERN to the other countries). Networking within each national research network was also expected to be adequately provisioned, but no assumption could be made on any other path. For this reason, the computing model was very strict on the kind of traffic that could be allowed, and international traffic was designed to happen only through Tier-1s and the Tier-0.

Developing the computing model early had a strong motivation: ATLAS's need for computing capacity long predated the first data. In 2003, ATLAS had already developed a system to produce simulated events for physics studies on Grid distributed resources, which were provided by three main flavours of Grid infrastructure: the EDG/EGEE, the Open Science Grid and NorduGrid. While all infrastructures relied on a common set of baseline services and protocols based on the Globus toolkit, the middleware to deliver compute and storage capacity built on top of the baseline varied considerably, and ATLAS was faced with the challenge of federating the different Grids. The ATLAS Production System (Prodsys) was developed with this modularity in mind: jobs would be defined in a central repository (an Oracle database) and orchestrated by a supervisor. Various executors with different implementations depending on the Grid flavour would interface with the specific underlying infrastructure. Initially, all executors worked in "push mode"; payloads were brokered to the most likely available site and then waited in the site queue for execution. The initial experience with the grid model had rough edges — the underlying infrastructure was very heterogeneous in terms of hardware, policies, performance and availability, and the efficiency was inherently lower than that of a typical batch cluster. ATLAS therefore switched paradigms and implemented a "pull mode" job dispatch mechanism, in which payload-agnostic jobs called pilots were submitted at the sites. Only once the pilot validated the resource would it pull from a central dispatcher the highest priority payload that had been brokered to that resource based on its capabilities and data availability. The system was able to hide Grid inefficiencies from users and remains the pillar of today's workload management. PanDA, the ATLAS Production and Distributed Analysis system was initially developed to leverage US ATLAS's computing resources and was quickly extended to replace all executors, so that it became and still serves as the core workload management engine of job brokering and dispatching for both organised and user processing, tightly integrated and orchestrated with data management.

Data transfer and access tools were sufficiently interoperable across Grid flavours that ATLAS was able to implement a central engine to manage data across the whole infrastructure. "Don Quixote" (DQ soon followed by DQ2), the ATLAS data management system, was initially implemented to transfer data between the Tier-0 and the Tier-1s and later between any ATLAS source and destination, evolving over the years to become a complete data management system. At the time of the first LHC beams in 2008, DQ2 offered all functionalities for data location replication and deletion. ATLAS files are logically organised in datasets, and in DQ2 a dataset represents the quantum of data management. Datasets could be replicated through a subscription mechanism: the placement policy for a given site is registered in the system and enforced through a set of agents, interacting with the underlying middleware and storage services. DQ2 did host a limited set of file and dataset level metadata, but the core component of metadata management was and remains the ATLAS Metadata Interface (AMI), aggregating meta-information from many sources into a central Oracle database and displaying them through a web interface, a Python interface and a command line tool.

The ATLAS distributed computing system went through a series of commissioning exercises during its development. The WLCG organised data challenges in 2006 aimed at testing the core functionalities with a focus on data transfers, with the final challenge involving direct participation of the experiments. In 2008, roughly a year before the first LHC data, ATLAS participated in a common readiness challenge together with the other experiments to validate the global performance of the data transfer services at the scale needed for first data. A more complex exercise was performed in 2009 covering all aspects of distributed computing, from data management to data processing and distributed analysis. At this time ATLAS developed the "HammerCloud" testing and commissioning framework, still at the core of computing operations today.

The ATLAS Distributed Computing (ADC) activity was formally established in 2006, encompassing all development and operational efforts under a single leadership with a tiered organisational model, greatly improving computing operations over the previous compartmentalised organisation. At the first level, the Distributed Analysis Support team acts as a front end for user requests, while the ATLAS Distributed Computing Support team identifies and reports issues with the infrastructure. Dedicated second level support units address issues with specific systems. Developers serve as third level support, with development governed by ADC's credo that operational needs drive development.

The ATLAS computing model has been reviewed and iterated several times in light of the data taking experience and evolution of the infrastructure and services. In 2010, the first year with a considerable amount of collision data, it was soon realised that the generous data replication policy foreseen in the computing model would not scale sufficiently with the increasing data volume. For example, one copy of the AOD (analysis format) dataset in each Tier-1, plus many additional copies at the Tier-2s were hard to sustain, in terms of total volume and transfer rates. Other formats such as complete reconstruction records, vitally important in the early stages of the experiment to understand the detector and its performance, became less critical over time. At the same time, the introduction of reduced data formats such as analysis ntuples was not initially foreseen and required additional disk space. The data distribution and disk copy retention policies were tuned to take this experience into account, and dynamic data management was introduced. A popularity service collected information about all datasets being accessed through production or analysis jobs. The system automatically triggered replications of frequently used datasets to more sites, while reducing the number of copies of less frequently used data. The system allowed the reduction of the amount of data being replicated and made sure the data on disk was that most likely to be used by analysts.

The hierarchical organisation of Tiers dating back to the MONARC model was also reviewed and updated. Thanks to the explosive growth of internet use in the early 2000s, powerful networks became widely available and economical. Restricting heavy data traffic exclusively to higher Tiers became an unnecessarily limiting restriction, so more direct connections between Tier sites on different research networks were utilised. Transfer efficiencies were monitored in a dedicated service and fed back to DQ2 so that the most efficient data transfer path could be dynamically selected. By the end of LHC Run-1, the hierarchical model was fully relaxed and data exchange was allowed among all sites.

The Run-2 running conditions presented major new challenges for computing. In particular the higher trigger rate and the higher pile-up implied at least a factor two more data in volume and a factor five more CPU. The long LHC shutdown between Run-1 and Run-2 (LS-1) provided a unique opportunity to review many of the distributed computing components, workflows and procedures, with enough time to re-engineer those most in need of revision in light of Run-1's experience and new requirements. The data management system that served the experiment very well during Run-1 would have difficulty scaling to the data rates in Run-2 and did not offer enough flexibility to leverage rapidly emerging technologies. A new data management system, Rucio, was therefore designed and implemented during the shutdown and commissioned in time for the start of Run-2. It built on the experience of the previous system and provided new capabilities in automation, scalability and modularity. It was conceived from the beginning to facilitate collaborative development work and assured quality of the code. Rucio implemented as built-in capabilities many of the features that were added as external features to DQ2, improving the overall robustness of the service.

The increased pile-up with respect to Run-1 required more memory in reconstruction, while the available memory on Grid nodes remained unaltered for reasons of cost. ATLAS therefore implemented multicore jobs using multiprocessing techniques that share memory between processes, enabling reconstruction to fit the available memory. Mixing multicore and single core jobs in multidisciplinary facilities required evolution of the workload management tools, tuning of operational parameters, and implementation of new monitoring at both the job and global levels. The analysis model was fully revised during the shutdown, drawing on the experience of Run-1 when individual groups used independent tools to select, collect and persistify the events and information relevant to a particular analysis in a format using the ROOT program. For Run-2, ATLAS organised this activity in a centrally managed shared workflow, the new "derivation framework" in which each physics group can define the selection criteria relevant for the specific analysis and what information to retain. The framework is executed regularly on AOD datasets and produces derived AOD formats suitable for analysis. The AOD format and its derived version were re-engineered to be readable through ROOT as well as Athena. The Prodsys production system had to be redesigned to support new workflows such as multicore jobs and the new analysis model. Prodsys-2 was developed during LS-1, preserving PanDA as the core component for workload management while re-implementing the higher layers dealing with user and production manager requests, definition of tasks, and task assignments to compute resources, to produce a more capable and scalable system with better automation. Prodsys-2 was complemented by the creation of JEDI, a new PanDA component able to flexibly and dynamically configure and assign workloads in order to process them optimally on currently available computing resources, setting the stage for extensive use of opportunistic resources by ATLAS during Run-2.

Several new components were developed to improve the way resources and systems were defined, organised and monitored. The ATLAS Grid Information System (AGIS) was developed to provide a centralised configuration repository for Prodsys-2, Rucio, and the resources and services at Grid sites. AGIS provided the flexibility to annotate and extend such information; it replaced several difficult to maintain precursor tools. Monitoring and accounting were also revisited, with Rucio, PanDA and Prodsys-2 developing new complementary monitoring systems that leverage current technologies and features of relational and non-relational databases. More recently a new suite of leading-edge data analytics tools was implemented to mine information from many distributed computing services and offer unprecedented monitoring capabilities. Finally, the deployment of core ATLAS computing services also went through a major revision, moving to virtualised hardware. This migration to standard tools widely used in the open source world was an important step in system simplification and long-term sustainability.

After a year of LHC commissioning in 2015 with steadily increasing machine performance, the accelerator delivered an impressive 20% more integrated luminosity than expected in 2016. This unexpectedly large data volume and the particularly long runs had to be handled in quasi-real time by the Tier-0 Management System responsible for reconstruction of the express stream, calibration, and calibrated reconstruction of the physics streams. The reconstructed data must be available within a few days after the data have been collected so that data quality can be assessed. The Tier-0 system leverages resources at CERN, but in 2016 some of the processing was spilled over to Tier-1 centres via the production system to ensure prompt data delivery during critical peak data taking periods. Because of this, ATLAS was able to present at the ICHEP conference, results that were based on data collected merely 15 days earlier — a proud achievement for distributed computing as well as the analysis teams.

The larger than expected volume of data and integrated luminosity in Run-2 required the complement of an increased amount of simulated data. Figure 7.9 shows the number of compute units processing ATLAS events in Run-2 as a function of time, showing the processing capacity almost doubling since the start of Run-2, with the majority of the resources going to simulation, while 30% are used to reconstruct and analyse the data. ATLAS constantly utilises all available processing and storage resources, with the production system prioritising requests based on inputs from the physics community. Compute resources pledged to ATLAS remain scarce nevertheless, and the physics programme benefits greatly from opportunistic usage of non-pledged resources. A fraction of such resources is provided in the form of CPUs at Grid sites and can be integrated in the production system with little or no effort, particularly if the extra capacity comes as expansion of the pledged one. Many diverse, opportunistically available resources are however not accessible via the Grid and present a challenge to use. ATLAS at the end of Run-1 started looking into integrating CPU capacity from cloud computing. Today, resources from several academic cloud resources are part of the ATLAS production pool, while several exercises have demonstrated

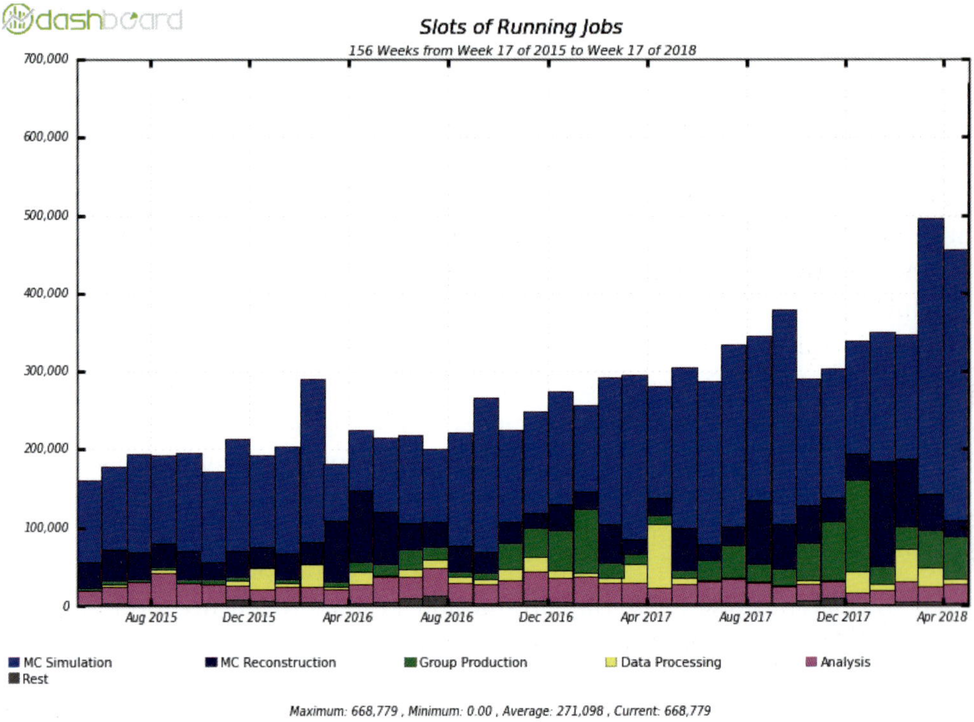

Fig. 7.9. ATLAS processing during Run-2.

the capability to leverage commercial cloud compute capacity in the United States, Europe and Asia-Pacific. The ATLAS High Level Trigger farm has also been instrumented with a cloud interface and is used for simulation in the non-data-taking periods. ATLAS has also integrated resources from many high-performance computing (HPC) systems (aka supercomputers) worldwide. While these large machines are built for use cases other than HEP, substantial CPU cycles are available to HEP at many facilities, and ATLAS has worked hard to leverage them either opportunistically or through grants. The usage of such facilities presents many challenges: the hardware can be different from the standard on the Grid, the access policies can be restrictive, data traffic capabilities can be limited, and in general each facility presents unique characteristics. ATLAS has successfully integrated about a dozen HPC facilities worldwide and expects to engage with more in the future. Finally, the ATLAS@Home project allows the general public, especially those with little computing knowledge but who wish to contribute to the ATLAS physics programme, to contribute spare capacity at home or office desktops and laptops for simulation of events. The system is implemented through the BOINC technology, the ARC middleware and its integration with PanDA, leveraging the technology used in production in NorduGrid since early 2000. Today, more than 25% of ATLAS production is run on non-Grid resources; this large contribution motivating the effort spent on integrating them.

ATLAS's large-scale CPU usage is complemented by large volume data flows. Figure 7.10 shows the volume of data transferred daily worldwide by ATLAS. Peaks correspond to intensive data management campaigns; overall the total traffic increased more than four-fold during Run-2.

Run-3 data taking conditions are expected to be similar to the last year of Run-2's and should therefore not present a big challenge. The next major challenge for computing will come a few years later in the high luminosity phase of the LHC. The data taking rate will increase ten-fold and the pile-up by a factor of five. The additional compute and storage capacity needed to digest such increases cannot be satisfied simply from anticipated hardware performance gains. Furthermore, substantial increases in funds for computing can in practical terms be excluded; planning assumes flat

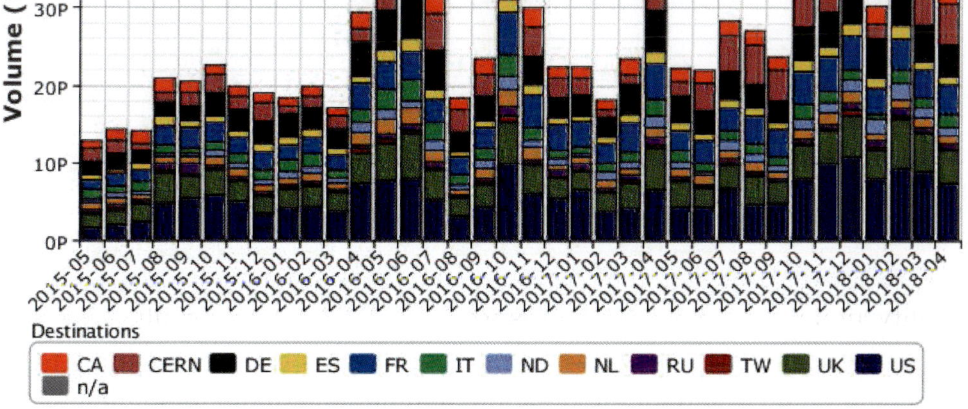

Fig. 7.10. The processing and data management activity during Run-2.

funding. The HEP community, with ATLAS in a leading role, is preparing to address this challenge through substantial evolution of the computing model, the offline software and the distributed computing systems, ensuring that the coming decade of ATLAS computing is sure to be as interesting as the last.

LHC Day One: In the ATLAS computing control room

Most ATLAS people know the control room and have seen the photographs of the first day of LHC data in 2008 taken from that room. What some people do not know is that in the same building, two floors above there was a much smaller, but still very crowded, control room to monitor computing activities on that same day. The ATLAS computing model relies on distributed resources and every institute in ATLAS participating with its computer centre was waiting for the first LHC data. The data volume of those few beam splashes was expected to be very small, and could have probably been sent around via email, but it was the first time the computing model was going to be exercised in real life. Obviously, the workflow had been tested several times, with challenges of increasing complexity, and the capability to handle the first year LHC data had been demonstrated with large contingency. When data arrived, the ATLAS Tier-0 system started recording its first run. The datasets were going to be exported only when the run was closed. That was how the data management system was working at that time. The data export started at 5PM, many hours after the data had been collected and the celebrations in the main control room had started. Most people at ATLAS Point 1 had left already when the computing people realised something was wrong. The export rate from CERN to the Tier-1 centres started increasing significantly, as was expected given the new data (see 17h in left figure) but with that increase, transfer errors started to appear as well (see right figure). Some sites worried more than others... "What is all that data you are sending us, guys? — Well, it is the LHC data..." However, the errors were affecting transfers to all destination sites, which implied the issue was at the source. An alarm was sent to the CERN computer centre, which promptly realised there was an overload of one disk server and fixed the issue. We were still uncomfortable. Why did that issue not show in the past when we had tried stressing the system orders of magnitude more vigorously? We then realised that the load was caused by a single user trying to access data with a rogue script. The irony was that the data access had nothing to do with LHC Day One; it was some test beam data collected years before. The user was not at all interested in the new data. The lesson learned: you should always separate chaotic user access from your production infrastructure.

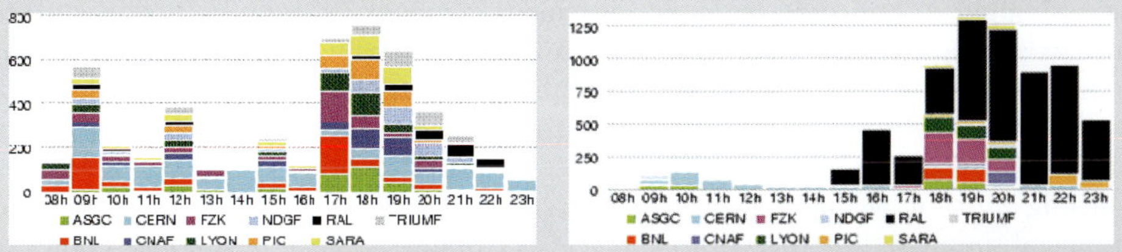

CERN to ATLAS Tier-1 computing sites on Day One of LHC beam-splash events. (Left) Average hourly transfer rate (in MB/s). (Right) Hourly number of transfer errors.

Earth, Wind, and Fire. And Water too.

The preparations for the ATLAS distributed computing infrastructure started in the early XXI century, as soon as it was realised that no single computer centre, not even CERN, would be able to purchase, deploy, power, and manage the enormous number of computing and disk and tape servers that would be needed to store, process and analyse ATLAS data. By 2007, one year before the start of LHC operations as foreseen then, the general organisation of computing centres supporting ATLAS was established, and infrastructure tests were well advanced. Computing management went on a tour of visits, or readiness inspections, of the ten Tier-1 sites around the world. The visits were very useful as they permitted the exchange of ideas and solutions and the free discussion of local conditions and constraints; for example, we learned that in Amsterdam they placed the computing equipment, despite its weight, on the first floor (above the ground floor) of their building to be safe from floods, as that is one metre above sea level. Other sites took different decisions, using refurbished or recycled spaces like a former nuclear shelter and an underground garage area.

We know that events that have low probability of occurring are actually likely to occur if a large number of units are concerned, and one waits long enough, and ATLAS computing centres are not immune from the risks due to geological, meteorological, and electromagnetic events. The Fukushima earthquake in Japan reduced suddenly the available power in all of Japan, including the University of Tokyo computer centre, but our Japanese colleagues managed with great efforts to keep enough power to sustain the data flow and processing of ATLAS data.

Hurricane Sandy in New York in 2012 blocked the Big Apple for several days; nevertheless, the ATLAS activities at the BNL Tier-1 centre continued unperturbed, thanks to the high level of automation of the computing systems and procedures, and the possibility for remote control.

Nuclear shelters are supposed to be the safest locations, except when a fire breaks out from the inside. In 2009, a power transformer caught fire in the Academia Sinica Tier-1 computer centre in Taipei; the fire was put out immediately but thick smoke invaded the computer room and deposited unwanted residue over the computing equipment. Despite this setback, this computer centre was able to get up and running again.

Finally, water... The basement of the Niels Bohr Institute in Copenhagen was thought to be safe from floods since Denmark is relatively flat and so any excess rain water runs directly into the surrounding seas. But despite this expectation, the NBI computer centre unfortunately was flooded a couple of years later. In 2017, a city water main pipe burst under the road in front of the CNAF computer centre in Bologna and water invaded the computer room, causing widespread damage. This is the only case of serious local data loss, and financial damage, but thanks to the ATLAS policy to have redundant copies of all important data in different locations there was no permanent damage to data processing or analysis.

Conclusion: the sun never sets over ATLAS computing operations!

Chapter 8

From Testbeams to First Physics

The road to stable data taking with the ATLAS detector began in the early 1990s, with beam testing of prototype designs for different detector subsystems. These were followed by similar testbeam campaigns for "module-0" prototypes of the final design, and then for the series production modules. Following the integration of all the detector subsystems, commissioning was done using the ATLAS calibration systems as well as with cosmic ray muons. The road was not always smooth; in 2008, the LHC experienced major mechanical damage that delayed the start of stable running for a year. However, for almost ten years, since 2009, the data delivery performance of the LHC and the data taking performance of ATLAS have been impressive. Table 8.1 lists a short timeline of beam tests, combined cosmic ray running and LHC running periods from 2004 to 2017.

8.1. From components to combined testbeams

The high luminosity of the LHC places significant constraints on the design of the detector components. Both the detector elements and the associated readout electronics must function at the planned instantaneous luminosity and survive the total integrated radiation dose that will be accumulated over the lifetime of the experiment. For the detector elements, these requirements affect the choice of both passive materials, such as the absorbers in the calorimeter subsystems, and straws in the Transition Radiation Tracker (TRT), as well as of active materials such as the liquid argon and scintillating tiles in the calorimeter, the gases in the TRT and the muon system, and the silicon-sensor technologies employed in the Pixel and SemiConductor Tracker (SCT) components of the Inner Detector (ID). Finally, the use of materials that are subject to significant activation must be avoided. Besides

Table 8.1. Timeline of beam tests, cosmic ray running and collision runs.

2002 to 2006	Combined testbeams with H6 and H8 SPS beam lines
2008 to 2009	Combined cosmic running
10 Sept. 2008	First splash events with single LHC beams
19 Sept. 2008	Incident at the LHC, resulting in mechanical damage
23 Nov. 2009	Splash events with single LHC beams
5 Dec. 2009	First stable colliding beams
2009	900 GeV pp collisions
2010 to 2012	Run-1
2010	7 TeV pp collisions, 2.76 TeV Pb-Pb collisions
2011	7 TeV pp collisions, 2.76 TeV Pb-Pb collisions
2012	8 TeV pp collisions, 5.0 TeV p-Pb collisions
2015 to 2018	Run-2
2015	13 TeV pp collisions, 5.0 TeV pp and Pb-Pb collisions
2016	13 TeV pp collisions, 5.0 and 8.2 TeV p-Pb collisions
2017	13 TeV pp collisions

the detector elements themselves, the electronics which read them out must also be qualified to operate in this high radiation environment. Radiation tests of both materials and electronics were done using, e.g., nuclear reactors for low-energy neutrons, or high-intensity gamma-ray sources for low-energy photons, along with high-intensity proton beams.

The relationship between signals from detector subsystems and the properties of the particles producing them are established in so-called "testbeam" studies in which production or prototype detector elements are exposed to beams of particles of known properties (type, energy, momentum, position, etc.). These studies provide basic information upon which the later fine-tuning of calibrations rely. The overall calibration strategy is discussed in the next section.

As already discussed in Chap. 1, early testbeam studies related to the development of the various subsystems of what would become the ATLAS detector were already being undertaken in the early 1990s, before the Collaboration formally existed. Dedicated R&D projects investigated design possibilities for the different subsystems of the inner tracking detector, the electromagnetic and hadronic calorimeters, and the muon spectrometer. As one example, the challenging and unusual design of the accordion electromagnetic calorimeter was driven in part by the goal to discover the Higgs boson via the decay two photons. Early prototypes were already being built and beam-tested by the RD3 Collaboration in the early 1990s, with a first combined test of the barrel electromagnetic (EM) and hadronic calorimeter concepts taking place in 1994. Beam testing of an early prototype in 1993 was also an integral part of choosing liquid argon as the technology for the ATLAS forward calorimeter, for which competing proposals were still under consideration at that time. Testbeam investigations of proposed TRT and silicon-detector technologies were undertaken over the same period. For the silicon tracker, the properties of both irradiated and un-irradiated silicon sensors, and the associated read-out electronics, were extensively investigated in regular beam tests at either CERN or KEK starting in the mid-1990s. For the TRT, testbeam studies of the particle identification capabilities of a small prototype were undertaken in 1990, followed by similar studies of a larger prototype, and investigations of the performance achievable with different choices of radiator material and gases, as well as at high rates. The final barrel TRT prototype, representing a 25-degree wedge was tested from 2002 to 2004, including as part of the combined testbeam program described below. For the muon system, testbeam studies were also used to investigate design issues, e.g., the use of different gas mixtures, for studies of performance and aging of the detector technology in an LHC-like environment (generated using an X-ray source in the X5-GIF facility at CERN) and for validation of the optical alignment system, which is critical to achieving the required momentum resolution.

Between 2002 and 2004, testbeam facilities were installed in the so-called H6 and H8 SPS beam lines in the north area of CERN, with the goal of gathering on the same beam line all prototypes of the various ATLAS subdetectors. Electrons, pions and muons with energies in the range of 10 to 350 GeV (for H8) were provided with a very good energy resolution, allowing detailed assessments of the detector performance in conditions which tried to mimic the final ATLAS environment, although each subdetector prototype represented only a small fraction of the final apparatus. For instance, the particles could be produced in bunches separated by 25 ns in time, as expected in LHC conditions; this was useful for testing the data-acquisition system and some features of the triggering system. The combined testbeam (CTB) programs also served the very useful purpose of gathering together the various subdetector teams to work together in close-by control rooms and on the analysis of data coming from several different subdetectors. A sketch of the H8 combined testbeam setup with the ATLAS subdetector prototypes aligned along the beam line is shown in Fig. 8.1.

For the ID, important information on the silicon detector's response to the passage of high momentum tracks (like the formation of pixel and strip clusters) was gathered and used to refine simulation programs. The ID

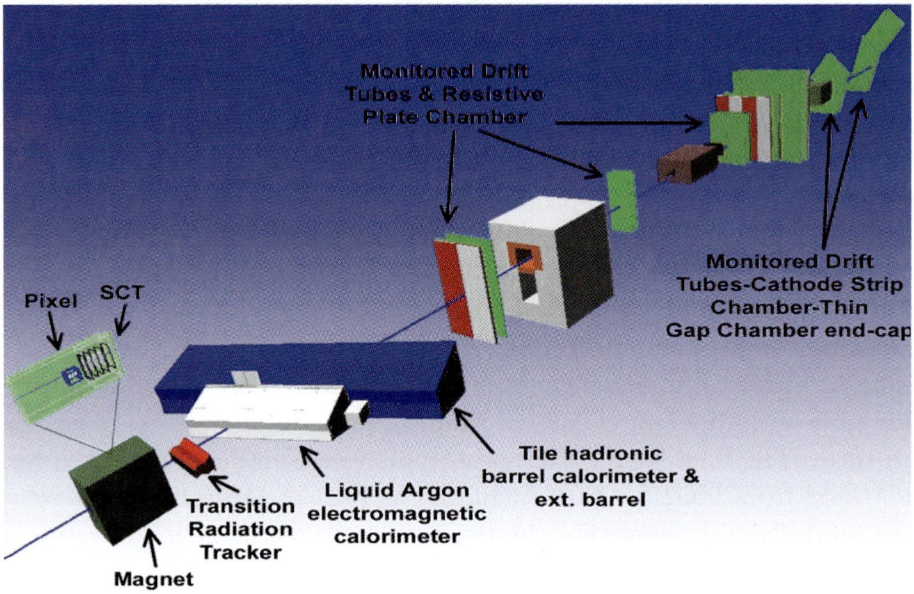

Fig. 8.1. Sketch of the H8 combined testbeam setup.

volume in the H8 CTB setup was divided into three containers, one each for the Pixels, SCT and TRT. Six Pixel and eight SCT modules were placed in their respective containers and located inside the bore of a dipole magnet used to mimic the field of the future ATLAS solenoidal magnet. The TRT setup consisted of two-barrel wedges, equivalent to 1/16 of the circumference of a cylinder. The number of tracking layers (3 Pixel and 4 SCT) provided enough redundancy to test the alignment procedure using curved tracks (see Sec. 8.2.1).

For the calorimeters, testbeams were used to establish the EM energy scales in the barrel region at H8 for each type of module of the EM barrel (EMB) and Tile and in the endcap regions at H6 for the EM endcap (EMEC), Hadronic Endcap (HEC) and Forward Calorimeter (FCal). These studies were followed by combined testbeams of the EMB+Tile, the EMEC+HEC and finally of the EMEC+HEC+FCal. The combined tests of the EM and hadronic layers of the calorimeter provided necessary information about the inter-calibrations and the response to hadronic energy deposits (mainly in jets). The EMEC+HEC+FCal combined testbeam additionally provided information needed for the calibration of particles and jets that fall into the overlap region between the EMEC/HEC and the FCal. The analysis of the electron data collected in EMB testbeams steered the development of the procedure for equalising the electronic response of the Liquid Argon (LAr) Calorimeter cells that is still used by ATLAS during the LHC data taking and proved that a long-range EM energy response uniformity within a calorimeter module better than 0.5% could be achieved. The procedure for correcting for the energy lost in the passive material upstream the EMB and EMEC was pioneered in the same testbeam studies and demonstrated that the EM calorimeter design specifications could be met, and even exceeded.

The Muon Spectrometer setup within the testbeam had two parts: a barrel stand and an endcap stand. The barrel setup reproduced 1/6 of an ATLAS barrel sector, with its Muon Drift Tube (MDT) and Resisitive Plate Chambers (RPC). It was followed by the endcap setup with six MDT chambers and three Thin Gap Chamber (TGC) units (one triplet and two doublets for triggering purposes) reproducing part of an endcap sector as well. An upstream dipole magnet located at a position mimicking the collider interaction point allowed bending of the muon trajectory in order to scan over the full sector area. Another dipole magnet was installed inside the endcap setup to allow a measurement of the particle momentum. Both setups were equipped with optical alignment systems, very similar to the ones foreseen for ATLAS. Controlled movement of the MDT chambers (translation and rotation) were performed to test the behaviour of the optical sensors over displacement ranges of several millimetres.

8.2. Calibration and alignment

The calibration of the response of each particle candidate in ATLAS begins with the calibration of the objects used to reconstruct them: e.g., charged tracks in the ID and EM showers in the EM calorimeters to build electron and photon candidates, ID tracks and energy clusters in the full calorimetric systems to build jets, tracks in the ID and the Muon Spectrometer to build muon candidates, etc. The basic calibrations for these objects are derived from testbeam studies and must be transferred to the final "as-built" detector. They are then refined and maintained using purpose-built calibration systems, as well as via in situ physics studies using collision data. The latter are based on classes of events where a well-measured object or event topology can be used as a reference, such as in $Z \rightarrow e^+ e^-$ events where the Z boson mass is well-measured from experiments at the LEP collider and therefore can be used as a constraint. In addition, for the tracking detectors that measure the particle momentum through a precise measurement of the reconstructed track curvature, a key feature is the alignment of the individual modules: Pixels and SCT sensors and TRT straw tubes units for the ID, and muon chambers, including their deformations, for the Muon Spectrometer.

8.2.1. *Calibration and alignment of the detector subsystems*

In the LAr calorimeters, the energy in each cell is reconstructed from the digitised samples of the current signals. The response of the read-out electronics is determined using the testbeam data but making use of electronic calibration data routinely acquired on the detector during periods where there are no colliding beams in the detector. Here, electronic pedestals are measured daily during the LHC data taking, and electronic gain and pulse response data are acquired on a weekly basis. The corresponding calibration constants are then re-calculated. The testbeam data are used to set the overall EM energy scale including an average correction for the detector sampling fraction. This scale is later refined using in situ studies. Over Run-1, these calibrations were observed to be stable at the level of 0.1%.

For the Tile Calorimeter, the overall response of the detector is probed using a radioactive cesium source that is inserted in the detector during periods with no beams and allows for the measurement of the gain response of the detector read-out. The light induced in the scintillating tiles is converted into an electrical signal by PMTs, which are routinely monitored with a laser system used to correct residual non-uniformities after the gain equalisation has been done with the cesium source. As for the LAr calorimeter, the initial conversion between the current signal and the energy deposit, calibrated at the EM scale, is based on testbeam data and later refined with in situ studies.

In the ID, the precision of the measurement of the parameters of the charged particle trajectories is limited by a combination of the intrinsic resolution, knowledge of the magnetic field and material distribution, and location of the detector elements. This information is initially estimated based on design and construction information and from installation surveys and is later refined using collision data and track-based alignment studies. At the end of Run-1, the positions of the individual active detector modules in all three ID subsystems were aligned to be better than 1 μm. To achieve this accuracy, a global fit involving about 36,000 degrees of freedom is performed on the residuals of about 10 million reconstructed helical tracks.

Whereas the ID alignment is almost entirely based on tracks, the Muon Spectrometer alignment relies on a combination of the use of straight tracks (data taken with toroidal magnets switched off) and the use of about 10,000 optical sensors. These sensors, which are accurately mounted on the MDT

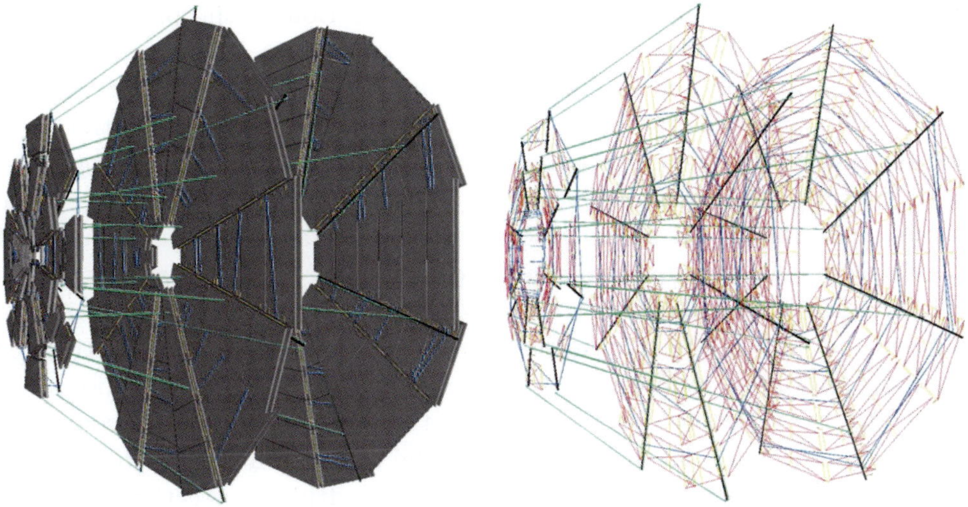

Fig. 8.2. Schematic view of one endcap Muon Spectrometer's alignment system. The figure on the right, with the drift chambers removed, shows all the optical lines connecting the alignment bars between themselves and with the chambers.

chambers and on intermediate structures (alignment bars in the endcaps are seen in Fig. 8.2, which shows four layers of muon stations along z, each containing eight controlled aluminium bars connected between themselves and neighbouring chambers with opto-electronical devices), monitor to within a few μm the relative movements of the chambers (a few mm when the toroids are switched on) and their deformations. A global fit is used to determine the position of the chambers, which can be more than 5 m apart, to a precision of a few tens of μm. For the MDT chambers, the relation between the measured drift time and the distance between the track and the central anode wire (r-t relation) must be carefully calibrated. The calibration procedure exploits the fact that each track traverses, on average, more than six tubes in a chamber, at various distances to the wires. This redundancy is used to fit the r-t relation for this chamber using a large number of tracks. To processes this large set of information, dedicated calibration centres located at several ATLAS institutes receive large data samples every few days, and monitor the changes of the r-t relation with time.

8.2.2. *Calibration of physics objects*

Particle candidates are reconstructed from the basic physics objects, e.g., tracks in the ID and the Muon Spectrometers and clusters of EM and hadronic energy deposits in the calorimeters. Their energy and direction are calibrated by exploiting the detailed simulation of the processes they induce in the detector systems.

Electron and photon candidates are built starting from clusters of cells in the EM calorimeters that may or may not be matched to tracks in the ID. Their energy calibration, which accounts for upstream energy loss, both lateral and longitudinal leakage and for the energy deposited in the passive material of the calorimeters, is based on a detailed simulation of the development of EM showers in the calorimeter. The calibration exploits a multi-variate technique to correlate the relative energy deposits in the different layers of the calorimeter with the true energy of electron or photon candidates, by optimising the reconstructed energy resolution. The procedure is separately optimised for electrons, unconverted and converted photons, and for the latter makes use of the conversion track properties to improve the energy resolution. A series of data-driven correction factors is applied to account for any difference between collision data and simulation, and in particular, for the inter-calibration of the response of the LAr longitudinal layers and for the imperfect simulation of the detector geometry. To achieve this, an overall scale adjustment of the electron and photon energy scale is performed by comparing the calibrated energy of electrons from $Z \rightarrow e^+ e^-$ events in collision data and simulation. The size of this residual correction is of the order of 1–2%. The energy resolution is simultaneously corrected in the simulation to match that observed in data, by adjusting the electron energy resolution such that the width of the reconstructed Z boson peak in the simulation matches that observed in the data.

Jets are reconstructed based on information in both the EM and hadronic calorimeter. The inputs to the algorithm are three-dimensional "topological" clusters of energy deposits in the calorimeter cells. Jets are initially corrected on an event-by-event basis, for energy deposits originating from pile-up, then their energy response is further calibrated by exploiting additional information about the calorimetric energy deposits, as well as that of the associated tracks in the ID and Muon Spectrometer, using a combination of simulation-based and data-driven corrections. These corrections are meant to account for the different response of the calorimeters to EM and hadronic showers, for the energy losses in the inactive regions of the calorimeters, and for the varying response in the detector depending on the jet particle composition. As for electrons and photons, a final in situ correction accounts for differences between data and simulation in the jet p_{T}, which originate from limitations in the simulation of the underlying event, physics of jet formation, pile-up activity, physics of EM and hadronic interactions in the detector, and description of the detector material. The corrections are calculated using γ/Z+jet and multijet processes, where the momentum ratio between a jet and reconstructed reference object can be built and compared between data and simulation.

8.3. Commissioning with cosmic rays in the pit

The commissioning of the ATLAS detector using cosmic ray particles marked a key transition from the building to the operation of the detector. Cosmic ray particles, which are mainly muons reaching the ATLAS cavern, were used for several purposes. First, cosmic rays were used to test the overall trigger and detector timing. Second, since muons typically traverse several subdetector systems, they were used to integrate subdetectors and test the full data-acquisition chain. Finally, these particles were used for the first studies of the performance of the detector and for important first comparisons of the data to the simulated model of the detector.

ATLAS was built in pieces over several years, with each subdetector having its own community of experts. Each subdetector required its own data acquisition, triggers, detector control systems and data monitoring. These systems were designed as modules of an ATLAS-wide system within a tree-like structure, with each subdetector representing a branch off the main trunk. These branches can be detached from the ATLAS tree, thereby giving control to the subdetector for standalone operations. In this way, the subsystems can be operated independently of each other (for example during calibration runs) or in a synchronised way for combined running using this top-level ATLAS DAQ infrastructure.

Running these subdetector modules as a single detector unit from the top-level DAQ infrastructure required that the timing and delay of the trigger and DAQ systems were precisely tuned between each subdetector. The data from the ATLAS detector are recorded at a rate of 40 MHz by on-detector pipeline memories and are stored until a decision is received from the trigger system to read out the data to the off-detector electronics. If no decision is received, the data eventually drop off the pipeline and are lost. For a given event, the time required for the electronics to receive a trigger decision must be precisely known. Otherwise, the wrong event information will be read out from the pipeline. This trigger delay or "latency" must be constant from event-to-event but depends on many factors such as the source of the trigger and the delays in the electronics systems.

During commissioning using cosmic ray data, several main triggers were available — triggers from the muon trigger system, a trigger based on energy deposits in the outermost layer of the Tile Calorimeter and a tracking-based trigger using the TRT. The trigger information from all sources is collected by the Central Trigger Processor (CTP). The function of the CTP is to take trigger decisions by comparing the collected information with criteria defined in a pre-programmed trigger menu. The trigger decision is dispatched to the front-end electronics to read out selected events for offline recording. By using the CTP in cosmic runs, the whole chain of first level trigger, including trigger menu functionalities, could be exercised and fine-tuned.

Understanding the trigger latency using cosmic data posed its own challenges. While the triggers from the muon system accepted the largest rate of cosmic ray muons, there was a large timing variation from one event to the next since the muon system covers such a large area. Therefore, only the trigger chambers on the bottom of the muon system were used. Tile Calorimeter-based trigger complemented the muon system triggers. Due to the sheer size of the calorimeters, the Tile-based trigger suffered from event-to-event timing variations similar to those described for the muon system triggers. To avoid this, only a small section of the calorimeter was used as a trigger. This provided a stable timing but had significantly lower rates of cosmic muons. Finally, a trigger based on the TRT system was employed in order to collect a clean sample of muons traversing the ID. At a later stage the High-Level Trigger was also tested, which included software tracking algorithms that searched for cosmic ray muon tracks. Figure 8.3 shows a cosmic ray track crossing the Pixel Detector volume, recorded in October 2008.

With multiple triggers, the system had several different latencies, each depending on the source of the trigger. Programmable delays were implemented into the trigger and data path to resolve these issues. The correct settings of these delays can only be achieved using real particles (i.e., not calibration runs) and therefore the cosmic ray data allowed us to estimate the trigger latencies before the beam was available.

Since muons traverse several detector subsystems, they were also used as the first integration tests. The first of such tests involved the two calorimeter systems: LAr and Tile. The latency of the integrated system was determined using signals from cosmic muons crossing both calorimeters, scanning through the pipeline memories to find the corresponding signals in each calorimeter. Eventually the muon and ID systems were also fully integrated into the top-level DAQ system.

These integration tests had several important by-products. Operations with cosmic ray data provided an opportunity to establish a routine of stable data taking, prior to the presence of LHC beams. For many ATLAS

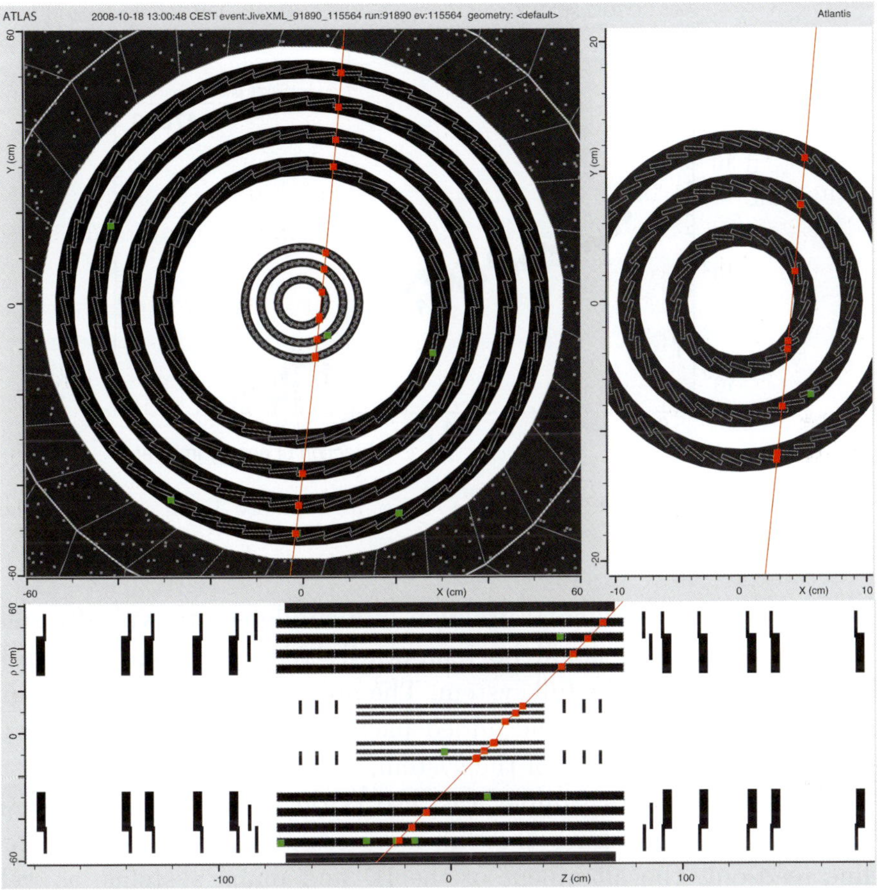

Fig. 8.3. Display of a cosmic ray muon going through the Pixel Detector taken on 18 October 2008. Shown are the XY view (of SCT and Pixels and of Pixels alone) and an RZ view.

physicists and technicians this was a cultural shift from the daily work over the previous decade, when each subdetector was independent and had its own daily run plan, to the new reality of a single experiment, with a detector-wide daily run plan and scheduled individual time for each subdetector. During this transition, short daily ATLAS-wide run meetings were started, to determine the priorities and procedures of the day so that there was no conflict between the ATLAS operation and the subdetector operations. The role of the shift leader was launched in order to oversee that program throughout the day. In addition, these integration tests also revealed the need for new, previously unforeseen features for the DAQ system. For example, to ensure smoother data taking, the development of the stopless recovery was

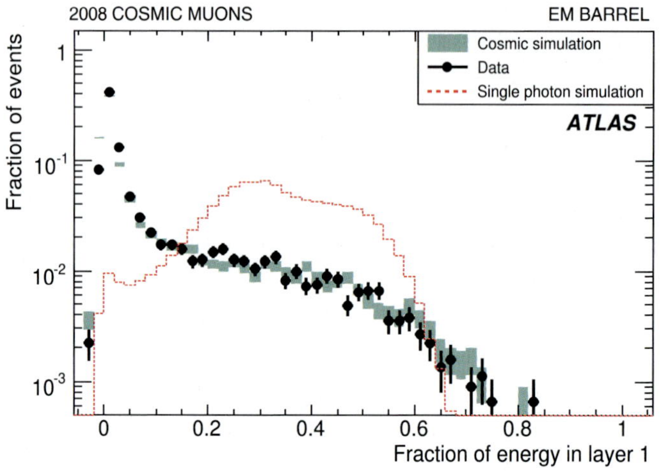

Fig. 8.4. Electromagnetic shower shapes using high-energy bremsstrahlung photons cosmic-ray data.

initiated, in which a very small fraction of a subdetector (typically smaller than 0.4% of channels) could be temporarily removed from the top-level DAQ without stopping the full system. The problematic subsystem channels could then be re-initialised and added back seamlessly into the data flow, without requiring a full stop of the system.

The cosmic ray data sample was also critical as a source of particles to be used for establishing the sub-nanosecond timing of the detector pipeline read-out, the alignment of the ID and muon systems, as well as the determination of dead and noisy channels and track reconstruction efficiencies. This data sample was also used for important first validations of the detector simulation. Occasionally, when a muon crosses the calorimeter, it radiates an energetic photon, which is then measured by the calorimeter systems. The energy of these photons as measured in data can be compared to the predictions of the simulation. Figure 8.4 shows the fraction of energy deposited in the first layer of the LAr EM calorimeter by a bremsstrahlung photon off a cosmic muon, in data and simulation, and for simulated single photons as expected in collision events. The simulation is able to predict the response as observed in the data.

8.4. First low energy collisions

The weeks leading up to first colliding beams in the LHC were a period of intense excitement with a flurry of final preparations. One concern was that the timing of the detector read-out systems would be incorrect and that we would fail to record the first data. This concern was accentuated by the fact that the control room during these times was packed full of reporters, recording our every move. Prior to the first beam circulation, the LHC sent a low intensity proton bunch, a so-called probe beam, step-by-step through each of the eight sectors of the accelerator. This probe beam was then purposely steered into a beam collimator located upstream of the ATLAS detector, thereby providing a large shower of particles travelling approximately parallel to the beam direction. The sheer number of particles produced by these collisions was enough such that every channel within the ATLAS detector had at least one particle passing through it. As a result, extremely high occupancy events were recorded and the timing of every channel could be determined. These data were then used to estimate the timing expected for colliding beams.

After the success of the "splash" events, ATLAS was ready for the colliding beams. However, a major concern at this point was the safety of the Pixel, SCT and Muon systems in the presence of any potential instabilities in the beam itself. As protective measures, the high voltages on the Pixel, SCT and Muon systems, and the Pixel pre-amplifiers were automatically turned off during the absence of stable beams. In this so-called STANDBY state, these systems can safely cope with non-stable beams and any LHC operations that might lead to an uncontrolled loss of the beams. However, during normal beam operations, the beam luminosity is always highest at the start of a fill and therefore, the ATLAS detector needs to be ready to take data as fast as possible once the LHC operators have declared that the beam is stable. To achieve this a "Warm Start" procedure was developed. All the front-end electronics are configured and data taking is started in STANDBY mode while the LHC is preparing the beam. The LHC provides a stable beam flag that automatically starts the turn-on of the high voltages on the Pixel, SCT and Muon systems. In this way, the ATLAS detector is able to seamlessly transition from STANDBY to RUNNING mode without the need to reconfigure all front-end electronics and the whole ATLAS data-acquisition system. After many months of preparation, the ATLAS detector received its first stable beam collisions on 5 December 2009. After 17 years of careful planning and construction, finally, colliding beams!

For this first run, the ATLAS detector was kept in STANDBY mode in order to ensure that the beam was in fact stable and safe for the Pixel, SCT and Muon systems. The systems that monitor the beam itself were vital here. Events in the detector can arrive from sources other than beam collisions,

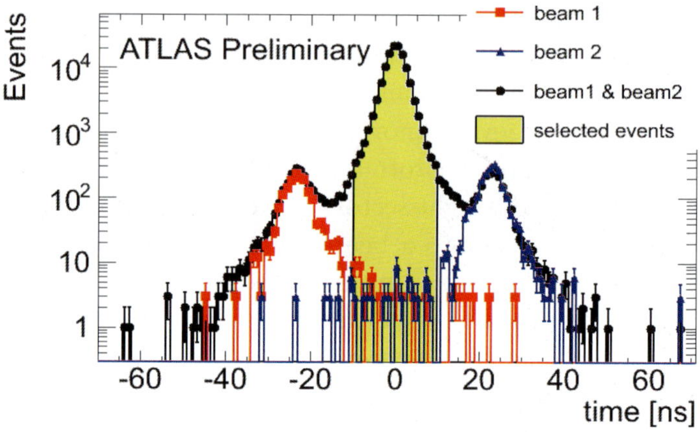

Fig. 8.5. Time difference between the two sides of the MBTS for the events recorded in December 2009.

such as particles from cosmic rays or particles induced from the beam itself, known as "beam halo" events. In this latter category, the particles cross the Minimum Bias Trigger Scintillators (MBTS), located on either side of the centre of ATLAS. This leads to a time difference between the two sides that is consistent with the time of flight and appears as two peaks in time. Figure 8.5 shows that the beam halo events from beam 1 (red squares) and beam 2 (blue triangles) peaked at 25 ns. All data events are shown by the black circles and the yellow area indicates events for which the time difference between the two sides is less than 10 ns.

The ATLAS Beam Condition Monitors (BCM) are responsible for quickly detecting abnormally high levels of radiation in the region of the silicon detectors, and if needed, triggering a dump of the LHC beam. Several tests were performed in which the LHC purposely provided higher radiation levels with ATLAS in STANDBY to check that the BCM did trigger the LHC beam dump as designed. For each of these tests, the level of the measured radiation was studied and thresholds, fine tuned. Using data from single-beam operations and first collisions, the beam monitoring systems were fully commissioned and were used to verify that the beam losses were acceptable, that the level of beam background was minimal and that the position of the beam itself was good.

During data taking, beam background events within the ATLAS detector are identified and monitored by searching for specific signatures in the muon, ID and calorimeter systems. A classic signature of a beam background event is an event with high energy deposits in a single calorimeter layer, with no associated tracks in the ID. These events, if not correctly identified,

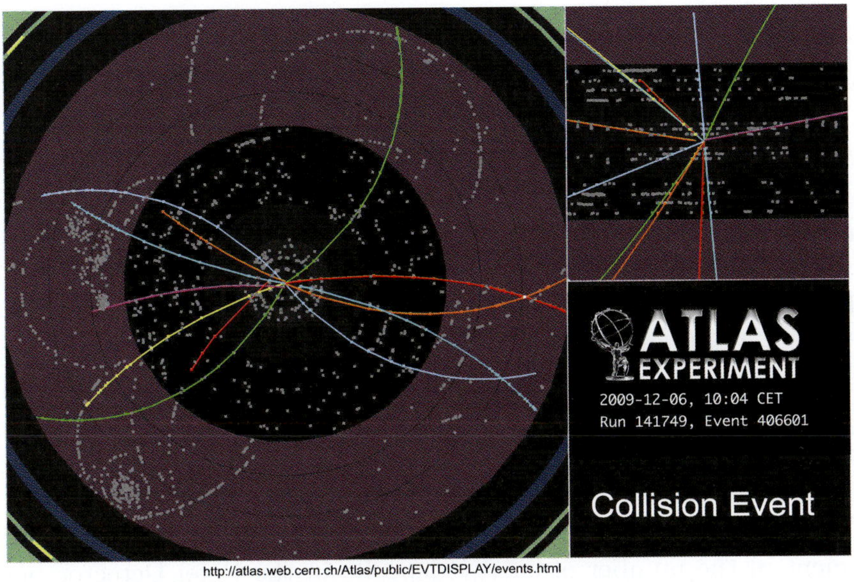

Fig. 8.6. One of the first proton-proton collision events recorded in ATLAS with stable beams.

can mimic exotic signals expected from physics beyond the Standard Model (SM).

Within a few days, confidence was established in the ATLAS beam monitoring and protection tools such that the high voltage of the Pixel, SCT and Muon systems could be turned on during stable beams. This step was taken very carefully with ATLAS in STANDBY mode, the LHC positioning the collimators in their normal position, and ATLAS experts carefully monitoring the beam conditions and the rate of hits in systems like the TRT to ensure that the beam losses were minimal. Tests of beam dumps were done to ensure that the protection systems would correctly respond. After these tests and with close communication between the expert teams in both the ATLAS and LHC control rooms, the high voltage on the Pixel, SCT and Muon systems were raised to nominal values. Figure 8.6 shows one of the first collision events including information from the ID system.

Once stable beams were established and checks that the protective systems worked as designed were completed, the ATLAS detector could start taking high-quality data to be used for refining the timing of the detector electronics, calibration of the systems, as well as for physics studies. An emotional milestone came only three days after the very first collisions, which was enabling the trigger system to reject events. The rate of collisions was so high that not all events could be saved. Therefore, the Level-2 trigger algorithms were activated for the first time to accept only the most interesting events.

8.5. First glimpse at the ATLAS performance

The first collision data from the LHC in 2009 allowed for detailed studies of the performance of the ATLAS detector and some early physics measurements. The early data provided an abundance of charged particles, which leave signals when passing through the ID. These hits were then reconstructed as tracks. These charged particles formed the heart of the first ATLAS physics paper: the measurement of the charged particle multiplicity. But first, key questions needed to be answered about the accuracy of the simulation, and in particular, the modelling of the material in the detector. Almost immediately, it was clear that the simulation described the data to a surprisingly high level of accuracy. Figure 8.7 shows almost perfect agreement in the number of energy deposits in the Pixel Detector between the data and the simulation. Such plots as these were the first published by the ATLAS Collaboration using collision data. These early data were used to measure the charged particle density as discussed in the "the road to the first minimum-bias publication" story.

The EM calorimeter was the next detector whose performance needed to be verified. This detector is vital for searches for new particles decaying to leptons or photons, which require good and well understood energy resolution. The observation of the abundant production of low mass SM

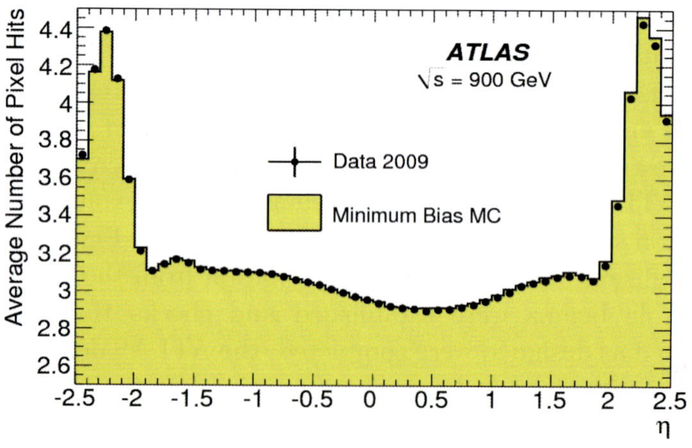

Fig. 8.7. The average number of Pixel hits as a function of pseudorapidity in data versus simulation.

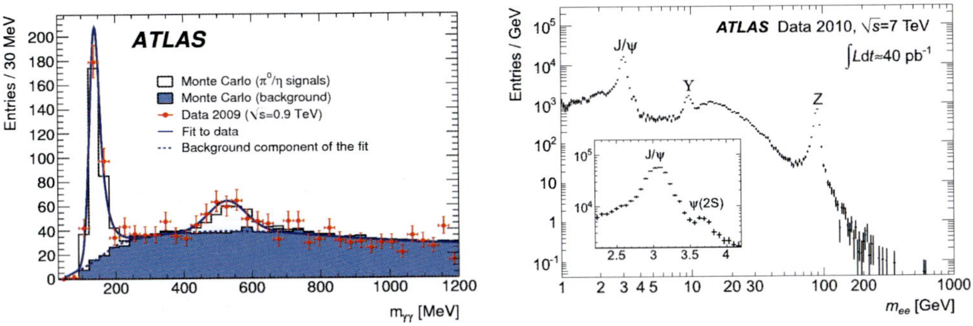

Fig. 8.8. Invariant mass distributions from di-photon events (left) and from electron-positron events (right) obtained with early proton-proton collisions data.

particles (Fig. 8.8, left) such as π^0 and η mesons decaying to pairs of photons confirmed our ability to reconstruct resonant peaks on top of a smooth background, opening the road to searches for the Higgs boson decaying to two photons. The observation of the $Z \rightarrow e^+ e^-$ peak using the calorimeter alone was a great early achievement but to fully validate the electron software algorithms, which matches tracks reconstructed in the ID with energy deposits detected in the calorimeter took more time. The invariant mass distribution of two electrons, which includes decays from J/ψ mesons, Υ mesons and Z bosons is shown on the right hand side of Fig. 8.8.

The data from the first collisions were used to improve the simulation of the observables involved in the selection of electrons, positrons and photons for physics analyses and to measure the corresponding efficiencies, profiting from these well-known physics processes. Such efficiency measurement techniques, referred to as "tag-and-probe", were also applied to particles decaying to pairs of muons. The very high efficiency for muons to be selected in data is well described by the simulation over a broad range of transverse momenta as seen in Fig. 8.9. These results, and others, allowed precision SM measurements, as well as searches for new particles with higher masses, to be made. These data were also used to validate and commission the trigger system. For both the Level-1 calorimeter and muon systems, excellent agreement between the objects measured "online" by the trigger system during the data taking and "offline" using stored data was found.

There were already a few thousand low momentum jets in this early dataset, so the performance was checked by comparing the transverse momentum distribution between data and simulation, as seen on the left hand side of Fig. 8.10. Excellent agreement was found, which validated the performance of the EM calorimeter, the jet reconstruction algorithms

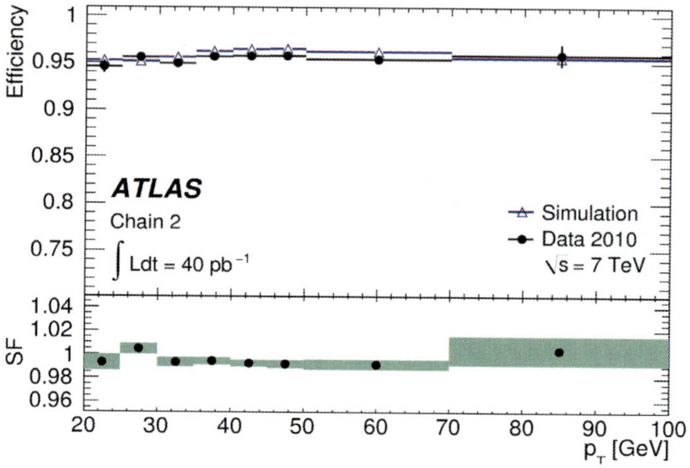

Fig. 8.9. Muon selection efficiency as a function of the muon transverse momenta.

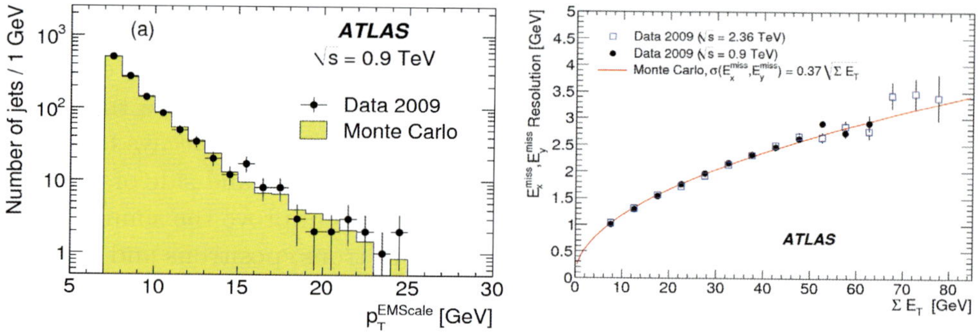

Fig. 8.10. (Left) The transverse momentum distribution of jets in data compared to simulation. (Right) The resolution of the missing transverse energy versus the total transverse energy.

and theoretical predictions. As the missing transverse energy is what is left once the energy of every other reconstructed object has been accounted for, it provided an early check of the full reconstruction chain. The missing transverse energy resolution in data is shown to be well-modelled by the simulation on the right hand side of Fig. 8.10.

8.6. Re-discovering the Standard Model

The first measurements with pp collisions at the LHC in 2010 were used to establish the validity of the SM at this new collider with an extended kinematic reach. In addition, the ATLAS Collaboration needed to prove that the fully commissioned detector could reconstruct the relevant final states emerging from the collisions with sufficient precision and acceptance to measure known SM particles, their properties and kinematic features with good resolution and high efficiency. This was an essential step towards using the detector for discovery physics.

Even with the relatively small amount of data collected in the first year of physics running, it was possible to measure and confirm several SM predictions, mostly with final states involving jets or the production of heavy particles like the W and Z bosons and top quarks. In addition, a set of measurements characterising the low momentum (soft) particle flow were conducted to provide input for the modelling of the underlying event and minimum bias physics in simulations. All measurements took advantage of the low proton beam intensities in 2010, which allowed for high precision reconstruction with only insignificant loss of signal quality due to pile-up generated by multiple proton-proton interactions in the same LHC bunch crossing.

The first SM analysis published by ATLAS was the measurement of the inclusive jet and dijet cross sections. This measurement not only established the use of a then new but now standard jet clustering algorithm, the anti-k_t algorithm, but also showed that in the LHC era, calculations of the hard processes driving the jet cross sections require precisions to at least next-to-leading order in perturbation theory. The left hand side of Fig. 8.11 shows

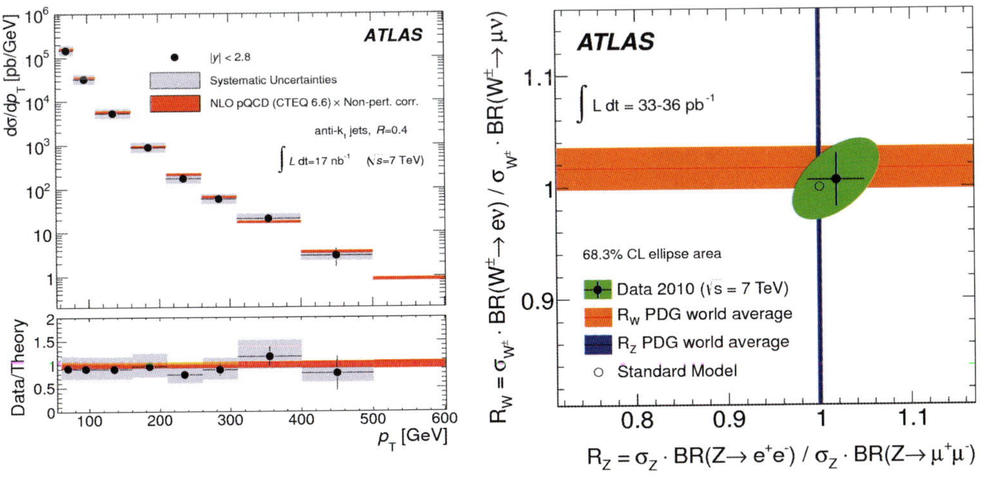

Fig. 8.11. (Left) The earliest measurement of the inclusive jet cross section with ATLAS. (Right) Measurement of the W and Z boson electron-to-muon decay ratios.

this first measurement of the inclusive jet cross section. Additional early measurements with jets investigated, e.g., jet shapes and spatial relations between jets. Among several others, those studies laid the foundation for the rich jet physics program in ATLAS.

The SM predicts the production cross sections of W and Z bosons, with state of the art theory predictions available in 2010. The measurements of these cross sections were an important confirmation of the validity of these predictions, and thus an important part of the re-discovery of the SM. The very first ATLAS publication of such measurements presented results achieved with a small collected data sample. The measurements focused on the leptonic decays of the gauge bosons. The same measurement using the complete 2010 data sample yielded precision results on, e.g., the correlated measurement of the electron-to-muon decay cross section ratios for W and Z bosons, as demonstrated in comparisons with previous world average results and SM predictions, as seen on the right hand side of Fig. 8.11. Other early ATLAS measurements exploring the electroweak sector of the SM included the measurements of the WW production cross section, the production of gauge bosons in association with jets, and the cross sections for Z and W bosons decaying into τ-leptons. These measurements represent only a beginning of precision electroweak physics in ATLAS. Similar to the early results for jet physics, their results indicated a strong need to control the systematic uncertainties in the experiment and to use predictions from the

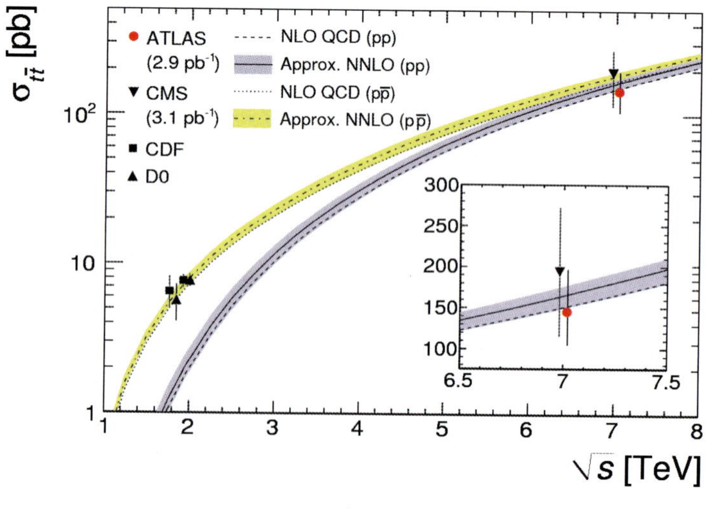

Fig. 8.12. First measurement of the top quark-pair production cross section in ATLAS.

most precise (highest order perturbation theory) calculations available when comparing to the SM.

The measurement of the top quark-pair production cross section in the early proton-proton collisions was another important part of the re-discovery of the SM at the LHC. It compared well to previous and contemporary results, when taking into account the expected difference between proton-proton and proton-anti-proton collisions in the evolution of the cross section from lower centre-of-mass energies. Figure 8.12 shows this result, together with a comparison to SM calculations employing next-to-leading order and next-to-next-to-leading order perturbation theory. This measurement laid the basis for extensive research in top quark physics. Several important features of the jet, electroweak and top quark physics in the SM were successfully confirmed in the first year of physics running at LHC.

Jet substructure in ATLAS

With the advent of the LHC, the hadronic physics community has undergone a revolution. At the Tevatron, hadronically-decaying massive particles such as electroweak bosons or the top quark were restricted to a p_T range below roughly 300 GeV. At the LHC, these particles frequently surpass 1 TeV, meaning that they are now sufficiently Lorentz-boosted for their decay products to overlap in the laboratory reference frame. In some cases, it is no longer possible to reconstruct the parent particle from its daughters. They now can be observed as a single jet with roughly the parent particle mass, and with an angular energy distribution consistent with a two- or three-body decay. Hadronically-decaying massive particles have transitioned from composite objects to single objects.

In preparation for the first LHC data, one of the earliest studies demonstrated the possibility of searching for the much-anticipated Higgs boson in an unconventional way, where its decay to a b-quark pair was sufficiently Lorentz-boosted to be contained within a single jet. They used the internal structure of the jet, the jet's substructure, to differentiate the Higgs boson from the Standard Model background. The authors included both ATLAS and theory colleagues, and this mix has proven beneficial to the field from its conception to the present.

Shortly thereafter, the BOOST conference series began, with a two-day kick-off event in 2009 attended by a mix of theorists and experimentalists. This has become the leading hadronic physics conference, and ATLAS is exceptionally well represented. The link between the ATLAS and theory communities remains strong, and it is common for theoretical techniques proposed at one year's BOOST to be followed up by ATLAS results the next year. The collaboration also goes in the other direction, where experimental constraints have ended up paving the way to new theoretical ideas. This strong two-way communication is a key piece of the initial and continuing success of the field.

Jet substructure techniques have changed how we search for new physics, measure Standard Model properties, and even how we trigger events. There has been an explosion of boosted jet taggers, supporting new analysis possibilities. For example, jet mass distribution has been used to probe hadronic resonances that have not been accessible since before the Tevatron. As such techniques have expanded, ATLAS has demonstrated its stability against high pile-up and increasing energies. Machine learning has proven to be important in a variety of areas, and its usage will likely grow. Jet substructure has also demonstrated the importance of both energy and angular information, inspiring developments of increasingly refined combinations of tracking and calorimeter inputs.

In the past decade, jet substructure has moved from initial conception to widespread usage. This rapid advancement was greatly aided by the strong support of ATLAS colleagues, close collaboration between the experimental and theory communities, and a very active annual conference series. Going forward, the community is continuing to grow and diversify, and jet substructure will remain a key piece of the ATLAS physics program into the HL-LHC era.

Taming pile-up

The desire to collect a lot luminosity for searches and measurements necessitates the collision of beams of protons with multiple interactions per bunch crossing. This implies that for any given event, there will be signals in the detector from the interaction of interest and from other colliding protons. These additional undesirable signals are referred to as *in-time pile-up*. Furthermore, the response times of some detector elements are longer than the interval between colliding bunches such that previous and subsequent colliding bunches also generate signals on top of the event of interest. These are referred to as *out-of-time pile-up*.

In the figure, an event with 25 simultaneous reconstructed vertices is shown. The blue lines are reconstructed charged particles and the coloured squares are the reconstructed interaction vertices where these originate. The two red lines represent reconstructed muons making the green vertex the one of interest. All the particles from other vertices contaminate this event of interest.

One of the main impacts of these additional particles on the performance is that they create undesirable effects in the calorimeter. For example, if several of these arrive in a similar place then they can create an additional pile-up jet. This first became an issue in 2011 and a simple approach based on the ratio of tracks coming from different vertices was employed. As pile-up increased, a multi-variate approach was developed to improve the rejection; this was used in the final Run-1 results. Further sophistications are now being developed as well.

A similar story has been true in many of the different areas where pile-up effects have been seen. For the extra energy falling inside areas of interest in the calorimeter, a series of corrections have been developed. Similarly, to improve energy resolution, techniques were developed during the shut-down period to enable the correction of these additional particles at the individual particle level.

Over the years, many people have worked on mitigating the effects of pile-up necessary to maintain the performance of the detector with increasing number of simultaneous interactions. These have allowed the physics program to continue with good performance into pile-up regimes above the design of the LHC.

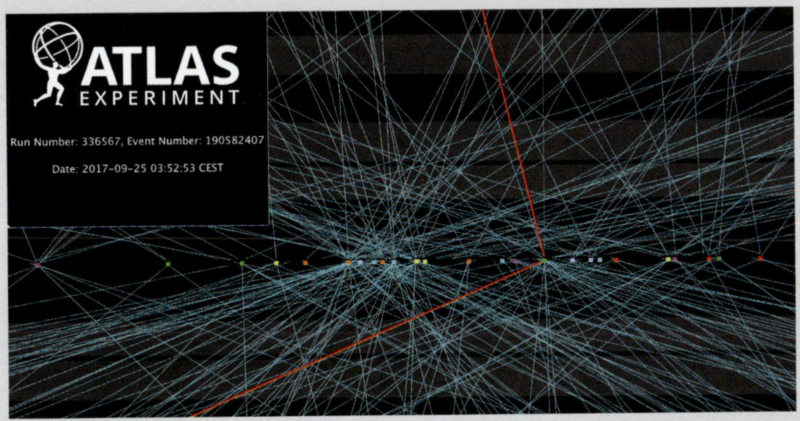

A view of the 25 reconstructed vertices in an event from 2017 data taking.

Chapter 9

Highlights of Physics Results (2010–2018)

9.1. The discovery of the Higgs boson

9.1.1. *A textbook discovery*

A key strength of the LHC physics programme was its "no-lose theorem". To ensure that the cross section for the scattering of longitudinal W bosons is well behaved, there should either be a Higgs boson no heavier than approximately 1 TeV or else "new physics" should occur near or below that energy scale. The LHC is sensitive to both cases. The search for the Higgs boson, over a wide mass range, was therefore a major goal of the LHC physics programme and the expected Higgs boson signatures served as benchmarks to optimise the detector design. The LEP collider and experiments had excluded a Standard Model (SM) Higgs boson with a mass up to 114 GeV, the precise electroweak measurements from LEP, SLC and Tevatron provided indirect upper limits on the SM Higgs boson mass of roughly 150 GeV and direct limits from the Tevatron excluded the range from 158 GeV to 175 GeV. To cover the full possible mass range of the Higgs boson, many different decay channels were studied. The four lepton $H \rightarrow ZZ^{(*)} \rightarrow 4l$ and the $H \rightarrow WW^{(*)} \rightarrow l\nu l\nu$ channels, which are sensitive to masses from approximately 120 GeV and above, and the diphoton channel $\mathrm{H} \rightarrow \gamma\gamma$, which is sensitive to lower masses as favoured by the precision electroweak data, were the discovery channels within their sensitive mass domains, which coincidentally overlapped between 120 GeV and 130 GeV. The key feature of these four lepton and di-photon channels was that both result in a clear peak in the invariant mass distribution of the final state particles, on top of a continuous background, and have an excellent reconstructed Higgs mass resolution. In the mass range below approximately 150 GeV, the fermion decay modes are also possible. The two main channels are the decays to a pair of tau leptons or a pair of b-quarks. These channels, as well as the WW^* channel, have a much worse mass resolution either due to the presence of neutrinos in the Higgs decay chain or the presence of jets.

When would we discover the Higgs? It was a question we were asked daily. The appraisals of the sensitivity for the Higgs discovery for the 1999 performance TDR foresaw that it would require a dataset corresponding to an integrated luminosity of approximately $100\,\mathrm{fb}^{-1}$ (which was only obtained by 2018) at a centre-of-mass energy of 14 TeV. The reappraisal of the sensitivities with updated predictions of the production cross sections showed, that for certain Higgs masses, a smaller amount of data would be sufficient, around $10\,\mathrm{fb}^{-1}$.

With barely $1\,\mathrm{fb}^{-1}$ of data analysed in summer 2011, Higgs searches had at least approximately two standard deviations sensitivity over a very wide mass range of 130 GeV up to 400 GeV. For intermediate masses around 160 GeV, the sensitivity was well above three standard deviations, in particular coming from the WW^* channel. By convention in particle physics, we only claim "discovery" at a significance of five standard deviations. However, towards the end of 2011 with approximately $5\,\mathrm{fb}^{-1}$ of data collected, the expectations were at their highest and every bump in the mass spectra was scrutinised in the greatest detail. Figure 9.1(a) shows the combined significance using the di-photon, four lepton and WW^* channels as a function of the Higgs boson mass as presented at the EPS-HEP conference in Grenoble in July 2011. At this conference, attention was more focused on the region around 145 GeV where both experiments had small excesses. More data were needed to draw definite conclusions, which fortunately, thanks to the splendid operations of the LHC, arrived promptly.

The discovery was timely and truly textbook! We were very clear that we did not want to combine the ATLAS and CMS results to achieve the discovery and wanted to see the evolution of any excess with the two experiments independently. In the 2011 dataset, a small excess had already started to appear in the EPS-HEP (summer 2011) mass spectra in the di-photon channel at a mass near 126 GeV that was joined by an excess in the four lepton channel. In December 2011, a special seminar was held on the update on Higgs search by both experiments. With the full 2011 dataset, the experiments had reached exclusion sensitivity over the entire mass range (from the LEP limit of 114 GeV to up to 600 GeV). The only un-excluded region for the SM Higgs boson was at low mass and the most significant excess, of three standard deviations, was at 125 GeV, as shown in Fig. 9.1(b).

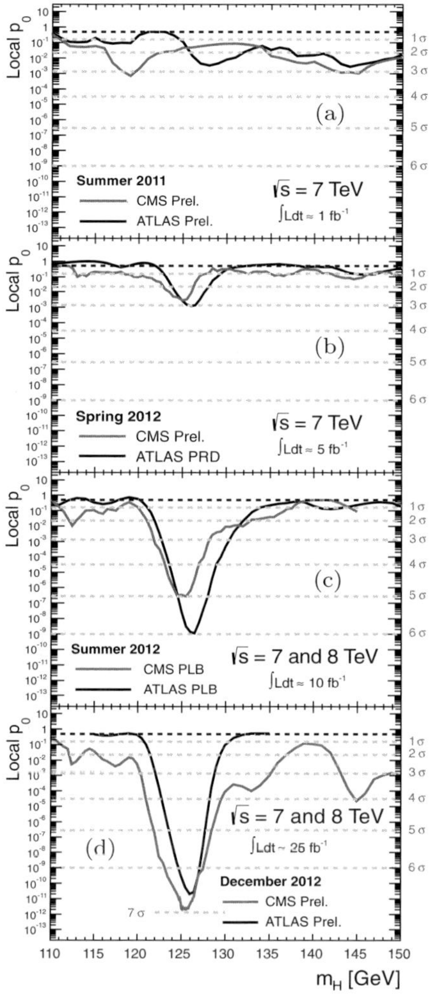

Fig. 9.1. Evolution of the combined significance of the signal from summer 2011 to December 2012.

The entire community had its eyes fully focused on the coming data in 2012 at a centre-of-mass energy of 8 TeV, for confirmation (or not).

By summer 2012, with a dataset of roughly 6 fb^{-1} (a little more than a third of the data that was going to be delivered that year), the excesses at a mass of approximately 125 GeV had grown significantly and the ATLAS and CMS experiments could claim the discovery of the Higgs boson. What made the discovery undisputable was that: (1) it was mostly made using the di-photon and four lepton channels, which featured a narrow signal over a continuous background, making it insensitive to variations in the background level, (2) it was made in three channels independently, (3) it was made in two different experiments independently, (4) the excess in each of the two experiments reached a significance of five standard deviations, (5) the excesses were compatible between different channels and between the two experiments in resolution with the expected signal, and (6) the excesses have independently evolved according to expectations with the growing amount of collision data. The excess and their evolution are illustrated in Fig. 9.1. With the complete 2012 datasets and the data taken since, this discovery has been fully confirmed.

9.1.2. *A gift of nature*

The discovery was announced on 4 July 2012 at a CERN seminar, jointly organised with a special event in Melbourne for the opening of the ICHEP conference, as well as broadcasted live worldwide. It was a memorable and historic event. The CERN Director General Rolf Heuer sealed the discovery with the conclusive statement that "as a layman, I think we have it" and that this was a "global effort and a global success". It truly was the achievement of an entire community. The seminar was followed by publications from both experiments. The new particle was then still cautiously referred to as the "Higgs-like new boson". A general concluding remark made by many was that, apart from theoretical prejudice, little was actually known about the newly discovered particle and that this was really "just the beginning" of its exploration.

The mass of the Higgs is at a unique point; there is only a tiny region in mass where the Higgs boson can be measured with a large number of channels that is precisely in the range from 120 to 130 GeV. Many of the further developments in Higgs physics at the LHC were possible thanks to the very specific mass of the Higgs boson that allowed it to maximise the number of channels through which it could be measured. That was how Fabiola Gianotti, the Collaboration's spokesperson at that time, described it as being truly a "gift of nature".

9.1.3. *The discovery channels*

To deliver a textbook discovery it was essential that both the ATLAS and CMS detectors, in spite of different technologies, would have similar sensitivity and that the choices made to build them were sound. In 2012, two decades of design, projections, construction, commissioning and operation in a vast number of areas including mechanics, electronics, software and computing had come together. The di-photon and four lepton channels served as important benchmarks for the design of the detectors and throughout the discovery these choices were put under a considerable amount of stress.

The key element of the ATLAS detector for the di-photon channel is the electromagnetic Liquid Argon (LAr) Calorimeter. In the range relevant for the Higgs decays to two photons, the energy response is linear to better than 0.3%. The resolution of the invariant mass for a Higgs boson of 125 GeV is 1.2 GeV, for Higgs bosons with large transverse momentum and both photons in the central region of the detector, and 1.9 GeV for events where at least one photon is in the forward region and at low transverse momentum. The lateral and longitudinal segmentation of the calorimeter allows the measurement of the shape of the electromagnetic shower and therefore to distinguish the narrow shower of a photon from the showers initiated by jets fragmenting to a leading neutral pion. Figure 9.2 (top) shows the photon identification efficiency measured with three independent methods in the 2012 dataset. The longitudinal segmentation also allows a measurement of the photon direction with sufficient precision that is robust in high pile-up conditions. The analysis classifies events into categories separating "better" events from "worse" events in terms of signal-to-background (S/B) ratio to enhance its sensitivity. Figure 9.2 (bottom) shows the invariant mass distribution (weighted by the signal-to-background ratio in each of the categories) at the time of the discovery. The signal can clearly be seen as a peak on top of a smoothly falling background.

The Higgs decay to four leptons (in this case muons or electrons) is a golden decay channel thanks to the large S/B ratio. It is the only decay channel with such a large S/B at the LHC for the inclusive production of the Higgs boson. A limitation of this channel is, however, the very small decay branching fraction: only about one Higgs boson out of ten thousand decays in this final state.

Key ingredients to observe the Higgs boson in this decay channel are the capability to reconstruct leptons down to small transverse momenta with good rejection of non-prompt leptons (not coming from a decay of a W or a Z boson), a hermetic detector to maximise the signal acceptance, and the good momentum resolution to minimise the amount of background under the resonance mass peak. The high granularity LAr calorimeter, which is relevant

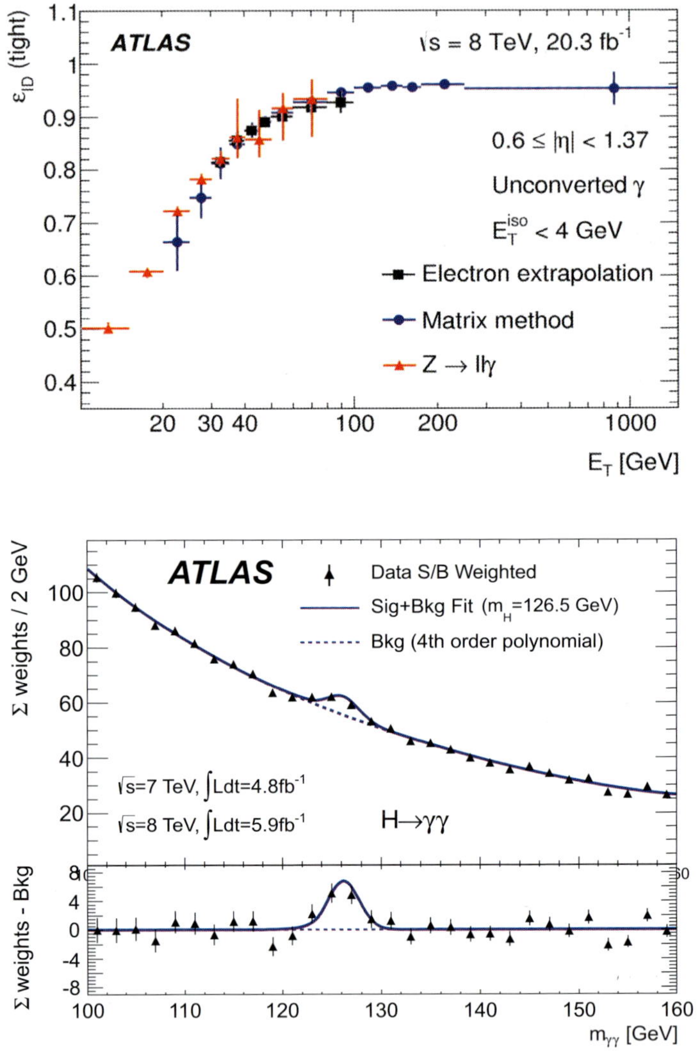

Fig. 9.2. Photon identification efficiency measured with three independent methods (top) and weighted distribution of the invariant mass of the di-photon candidates after all selections (bottom).

for electron measurement, was discussed above. The muon reconstruction uses both the Inner Detector (ID) and the Muon Spectrometer (MS). The MS plays a main role in terms of hardware-based trigger, muon identification and momentum resolution at high transverse momenta and in the forward region of the detector. The ID drives the momentum resolution for lower transverse momenta, below 80 GeV in the central part and below 30–40 GeV in the forward part of the detector. The two tracking detectors are therefore complementary and their combined performance is crucial to be able to observe the Higgs boson decaying into four leptons. A typical mass resolution of the order of 1.6 GeV for a 125 GeV Higgs boson (1.3%) was achieved in Run-1, before the Higgs discovery. The cleanness of this channel can be appreciated in Fig. 9.3, where the observed four lepton invariant mass is compared with the resonant signal. In a 5 GeV mass window around 125 GeV, 13 candidates were observed in the dataset analysed for the Higgs discovery, while the expected background amounted to about five events — a handful of events only, but essential to the Higgs boson discovery.

Another albeit more difficult discovery channel was the Higgs boson decay to a WW^* pair, where each W boson subsequently decays to an electron (or a muon) and a neutrino. Unlike the aforementioned channels, this channel does not provide a narrow mass peak on a continuous background, due to the presence of undetected neutrinos, but benefits from a larger branching ratio. Precise estimates of backgrounds from the production of non-resonant $pp \to WW^*$, $t\bar{t}$ and W + jets events are therefore necessary. A sizable S/B ratio in this channel can be achieved, for example, by making use of the spin-zero property of the Higgs boson. By angular momentum conservation the two leptons from the W bosons are emitted closer to each other in azimuthal angle than in non-resonant WW^* events. Nevertheless,

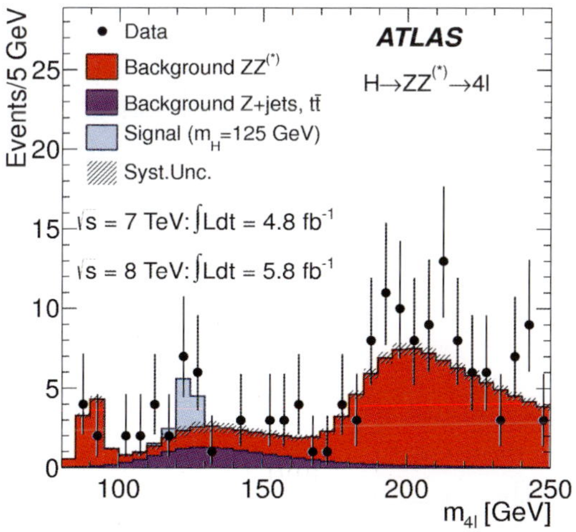

Fig. 9.3. The distribution of the four-lepton invariant mass, m_{4l}, for the selected candidates, compared to the background. The signal expectation for a SM Higgs with a mass of 125 GeV is also shown.

the backgrounds are complex and require a very accurate and state of the art simulation, or in some cases must be estimated from the data. Providing convincing evidence of the Higgs observation in this channel was therefore challenging. In a tour de force effort, this channel eventually achieved and even surpassed the sensitivity of the main discovery channels. Beyond its role in the discovery of the Higgs boson, the WW^* channel is also key in the measurement of its properties, as for instance comparing its rate to that of the ZZ^* channel, which provides a probe of custodial symmetry (the remaining symmetry of the SM electroweak sector after the spontaneous symmetry breaking induced by the Higgs mechanism).

9.1.4. *The precise measurement of the mass of the Higgs boson*

Prior to the discovery, the mass of the Higgs boson was a key missing fundamental parameter not predicted by the SM. All SM couplings of the Higgs boson to gauge bosons and fermions were known, determined by their masses and the vacuum expectation value of the Higgs field. The trilinear and quartic self-coupling strengths of the Higgs boson, its total decay width and all decay branching ratios are predicted once the mass of the Higgs boson is known.

Owing to their excellent mass resolution, the four leptons and di-photon channels dominate the Higgs boson's mass measurement. A refined calibration of the detector response to electrons, photons and muons was required to reach the highest possible precision. The ATLAS combined measurement yielded, $m_H = 125.36 \pm 0.37\,(stat) \pm 0.18\,(syst)$ GeV, with the first uncertainty being statistical and the second of systematic origin. The combination with the CMS measurement performed shortly after resulted in $m_H = 125.09 \pm 0.21\,(stat) \pm 0.11\,(syst)$ GeV with a 0.02% relative precision, which was a tremendous achievement.

9.1.5. *From "Higgs-like" to the Higgs boson*

The observation of the Higgs boson in the vector boson decay channels gave strong indications about its nature. It had to be neutral and it had to be a boson. Angular momentum conservation and spin-statistics arguments further constrained the range of possibilities for it to be either of spin-0 or spin-2 nature (higher spins were not strictly impossible, but they were not studied in detail). Given the agreement in production and decay rates, the question of the spin of the discovered particle was considered by the community to be mostly academic (and for some even rhetorical). By studying the angles of the decay products in the di-boson final states of di-photons, ZZ^* to four leptons and WW^* to di-leptons, and the invariant mass distributions in the ZZ^* channel, the spin quantum number of the Higgs boson can be probed. By combining these three decay channels, for most of the large number of models considered, the spin-2 hypothesis was rejected at more than 95% confidence level (CL) in favour of the SM spin-0 hypothesis.

The other outstanding questions about the nature of the discovered particle are its charge and parity (CP) symmetry properties. The observation of the decay into two vector bosons already gave indications of the Higgs boson being dominantly CP-even, since the coupling should otherwise proceed through loops and would be suppressed. A CP analysis could only be performed with the four leptons and WW^* channels. In the ZZ^* channel, the production and decay angles of Z bosons are available as well as the M_{Z*} threshold distribution, providing naturally CP-sensitive observables. The analysis was carried out and excluded at the 95% CL, a pure CP-odd Higgs boson. First studies of a possible CP-odd admixture were also performed, resulting in stringent limits on such contributions.

9.1.6. *Evidence of Yukawa couplings to fermions*

The Brout–Englert–Higgs mechanism of electroweak symmetry breaking establishes that the couplings of the Higgs boson to the electroweak gauge bosons are proportional to the square of their masses. In the SM, the Higgs mechanism is responsible for restoring gauge invariance in the fermionic mass sector and as a consequence the couplings of the Higgs boson to fermions is proportional to their masses. To further understand the electroweak symmetry breaking mechanism, the study of processes that involve interactions between the Higgs boson and fermions are therefore of fundamental interest. In the SM, the most abundant Higgs production is through the gluon fusion process, which involves an internal quark loop (predominantly the top quark), from which the Higgs boson is emitted. The observation during the discovery and subsequent measurements that the Higgs boson production's cross section was compatible with the SM predictions provided indirect evidence for the coupling of the Higgs boson to fermions. A direct measurement of the decay to fermions is not only interesting per se, it is even more so when it is compared with the SM loop responsible for the main production process. The Higgs boson decays to heavy fermions, b-quarks and taus, with predicted branching ratios of approximately 58% and 6%, respectively, and the Higgs production in association with top quarks are the most promising processes. These final

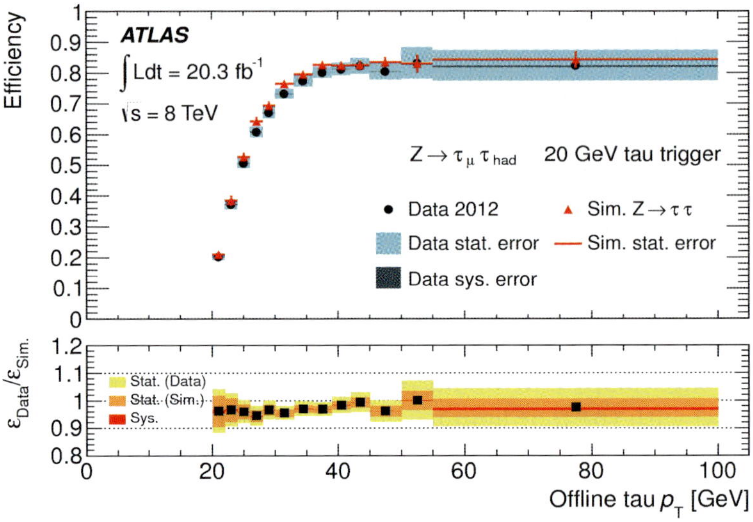

Fig. 9.4. The measured tau trigger efficiency in data and simulation.

states are typically more challenging to reconstruct and identify than the processes that played a role in the Higgs discovery. SM processes with similar signatures give rise to large backgrounds, which need to be carefully estimated either from simulations or from the data. Taus can, for example, decay to hadrons, thereby making it difficult to distinguish them from jets. Figure 9.4 shows the efficiency to trigger on a taus in data compared to that predicted by the simulation. In addition, due to the processes being reconstructed from hadronic signatures and having neutrinos in the final state, signals with taus or b-quarks do not have a very good mass resolution.

Finding evidence for the Higgs boson decaying to two taus was achieved in summer 2012 after a long painstaking analysis of the Run-1 data. The process was seen with a significance of 4.5 standard deviations. After combining the results from CMS, the two experiments were able to claim the observation of Higgs boson decaying to two taus in summer 2016. These results were also confirmed with the 2015 and 2016 Run-2 datasets by both experiments.

9.1.7. *Does the discovered boson interact as predicted by the SM?*

To further probe the nature of the discovered particle, all the channels can be analysed in combination to extract useful information about the properties of its couplings. The state of the art so far in terms of combined results was achieved in the ATLAS and CMS full Run-1 combination. This combination is based on the five baseline decay channels discussed above in the two experiments, where each is subdivided in event categories, enriched in specific production processes. Altogether, the combined result had to fit 600 categories with more than 4000 varying "nuisance" parameters to account for systematic uncertainties. The combination resulted in an unambiguous observation of the di-photon, ZZ^*, WW^* and di-tau decay modes of the Higgs boson, and evidence for the bb decay. In addition, it led to the observation of inclusive and vector boson fusion production and offered evidence for the production of the Higgs boson in association with a vector boson.

The results of the combination can be further interpreted in terms of constraints on the Higgs boson couplings to gauge bosons and fermions. At the LHC, when measuring cross section times branching ratio, the couplings of the Higgs boson to the other SM particles cannot be directly determined without assumptions, in particular on the total width of the Higgs. However, the couplings of the Higgs boson can be probed under assumptions on non-SM contributions to the total width or on couplings of the Higgs boson to some SM particles. Several approaches have been developed in this context. To perform the Higgs couplings' fit, the community unanimously adopted a simplified approach that is often referred to as the "kappa" framework. This framework is essentially a simplified effective Lagrangian that assumes that the Higgs boson couplings are CP-even. In this framework the strength of the interaction of the Higgs boson with a given particle is assumed to scale as the square of a coupling modifier, kappa (κ), with κ being unity in the SM. The advantage of this approach is that it allows one to test the SM using expected relationships between the production and decay processes. The disadvantage is that it assumes SM-like kinematics and in addition makes assumptions on the total width of the Higgs boson. The interpretation of the combined results in this framework is illustrated in Fig. 9.5. These measurements rely on precise determinations of the SM's signal expectations. Significant improvements in theory predictions were made at the time but were not yet included in this combination. Within the given precision of the data and theoretical predictions, the measured properties were found to be fully consistent with the SM Higgs boson. The newly discovered particle was henceforth referred to as the Higgs boson.

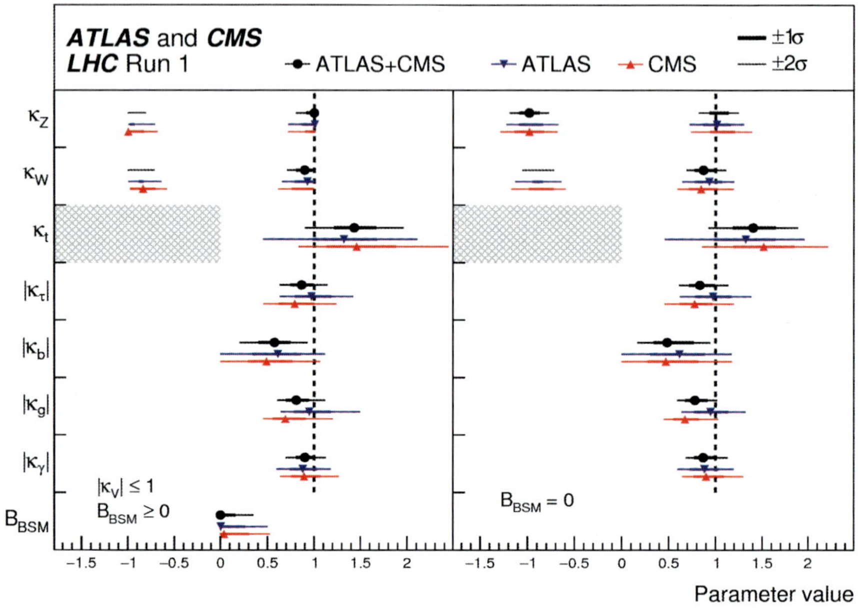

Fig. 9.5. Best fit values of the coupling strength modifiers allowing non-SM loops and in two different assumptions on non-SM contributions to the total width: (left) no assumption on the total width but that the coupling to vector bosons should be smaller than one, and (right) that there is no non-SM contribution to the total Higgs boson width.

9.1.8. *A flurry of new ideas redefining the landscape of Higgs physics*

With the discovery of the Higgs boson and in a very short amount of time, the landscape of Higgs physics became much wider and richer than anticipated. A flurry of interesting new ideas was proposed. For example, could the top quark decay to a Higgs boson and a lighter quark? Such flavour-changing neutral currents were immediately probed in the di-photon channel and soon after in other channels. The detailed study of the interference between the continuum background and the signal in the di-photon channel led to the interesting observation of a predicted mass shift of approximately 40 MeV with respect to the four-lepton channel. The mass shift depends on the Higgs boson's total width and a precise measurement in the di-photon channel could lead to constraints on the width that are model independent. New decay channels probing couplings to light quarks were also proposed, such as the Higgs boson decaying to quarkonia and a photon, or lepton flavour violating decays that were probed in electron-muon and tau-muon channels.

ATLAS also performed a large range of searches for additional Higgs bosons. Many models of new physics predict the existence of more than one Higgs particle. The most prominent example is supersymmetry. It has motivated a rich programme of searches for heavier scalars decaying to a pair of bosons (Z, W or photons) or fermions (preferably top, bottom or taus). Another important direction represents searches for resonant and non-resonant pair production of Higgs particles. Here, various channels were investigated such as $HH \rightarrow bb\tau\tau$, $\gamma\gamma WW^*$, $\gamma\gamma bb$ or $bbbb$. In addition, the Higgs programme includes searches for scenarios where the Higgs serves as a mediator to new physics and potentially acting as a portal to dark matter, such as the Higgs boson decaying to dark matter particles or into "dark Z". Searches for these kinds of events are similar to those described in Sec. 9.3.5.

Direct searches as well as studies of the Higgs boson couplings and invisible Higgs decays all confirmed the SM and led to constraints on new phenomena. However, a surprise came from the search for a heavy neutral particle decaying into two photons. In the data collected in 2015 with an increased proton-proton collision energy of 13 TeV, there appeared to be a resonance with a mass of about 750 GeV with a local significance of more than three standard deviations. This significance is much reduced when considering that such resonances may occur at any mass value (called global significance) and even in other distributions (which represents an additional, but hard-to-estimate, look-elsewhere effect). This result, together with a similar excess seen by the CMS experiment, triggered a flood of theoretical papers. In just a few weeks, physicists came up with a few hundred papers with possible explanations beyond the SM. No excess was found in the much larger 13 TeV dataset taken in 2016 and it was concluded that the events in 2015 were due to a statistical fluctuation. Occasional excesses are expected on a statistical basis.

9.1.9. *Higgs physics at Run-2 and beyond*

With the increased LHC centre-of-mass energy during Run-2 and the sizably larger dataset, searches for the Higgs boson decaying to two b-quarks and the Higgs boson production in association with a top quark pair have become more sensitive. While only approximately half of the Run-2 dataset have been analysed, the Higgs boson production in association with two top quarks was observed with a significance of 6.3 standard deviations in spring 2018, providing the first direct indication for the Higgs boson coupling to the top quark. This result required the combination of analyses aiming at

Higgs decays to two Z bosons, two W bosons, two photons, two taus and two b-quarks. In summer 2018, the Higgs boson decaying to two *b*-quarks were observed with a significance of 5.4 standard deviations, predominantly through Higgs production in association with a vector boson. The measured yields for these processes were found in agreement with the SM predictions. These observations represented a huge step towards a better understanding of the origin of fermion masses.

These analyses of highly complex final states have required the development of original techniques, often based on machine learning algorithms, and state of the art simulations of intricate background processes. The reach in sensitivity significantly surpassed the expectations prior to the start of the LHC.

The Collaboration is also preparing results expected to bring substantial sensitivity with larger datasets, in rare processes such as the Higgs boson decaying to two muons and $Z\gamma$, or the study of fiducial and differential cross sections. The Higgs boson to two muons decay plays a special role in future goals: up until now only the third fermion generation has been measured at the LHC, and only upper limits have been provided for processes involving other fermion generations. The observation of the Higgs boson decaying to two muons and the measurement of its branching ratio would shed light on whether the Higgs mechanism is also responsible for the masses of other fermion generations.

Finally, a concerted effort to understand how the Higgs boson trilinear coupling (an interaction point involving three Higgs bosons) can be constrained is currently ongoing, both directly from the di-Higgs bosons production measurements and from their indirect impact on differential Higgs production distributions.

9.1.10. *Outlook*

The Higgs boson discovery is a landmark result in the field of particle physics. It was an immense success for the entire community. It fully validated the design choices made for the ATLAS and CMS detectors, and for the LHC whose performance has been and continues to be outstanding, by far surpassing expectations. The discovery is a result of fundamental importance. Arguments of "naturalness" with respect to the hierarchy of energy scales suggest that new physics should exist near the TeV scale. This is a major motivation for direct searches of new physics at the LHC, but it is also intriguing because no significant deviations from the SM have been observed so far. The mass of the Higgs boson also has far reaching consequences; not only does it allow the SM to be extrapolated to very high-energy scales (equivalent to very small distances), but also within the SM its value suggests that the vacuum is metastable. The discovery also brought a rapid development of the programme of Higgs boson precision measurements and a host of new searches using the Higgs boson as a tool. For these reasons, the discovery of the Higgs boson, which is the culmination of half a century of refinement of the SM and the development of collider physics also heralds an exciting and challenging future programme at the LHC and beyond.

9.2. Measurements of Standard Model processes

When protons are accelerated to very high energies and collided inside the ATLAS detector, a wide range of processes may occur. One possible strategy to analyse the data is to focus on measurements of single processes that are expected to occur through interactions of the known particles and forces as described by the SM. Such measurements have been of major importance for many aspects of the ATLAS experiment. In the initial phase of the experiment, it was speculated on how well pp collisions at the new energy frontier could be predicted by the SM. After establishing broad agreement, the studies moved on to challenge the predictions of the SM or measure the parameters of the SM.

The two fundamental ingredients to these studies are the excellent understanding of the performance of the ATLAS detector to achieve a high experimental precision and a systematic framework to calculate and improve predictions within the SM that can be compared to the measurements. The former is the daily work of an experimental physicist in ATLAS and includes, for example, to measure the efficiencies to observe charged-particle tracks, electrons, muons or jets originating from b-quarks and to ensure that the calibration of the momentum of the detected leptons and jets is correct within the uncertainties. Similarly, working out reliable and precise predictions for the frequency with which specific SM processes occur in complex interactions of the composite protons, the cross sections, has taken decades of work by the theory community. In the last few years many so-called higher order calculations were completed, allowing one to confront the precise measurements with equally precise predictions. Such a procedure may eventually reveal shortcomings of the underlying theory and thus help to establish evidence for new physics.

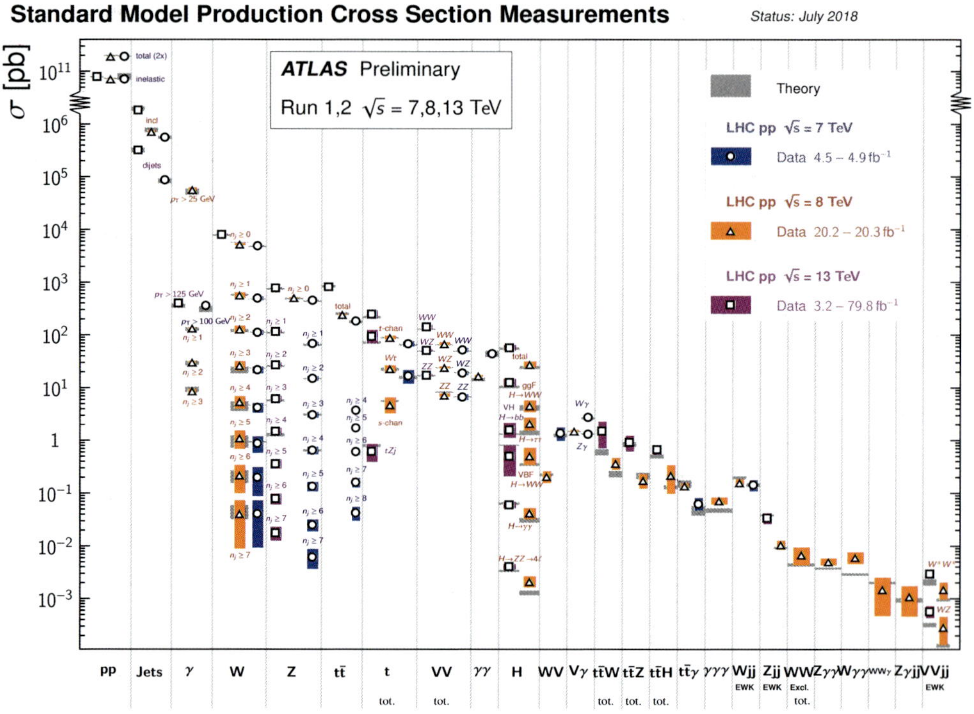

Fig. 9.6. Summary of the cross section measurements of SM processes compared to the theoretical expectation.

By the end of 2017, a total of about 270 publications on SM measurements have been completed by ATLAS and a small selection of these is presented in this chapter. Figure 9.6 gives an overview of a selection of cross section measurements and compares them to the respective SM predictions based on state of the art calculations involving both the electroweak and strong interactions. Shown are the cross sections for inclusive pp processes, jet and photon production, production of W and Z bosons (collectively referred to as V) inclusively and with jets, pair and single production of top quark, pairs of bosons (VV, $\gamma\gamma$, WV and $V\gamma$) and various processes involving a Higgs boson. Also shown are the more rare processes of top pair production processes in association with a boson (ttW, ttZ and ttH), vector boson scattering and vector boson fusion processes (Wjj, Zjj, $Z\gamma jj$, $VVjj$), and triboson processes ($Z\gamma\gamma$, $W\gamma\gamma$, WWW). The figure not only highlights the impressive agreement of the theory predictions with the data within the uncertainties, but that extremely rare processes (i.e., smaller and smaller cross sections) can also now be studied.

9.2.1. *Minimum bias, jet and single gauge boson processes*

The most abundant SM process at the LHC is the inclusive interaction of the highly energetic protons of $pp \to X$, where X is an unspecified set of final state particles. While the total cross section for this process has been measured at numerous pp and $p\bar{p}$ colliders since the 1970s, little was known about the interactions at the highest energies of several TeV. Thus, ATLAS has measured this process at each new centre-of-mass energy with different techniques. The most precise technique uses the so-called Absolute Luminosity For ATLAS (ALFA) detector, which consists of a set of devices installed close to the LHC beam about 200 m downstream from the centre of the ATLAS detector. In fact, ALFA measures the elastic scattering of $pp \to pp$, where the two incoming protons interact but remain intact, by detecting the outgoing protons and measuring their momenta. The cross section for inelastic pp interactions, where the protons break up, is measured indirectly by ALFA by a fundamental physics law called the "optical theorem". Another measurement of the inelastic cross section has been obtained by directly counting inelastic events by detecting the particles produced by the proton interactions in large scintillation counters, the Minimum Bias Trigger Scintillators. Both measurements are shown in Fig. 9.7, together with various theoretical predictions and other measurements. The figure shows that some predictions can describe the rise of the cross section with the centre-of-mass energies to a reasonable accuracy.

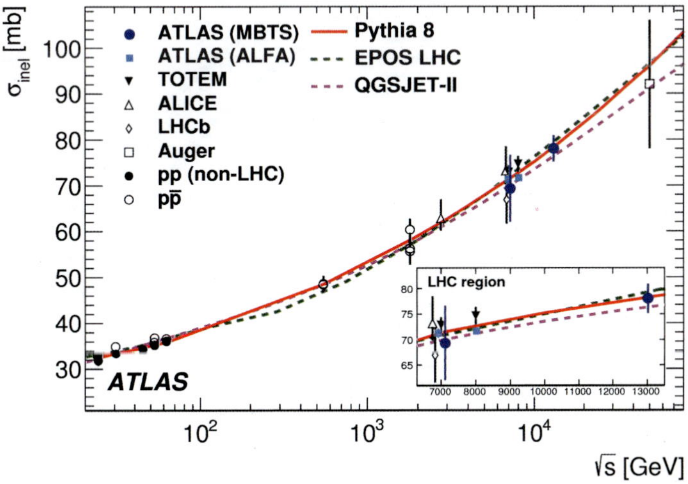

Fig. 9.7. Summary of the inelastic pp and $p\bar{p}$ cross section measurements.

As all measurements of absolute production rates, these analyses require a precisely determined integrated luminosity. Using van der Meer's beam-separation scan technique, this luminosity is known to an uncertainty of about 2%. See the story on the van der Meer scan for more details.

Another abundant and important process at the LHC is the production of energetic jets, which are collimated sprays of hadrons produced through the interactions of the proton constituents, quarks and gluons. The production of jets is one of the most copious processes at the LHC and therefore jets are present in many events recorded by the detector. The cross section for jet production is shown in Fig. 9.8 as a function of the transverse momentum, p_T, of the jet in different rapidity regions ranging from the most central region ($y < 0.5$) up to the forward region ($2.5 < y < 3.0$) of the detector. The measurement covers a wide range of p_T values and the cross section falls by about ten orders of magnitude within this range. It is a major challenge to provide an accurate theoretical prediction over this wide range of both jet energies and cross sections.

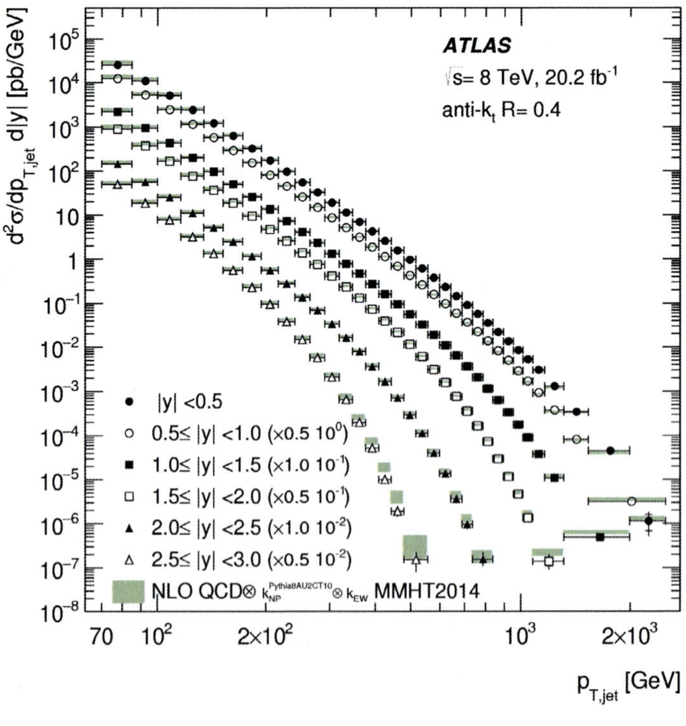

Fig. 9.8. Differential cross section for jet production as a function of the transverse momentum, p_T, for various ranges in rapidity, $|y|$.

The theoretical calculations shown here are performed in terms of the fundamental proton constituents interacting through the strong interaction, known as quantum chromodynamics (QCD). During the past decade the precision of these calculations has improved significantly and they provide a good description of the data in the whole range. The experimental precision of this analysis is also impressive and allows us to probe the fine details in the theoretical modelling. The precision of this measurement is only possible with a good understanding of the energy of the jets, the jet energy scale.

The calibration of jets relies on a detailed modelling of the detector's response to the particles that make up a jet as well as measurements from data. The latter can be performed in events, where a less complex object such as a single photon or a Z boson decaying to two electrons or two muons with a well-known energy scale is balanced against a jet and therefore provides a reference scale to estimate the jet energy. The relative uncertainty on the jet energy scale is seen in Fig. 9.9. In the central regions, the uncertainty in the jet energy calibration is better than 1.5% in a wide range, which is a large improvement compared to detectors at previous colliders.

The production of the heavy W and Z bosons through the electroweak interaction of quarks and anti-quarks and their decay to leptons provides opportunities for precision measurements and interpretations that are otherwise hard to achieve at a hadron collider. Experimentally, this is

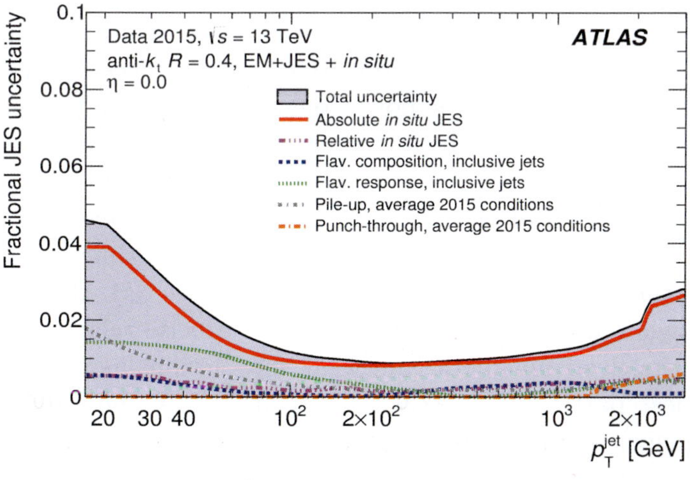

Fig. 9.9. Uncertainty on the jet energy calibration as a function of the jet transverse momentum.

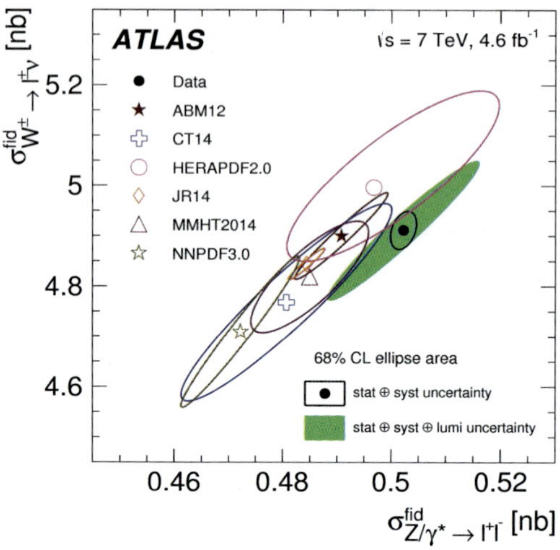

Fig. 9.10. Measurements of the W and Z boson cross sections, compared to theory predictions using various PDF models.

because the detection of electrons and muons is relatively simple and can be understood to a very high accuracy using the large sample of $Z \rightarrow ee$ and $Z \rightarrow \mu\mu$ decays. The most precise measurements so far have been achieved in the 7 TeV data, where the precision, excluding the common luminosity uncertainty, has reached between 0.3% and 0.5%. Figure 9.10 compares the measurements of W and Z boson cross sections to SM predictions, based on different sets of parton distribution functions (PDFs) that describe our understanding of the composition of protons in terms of elementary quarks and gluons. Such data are clearly sensitive to differences between PDF sets and have been shown to provide novel constraints, e.g., on the strange-quark contribution to the proton, which is found to be larger than previously often assumed.

Whereas the mass of the Z boson was precisely measured at the LEP electron-positron collider to be 91188 ± 2 MeV, the same was not true for the mass of the W boson. Within the theory of the SM, these masses arise from the Higgs mechanism, which introduces precise relations between the strength of the electromagnetic and weak interactions as well as the masses of the heaviest particles in the SM. Experimentally the measurement is a major challenge, as the W boson used in the analysis decays to a charged lepton and a neutrino, of which only the charged lepton is measured in the detector. To infer the mass of the W boson from this, incomplete information has required an experimental and phenomenological programme of several years, as all aspects of the production and decay of W bosons have to be modelled to an extreme precision. The ATLAS experiment has completed the first measurement of the mass of the W boson at the LHC with a result of $80370 \pm 19\,\text{MeV}$, in agreement with the prediction of the SM. While this result is still almost ten times less precise than the mass of the Z boson, this measurement and the measurements performed at the Tevatron collider represent the highest precision reached so far. To achieve this relative precision of 0.02% requires, for example, a corresponding extraordinary understanding of the calibration of the leptons from the W boson decay. The electron and muon momentum measurements are calibrated using the known Z-boson mass. The time stability of the calibration is illustrated in Fig. 9.11 for electrons for two different methods: m_{ee} compares the measured di-electron mass to the known Z mass value and E/p compares the calorimetric energy measurement of the electrons to the track momentum. The calibration is stable over time to one part in a thousand.

9.2.2. *Diboson and triboson processes*

Interactions of electroweak bosons (γ, W, Z) can be very rare but are of major interest in order to test the couplings of the bosons to each other. For example, the production of a W and a photon probes the coupling of $WW\gamma$.

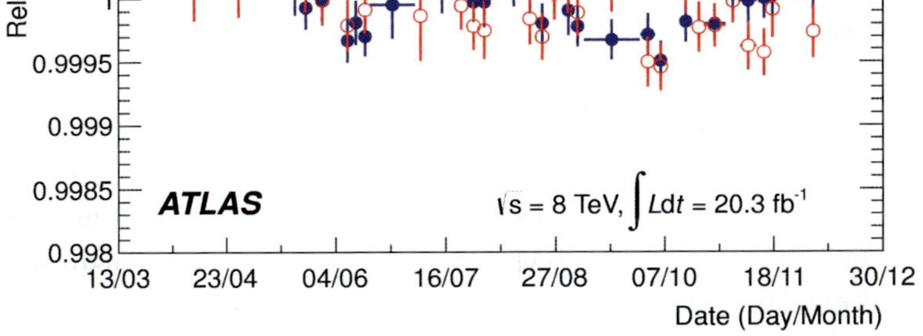

Fig. 9.11. Relative response of electrons as function of date for 2012.

These couplings are predicted precisely within the SM and deviations could be a sign of physics beyond the SM. The rate of these processes is about a factor of a thousand smaller than that of single boson production. In many cases the rates are measured for the first time differentially, that is, as a function of an observable at the LHC. ATLAS has measured diboson production in all possible combinations, inclusively and differentially, i.e., $\gamma\gamma, W\gamma, Z\gamma, W^+W^-$, WZ and ZZ, many of which are shown in Fig. 9.6. The theory predictions of these processes are estimated up to the next-to-next-to-leading order (NNLO) accuracy. Calculations at that precision are computationally very challenging and were only finalised about 2014, after ATLAS had observed tensions in the $W\gamma$ and WZ production cross sections with the less precise next-to-leading order (NLO) predictions. Since 2014, all relevant NNLO calculations were completed and they describe the data well. This represents a major theoretical advancement of the last few years. The measurements of these processes are used to investigate if the couplings between three gauge bosons have anomalous values, as compared to the SM predictions.

In addition to couplings involving three gauge bosons, the SM also contains direct interactions among four gauge bosons through quartic couplings. Final states with dibosons can be produced from three or four gauge boson interactions. The four gauge boson couplings are measured using electroweak production of dibosons, where two gauge bosons are radiated from the incoming quarks and interact to produce again a pair of gauge bosons in the final state. This process is known as vector boson scattering and is closely related to the vector boson fusion process studied in Higgs production. These processes are of particular interest since their cross sections would grow with energy if there were no Higgs boson. At large enough energy, the number of predicted diboson events would eventually exceed the number of incident pp collisions, which is unphysical. The Higgs boson modifies the cross section such that this does not occur and "unitarity" is not violated. The first measurement of the $W^{\pm}W^{\pm}$ + 2 jets production process, which has relatively low background from other processes, is based on selecting events with the two leptons with the same electric charge, from the two W bosons, and two well-separated, high-energy jets. The measured cross section agrees with the SM prediction. The difference in the angle along the beam line between the two high-energy jets is shown in Fig. 9.12. It is seen that the data agree well with the prediction. The data are used to exclude large anomalous values for the quartic gauge boson couplings.

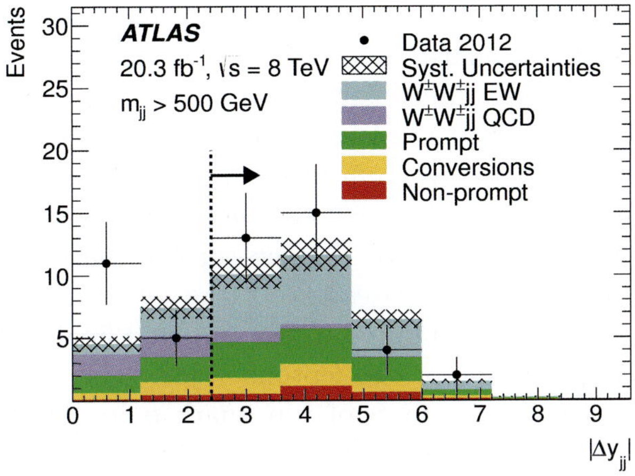

Fig. 9.12. Absolute value of the angle along the beam-line between the two high-momentum jets in the pp→$W^{\pm}$ $W^{\pm}$+ 2 jets analysis.

9.2.3. *The bottom quark*

Precision tests of the flavour dynamics apply challenging constraints on the SM and are therefore an important test of the SM itself, as well as a window into possible new physics. The large production cross section of bottom quarks at the LHC allows for sensitive tests of any possible new physics contributions. These tests include measurements of the charge-parity violating phase, searching for anomalous rates of rare decays of B hadrons as well as precise tests of QCD by studying, for example, the production mechanisms of b-quark pairs.

One exciting discovery was that of the χ_b (3P) boson. The χ particle consists of a bottom and anti-bottom quark that are bound together. Although this excited state had been theorised by the SM, it was only first observed in 2009. Measurements of this state probe our understanding of strong interactions in the SM.

9.2.4. *The top quark*

The heaviest known quark, the top quark, was discovered at the Tevatron collider in 1995. The LHC is often referred to as a top factory, where at the highest luminosity conditions achieved so far, about ten top-quark pairs were produced every second. With this plethora of data, ATLAS is able to produce some outstanding, high precision measurements, many of which were not possible before.

The top quark decays almost instantly with a lifetime of less than 10^{-24} s to a W boson and a b-quark. The presence of the latter can be inferred by identifying a jet containing a long-lived b-hadron (b-tagging), while the decay of the W boson may yield either two jets or a charged lepton and a neutrino. Thus, in events where a top-quark pair is produced, one of the challenges is the complex final state consisting of no leptons and six jets, one charged lepton and a neutrino and four jets, or two charged leptons and two neutrinos and two jets. In all cases two of the jets are typically b-tagged. Analyses are typically differentiated in the targeted final state as "all jets", one (charged) lepton + jets, or two leptons + jets. The measurement of the inclusive production cross section of top quark pairs has an experimental precision of about 4%, which is more accurate than the theoretical prediction at the NNLO level.

While the precision of the inclusive top cross section measurements is noteworthy, one of the biggest excitements has been the ability, with the ATLAS detector, to precisely measure differential cross sections, especially up to high energies of several TeV. In addition, many of the latest measurements have followed an approach pioneered by ATLAS where the cross sections are measured in well-defined kinematic regions using final state observables such as leptons and jets rather than using the simulation to extrapolate to stable top quarks. This technique has the advantage that the measurements are less sensitive to theoretical aspects such as the details of the top quark decay. These measurements are then compared to state-of-the-art generators but can also in the future be compared to new predictions that could, for example, directly include the calculation of the top-quark decay at higher accuracy.

Another key advancement of the past few years is the use of algorithms to select the decays of high momentum top quarks, where the decay products are not resolved as single objects but rather "boosted" and overlapping in a relatively small area of the detector. With the help of algorithms to reconstruct these boosted top quarks, ATLAS is able to measure the differential cross section and properties of top quark pairs up to the highest momenta. An example of the measurement of the fiducial cross section as function of the leading top quark transverse momentum is shown in Fig. 9.13. Theoretical predictions describe the data in the entire measurement range up to 1 TeV.

The mass of the top quark is one of the fundamental parameters of the SM. Its precise value has implications on the overall consistency of the SM, similar to the mass of the W boson discussed above. These measurements

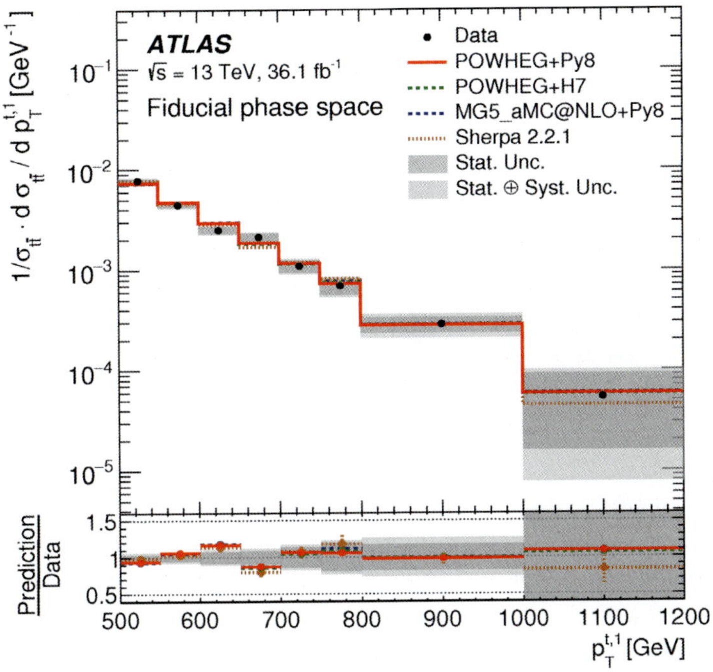

Fig. 9.13. Normalised differential top pair cross section as a function of transverse momentum of the leading boosted top quark jet candidate.

are complicated by the complex final state of top quark pair production and thus numerous approaches have been studied over the years by the Tevatron and LHC experiments. When determining the mass, two different event signatures typically give the most precise results: the one lepton + jets and two leptons + jets final states. Two major improvements to these analysis techniques have been developed by ATLAS to enhance the precision of the measurement. First, the two final states were used together to extract the mass. By doing so, anti-correlations in the uncertainties between the two final states can be exploited, resulting in a reduction of the overall uncertainty on the measurement. Second, the event selection, such as the energy of the leptons in the event, was optimised to reduce the total uncertainty in the measurement. This results in a 200 MeV improvement on the top mass uncertainty. Figure 9.14 illustrates the impact of the first of these improvements. Successive combinations of top mass measurements are shown, with the resulting total uncertainty indicated by the blue line. The method discussed here is highlighted in red and results in a 41% improvement in the total uncertainty over the di-lepton measurement alone. The absolute

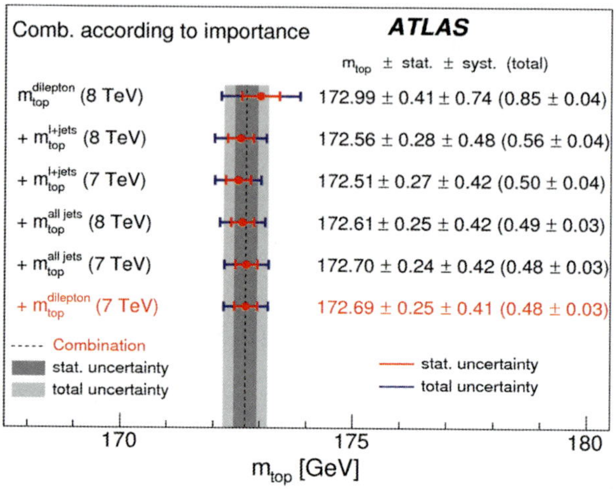

Fig. 9.14. ATLAS's top mass combination by successively adding results to the most precise one. Each line of the figure shows the combined result when adding the result listed to the combination, indicated by a "+".

precision on this combined top quark mass measurement is now 0.50 GeV (less than 0.3%).

With the large data samples from the LHC, many more properties of the top quark and its production have been measured. This includes measurements of the charge asymmetry, the polarisation, the charge, the spin correlations and searches for strongly suppressed decays of the top quark due to possible flavour-changing neutral currents (if seen, such decays would indicate presence of physics beyond the SM). One example is the measurement of the polarisation of top quarks originating from single production via electroweak interaction. In this analysis, five fundamental constants are probed governing the interaction between the top quark, the bottom quark and the W boson (the Wtb vertex). In the SM, all couplings (at leading order) are expected to be zero except for one, V_L, which is equal to the CKM matrix element V_{tb}. Thus, the observation of any non-zero component to any of the other interactions or deviations of V_L from V_{tb} would be evidence for new physics. Limits on these parameters place direct constraints on the nature of possible new physics couplings occurring in the Wtb vertex. By studying the full triple-differential decay patterns of the top quark using angular information, ATLAS measures many properties of the Wtb interaction simultaneously, without some of the assumptions that have been made in the past. Figure 9.15 shows an example of the simultaneous constraints placed on the real versus imaginary part of the ratio of two of the

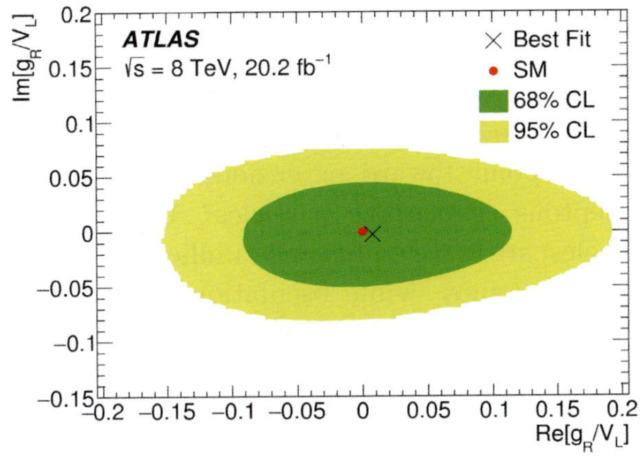

Fig. 9.15. Joint likelihood contour of the imaginary versus the real part of the anomalous coupling ratio g_R/V_L.

five fundamental parameters (g_R/V_L). The data are consistence with zero for both parameters, indicting that it is consistent with the SM prediction.

9.3. Searches for new phenomena beyond the Standard Model

9.3.1. *The hunt begins*

On 13 August 2010, ATLAS published a paper on the search for new strongly interacting particles in the production of a pair of jets (dijets). This was the first published paper by the LHC experiments based on the 7 TeV run, which had just begun shortly before. Despite having 3500 times less data than collected by the two experiments at the 2 TeV energy Tevatron collider, ATLAS was able to set exclusion limits on the existence of representative models of new physics beyond that of the SM, superseding the Tevatron bounds, thanks to the higher LHC *pp* collision energy. This paper marked the beginning of a new era, following upon a long period of preparation: ATLAS was finally exploring new territory at the energy frontier.

9.3.2. *Motivation to search for extension to the Standard Model*

The field of particle physics is at an interesting junction. The SM was very carefully examined by earlier e^+e^-, proton-proton, νN and low-energy precision measurements and searches for new phenomena, as well as measurements and searches by ATLAS and other LHC experiments. No significant deviation from SM predictions has yet been detected. However, there are strong reasons to believe that the SM does not provide the complete picture. Several questions remain unaddressed by the SM in its current form and these may point toward extensions of the model. For instance, are quarks and leptons elementary particles? There are three families of fundamental particles; are there additional families? The SM was constructed assuming massless neutrinos, while oscillation between different types of neutrinos shows unambiguous evidence of mass, albeit very small ones. Why are the neutrinos so light and how do neutrinos acquire mass? Various astrophysical and cosmological measurements show that the visible matter in the Universe is not enough to explain the gravitational features observed. Is there additional dark matter that is responsible for these effects? What is the reason for the observed asymmetry between the number of baryons (matter) and anti-baryons, exceeding the SM prediction by many orders of magnitude? Why is there such a gap between the electroweak unification scale around 100 GeV and the Planck scale around 10^{19} GeV? The discovery of the Higgs particle completed the SM, but without new physics the observed mass of the Higgs boson would require an extremely finetuned SM to keep its mass near the electroweak scale in the presence of other, much larger energy scales (a mystery also known as the "hierarchy problem"). Most physicists believe that the explanations of some or all of these puzzles lie beyond the SM.

While the experimental searches are often motivated by proposed SM extensions, the goal of the ATLAS search programme is to detect any new phenomena at the energy frontier. Since new physics can appear anywhere, and in any process, it is a challenge to know where to search. Therefore, theories predicting new physics are used as a guide, while designing the analysis to be as general as possible so as to not miss a signal. The search strategy is a balance between these two poles, with general searches for the unknown such as pair production searches and optimised searches for a set of new physics models, such as extra dimensions or supersymmetry (see Sec. 9.4). However, theories of new physics evolve over time, changing based on the latest experimental results and reflecting new developments in theory; therefore, the search programme is constantly refined and redefined. This often requires improvements in the triggers and event reconstruction performance as, for example, in the search for long-lived massive particles. Improved modelling of background processes with state of the art theory predictions and making use of machine learning techniques in object identification and physics analysis further enhance the sensitivity of a search.

9.3.3. *Search techniques*

A powerful way to look for new physics in the ATLAS data is to search for the decay of new heavy resonances into pairs of lighter particles. Various theoretical models such as grand unified theories, little Higgs, or extra dimensions predict the existence of heavy neutral (Z') or charged (W') gauge bosons. The simplest channels to examine these are via their decays into a pair of SM particles. These channels are easy to detect, have high cross sections, but suffer from high levels of background. Key challenges are therefore distinguishing between signal and background and understanding the selection acceptance and efficiency, which are model dependent, and in many cases, estimated using simulation. The estimation of the backgrounds is in most cases done by studying the data in control regions where signals of new physics are not expected. For all of these, a precise understanding of the detector and the reconstruction of objects such as electrons, muons and jets are critical. The selected data are then compared to the SM expectation. In the absence of any deviation, a limit is set, excluding a given model of new physics in the channel examined. The exclusion can either be for a specific model or for a more generic result, which can be later expressed in terms of concrete new physics models.

In searches for the unknown, experimentalists are concerned that personal prejudices may introduce biases in the search itself, for example, through the event selection. Therefore, protocols are developed to minimise this possibility. In many of the analyses, the selection criteria are fixed, and the efficiency and backgrounds are estimated *before* uncovering the region where the signal is expected to appear.

9.3.4. *Pair production searches*

Since new physics can hide anywhere, a common practice is to look for pair production of SM particles. When protons collide, the most probable final state is jets of hadrons. Hence the flagship of the exotics searches is to look for an excess in the distributions of hadronic jets produced in the interaction. With any increase in the LHC collision energy or significant addition of collision data events with high transverse momentum, jets are selected. Such an energetic dijet event is shown in Fig. 9.16. The mass distribution of the two highest momentum jets is statistically examined to search for a

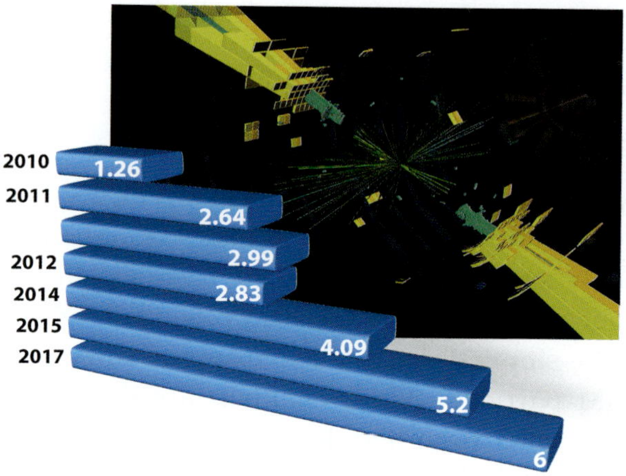

Fig. 9.16. The evolution of exclusion limits set by ATLAS on excited quark model-based searches from 2010 to 2017. The background image shows a display of a high-mass ATLAS dijet event.

"bump" above the SM prediction. In order to search for a deviation from the SM expectation, that expectation has to be well known. The ATLAS dijet resonance search uses a smooth functional fit to the measured data spectrum to determine the SM expectation. The function cannot account for narrow peaks in the spectrum, which would then stick out as deviations from the background fit. ATLAS repeatedly exercised this search starting with a first dataset of $315\,\mathrm{nb}^{-1}$. The sensitivity of this search can be compared to the most recent dijets result based on about 120,000 times more data ($37\,\mathrm{fb}^{-1}$) and larger *pp* collision energy. Figure 9.16 demonstrates how the exclusion boundary on the mass of a hypothetical excited quark, q* (indicating quark substructure), improved from 1.26 TeV in 2010 to 6 TeV in 2017.

Similarly, to jet pairs, new physics is searched for in events with pairs of leptons. The first search candidates were pairs of muons or electrons, as their signature is simple and the reconstruction of the particles is straightforward (albeit the accurate measurement of very high-momentum muon tracks is challenging as they appear as almost straight tracks in the detector). Some new physics models predict a Z' boson, which decays into a pair of leptons or a W' boson, which decays into a charged lepton and a neutrino. The neutrino escapes detection and is observed as large missing transverse energy in the event. The sensitivity of these searches requires excellent understanding of the characteristics of high momentum electrons and muons in the ATLAS detector. The precise modelling of the muon reconstruction is demonstrated

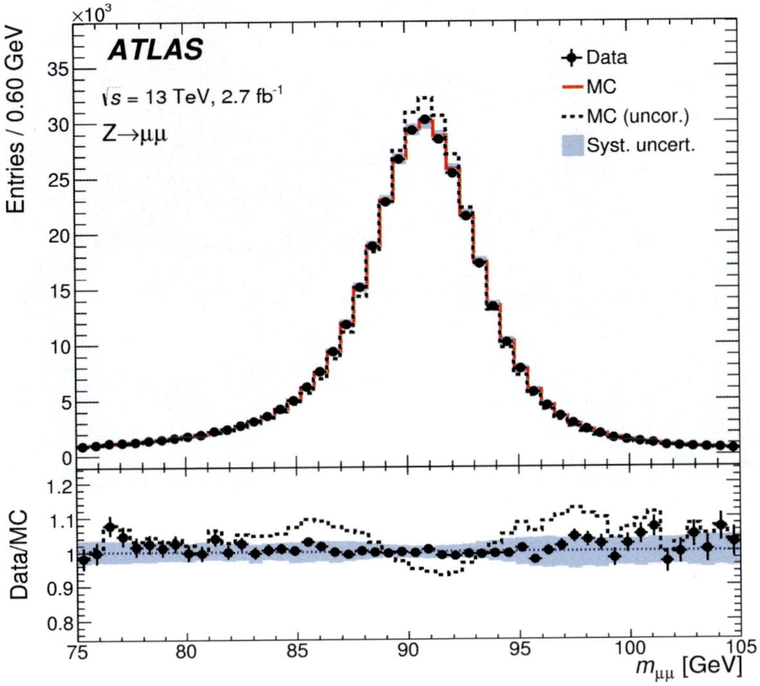

Fig. 9.17. The di-muon invariant mass in Z boson events for data and theoretical predictions.

in Fig. 9.17, where the invariant mass of muon pairs originating from Z boson decays is compared between data and simulation before and after applying corrections.

The top quark is one of the six fundamental quarks but its nature is unique. Its large mass of about 170 GeV, which is similar to that of a gold atom, may be a hint that this particle could play a special role in the phenomena of physics beyond the SM. Searches for new heavy particles decaying into a ditop ($t\bar{t}$) pair are therefore highly motivated probes of new physics. When a new heavy particle is produced and decays into $t\bar{t}$, individual top quarks are strongly boosted with large momentum. Since the top quarks promptly decay into a W boson and a b-quark, the boosted top quark can be observed as a narrow jet composed of three "sub-jets" associated with the b-quark and the $W \to q\bar{q}$' decay. The identification of such a topology of jets with internal "substructure" provides a very powerful tool to isolate boosted massive particles such as top quarks, W/Z or Higgs bosons originating from the decay of a new heavy particle. Many ATLAS

Fig. 9.18. Display of a top quark pair production event.

searches and measurements have begun to exploit substructure methods. Figure 9.18 depicts an event selected by the search for a resonance decaying into $t\bar{t}$. Here, one top decays into a b-quark and a W boson that further decays into two jets, and the second top decays into a b-quark and a W boson, which then decays into a muon and a neutrino. In the figure, the red line shows the path of a muon with transverse momentum and the dashed line shows the direction of the missing transverse energy. The green and yellow bars indicate energy deposits in the Liquid Argon and Tile calorimeters; from these deposits four small-radius jets are identified. Three of these jets are re-clustered into a large-radius jet (not shown explicitly in the figure) with a jet mass near the top quark mass. One of these three jets and a fourth jet are identified as having originated from b-quarks. Similar substructure techniques were implemented in the searches for resonances decaying into boson pairs.

9.3.5. *Dark matter searches*

Production of the elusive and much sought-after dark matter particle at the LHC would be observed in the detector only via a signature of missing transverse energy (similar to a neutrino). ATLAS detects such signatures by searching for a SM particle "X" recoiling against missing energy. These signatures, collectively called "Mono-X" searches, are powerful probes not only for dark matter, but for the many non- or weakly-interacting particles in association with SM particles as well. ATLAS has performed a number of Mono-X searches with X being jets, photons, heavy flavour quarks, W/Z or Higgs bosons, all of which are consistent with the SM expectations. Backgrounds events can stem from Z+X processes, where the Z decays into two neutrinos. Searches with this signature are challenging, as the missing transverse energy measured in the detector must be well understood. Since the missing transverse energy is estimated from other measured objects in the event, any uncertainty on the measurement of these objects is compounded into the missing transverse energy measurement. Ideally the measurement of the missing transverse energy should not be sensitive to the amount of pile-up in the event. Figure 9.19, which shows the x- and y-components for the missing transverse energy as a function of the number of collision (reconstructed vertices) in the event, indicates however that there can be a strong sensitivity. The different coloured lines indicate different algorithms used to estimate the amount of missing transverse energy. ATLAS has spent significant time to develop algorithms that have little to no dependence on the amount of pile-up in the event. An example of the missing transverse energy spectrum is shown in Fig. 9.20. The SM processes are shown in the solid colours, possible New Physics models are shown by their dashed lines and the data are indicated by the black points. Thus far no indications for dark matter at the LHC have been observed.

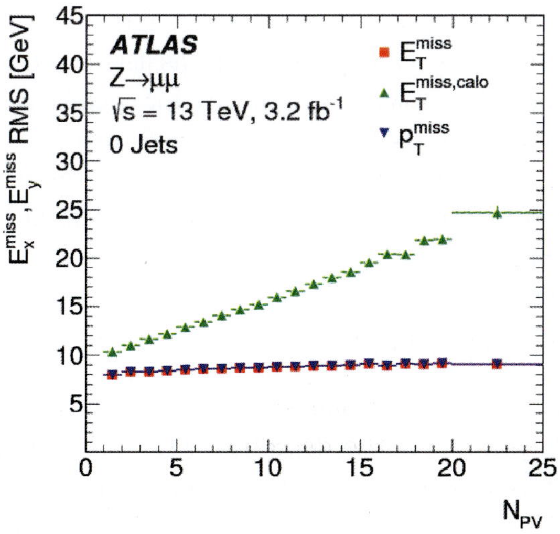

Fig. 9.19. The RMS of x- and y-components of the missing transverse energy for different algorithms as a function of the number of reconstructed vertices (pile-up).

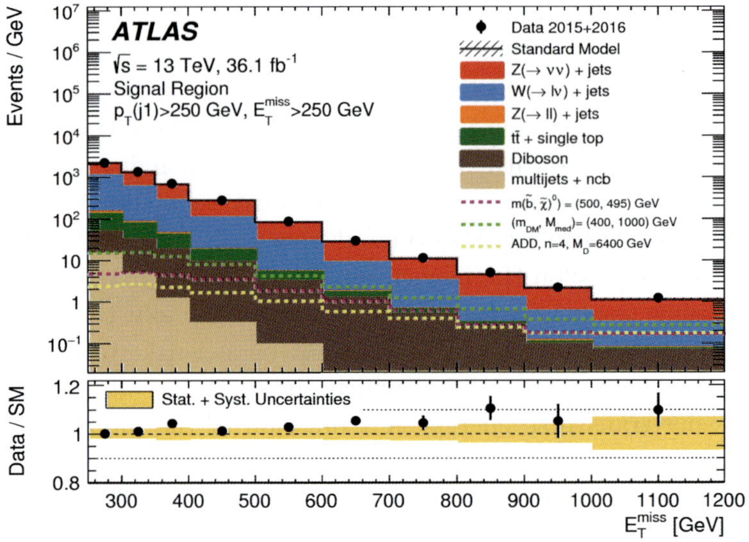

Fig. 9.20. Missing transverse energy distribution in events with at least one energetic jet.

9.3.6. *Searches for extra dimensions*

Theories of extra dimensions, which are introduced in various SM extensions to resolve the gap (or hierarchy) between the strength of the SM forces and that of gravity, can result in similar detector signatures as dark matter particles. Assuming that gravity is diluted in the Universe by propagating into "large" extra space dimensions, the strength of fundamental gravity in extra dimensions space can be as large as the electroweak force, thus eliminating the hierarchy. A typical signature for such large extra dimensions at the LHC is large missing transverse energy associated with the momentum escaping into the extra dimensions. Theories of extra dimensions also predict microscopic black hole production at the LHC. In March 2008, a lawsuit in Hawaii that requested to delay the start of the LHC caused a sensation when it claimed that the LHC could create a black hole that would destroy the Earth; the claim was unfounded since macroscopic black holes cannot be produced in LHC collisions. ATLAS has a rich programme of searches for microscopic black holes. If there are extra dimensions at an energy scale of a few TeV, two protons colliding at the LHC could get close enough to collapse and form a microscopic black hole. That black hole would have a very short lifetime and it would evaporate via what is called Hawking radiation before it could pull in any additional matter. These black holes are expected to decay into multi-object states including multi-jets, photons, leptons or even Higgs particles. If detected it could tell us that the Universe has more dimensions than the one time and three-space dimensions, and thus make the LHC a fascinating laboratory for studying the microscopic gravitational phenomena. ATLAS has been looking for various signatures with sensitivity to black hole masses of up to about 10 TeV. No signal has been observed so far. Similar to searches in events with pair production of jets, the main challenge in these analyses is to accurately estimate the background from the SM jet production. Theoretical predictions are generally not precise enough, so the number of background events is determined from the data.

9.3.7. *New types of quarks*

Are there more generations of quarks beyond the three known from the SM? The discovery of the Higgs boson sheds new light on this important question. Since the Higgs boson couples to mass, it prefers to interact with the heavy top and bottom quarks. If additional heavier quarks existed, the Higgs would be produced at a rate much larger (roughly a factor of nine) than that predicted in the SM. Since the measured Higgs boson production rate agrees with the expectation within an uncertainty of about 10%, it rules out the presence of a simple SM extension by an additional "4th" generation of quarks with properties similar to those of the first three generations. However, ruling out a fourth generation points to new directions in searches for heavy quarks, in particular with so-called vector couplings. These new particles do not acquire their masses by the Higgs mechanism and hence are not constrained by the Higgs production rate. Those hypothetical "vector-like" quarks can play an important role in stabilising the Higgs mass, the so-called mass-hierarchy problem, without significant fine-tuning. Pair production of vector-like quarks leads to complex final states including multiple heavy quarks and electroweak/Higgs bosons. These new quarks have been searched for with ATLAS in various final states, excluding an extensive mass range. Furthermore, the similarity between the structures of the lepton and the quark sectors may indicate that there is some unknown mechanism that relates the two. Such a possibility, predicted by Grand Unification scenarios, may also lead to the existence of new states that transform leptons and quarks, generally called "leptoquarks". These states can be used to explain some the recent anomalies observed in the flavour sector at LHCb and B-factories and have therefore gained considerable interest. The pair production of leptoquarks results in a wide range of final states composed of multiple charged or neutral leptons and jets with all different flavours.

ATLAS searches have primarily focused on pairs of charged leptons and light-flavour quark jets and are extending the search towards more complex final states, including the third-generation fermions.

9.3.8. *Extreme exotica*

All the searches above involve final states with objects like electrons, muons and jets. But what if new physics involves much more exotic final states? Could it still be detected and discovered? One good example is the hidden valley scenarios, which assume the existence of a *hidden* set of particles that interacts only weakly with SM particles. The hidden particles may have a relatively long lifetime, producing in their decay vertices away from the interaction point. Such decay vertices may appear in the inner detectors, calorimeters, or even as far away as the muon spectrometers. ATLAS has developed unique trigger and reconstruction tools in order to explore all these options. The ability to search for particles that decayed far from the interaction point was not initially foreseen and the success of these results highlights the flexibility of the ATLAS detector.

Exotic particles like magnetic monopoles or massive particles with multiple or fractional electric charges are predicted in various new physics scenarios. Searches for those particles usually require the extension of the standard reconstruction techniques. Magnetic monopoles were predicted as a solution to the electric charge quantization problem. If they are produced at the LHC, they could have the signature of a highly electrically charged massive particle traversing the ATLAS detector, leaving large energy loss due to ionization. This was searched for using the Transition Radiation Tracker and electromagnetic calorimeters. Evidence of such highly ionising massive particles was not observed, but it demonstrates that ATLAS is a versatile detector that provides a wide range of opportunities for new physics searches, including those with non-standard signatures.

9.3.9. *Outlook: Searches in the next decades*

With the opening of the TeV range and the wealth of high-quality data provided by the LHC, ATLAS has searched for an extremely broad range of signatures. The absence of deviations from the SM has been used to put severe constraints on new physics models and the number and variety of studied final states bear witness to both the excellent performance of the detector and the ingenuity of the analysers. Questions that are frequently asked on the basis of these impressive results are: Is there no new physics?

Should experiments just stop looking? The answer to both questions is a clear no. The data collected up to now have only allowed the exploration of the *easier* final states, with high cross sections, and clear handles for separating the signal for the backgrounds. Much larger data samples are necessary for extending the exploration beyond what has been already achieved. Examples are very rare processes or processes that resemble the SM background. For these and for high mass particles, larger datasets will be needed and the increase in centre-of-mass energy will be beneficial. The invention of new experimental techniques to push the reach of these searches well beyond expectations is also being undertaken.

9.4. Supersymmetry

Supersymmetry (SUSY) has been for decades one of the most compelling theories to address known problems in the SM. SUSY introduces new partners (superpartners) of the known bosons and fermions that share with them the same mass and internal quantum numbers but have spin different by half-unit. As shown in Fig. 9.21, quarks have *squarks* as superpartners, while leptons have *sleptons*. Force carriers also have their counterparts: *gluinos* are superpartners of the gluons, whilst *neutralinos* and *charginos* are a mixture of the W, Z and Higgs bosons superpartners (higgsinos, winos and binos). The discovery of SUSY would signal the restoration of fermion-boson symmetry and could solve the long-standing hierarchy problem, arising from the fact that the Higgs mass is sensitive to quantum corrections from physics at very high-energy scales. SUSY could also possibly explain the

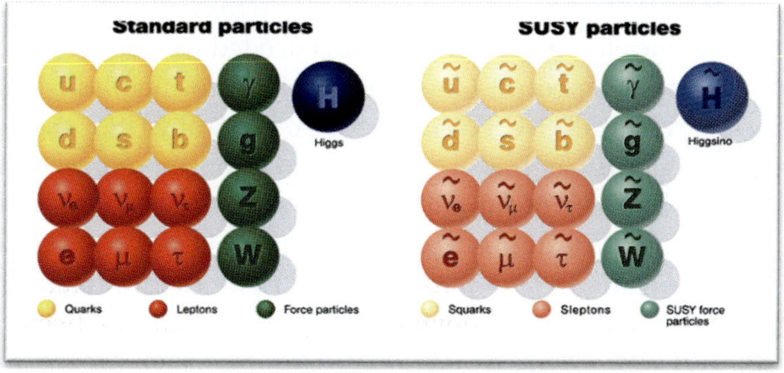

Fig. 9.21. An example supersymmetry spectrum: for each SM particle (quarks, leptons and force particles), a SUSY partner is associated (squarks, sleptons and SUSY force particles).

nature of dark matter in the universe if R-parity, which relates the baryonic and leptonic numbers of a particle to its spin, is conserved. In R-parity conserving SUSY models, superpartners of the SM particles are produced in pairs and the lightest supersymmetric particle (LSP) is stable, providing a possible candidate for dark matter.

Since no light superpartners have been discovered, the masses of the superpartners are not equal to the SM particle, thus SUSY is a broken symmetry. The mass scale of the SUSY particles is no longer determined after symmetry breaking and the experimental signatures depend mostly on the breaking mechanism and on the nature of the lightest and next-to-lightest supersymmetric particles. In a large variety of scenarios, the LSP is the lightest neutralino. In general, the nature of the LSP is defined by the mechanism that breaks supersymmetry and by the details of the chosen theoretical framework. If R-parity is not conserved, the LSP can decay into SM particles and a variety of additional signatures might arise depending on the R-parity violating terms. This very open theoretical framework provides a plethora of possible SUSY models to be explored, and experimentalists have been searching widely and strenuously for several years.

9.4.1. *A brief history of SUSY at ATLAS*

SUSY searches at ATLAS have evolved over the years. Before the turn-on of the LHC, the initial approach was based on targeting a very specific minimal supersymmetric Standard Model (MSSM) and the focus was on devising a set of analyses that would allow a quick discovery of squarks and gluinos in the TeV mass range. With the arrival of the data, and the negative results of the first searches, the uncovered space for constrained models was severely restricted; it became clear that guidance was needed to identify investigation-worthy SUSY particle spectra. One of the guiding principles is *naturalness*. What defines a "natural" SUSY scenario? A natural SUSY scenario solves the instability of the Higgs boson mass due to quantum corrections. This is only possible if the masses of the superpartners are not too far above the TeV scale. In particular, naturalness arguments suggest that the lightest of the charginos and neutralinos may have masses at the electroweak scale, third-generation squarks (stop and sbottom) should be expected with masses of order 1 TeV, and gluinos with masses just above. The masses of sleptons and first- and second-generation squarks are less constrained but their role could be relevant to determining the phenomenology that drives the experimental searches. Hence, the emphasis was shifted from specific SUSY breaking models to a more generic approach, based on simple SUSY decay chains and searching for particles participating in them in terms of their role within natural scenarios. The production and decay of squarks of the first two generations and gluinos, of third-generation squarks, the direct production and decay of sparticles carrying electroweak quantum numbers were investigated with dedicated searches. The development of the SUSY studies in ATLAS is timed by the evolution of the LHC energy and by the increase of available data uncovered by the previous round of searches, either because of the experimental difficulty, requiring the development of new analysis techniques or because of the too small rate of the processes involved. In this framework there has been a progressive growth of interest and effort on more general SUSY models yielding exotics signatures like long-lived particles, which require dedicated modifications of the reconstruction software of the experiments (see for example, Sec. 9.4.5).

9.4.2. *Searches for first- and second-generation squark and gluinos*

The initial approach to SUSY searches was aimed at covering the largest possible spectrum of SUSY models, by addressing simple final states with high cross section and distinctive signatures. The highest SUSY production cross sections are expected for squarks and gluinos that have strong couplings to the initial state partons. In R-parity conserving scenarios, the first main target of SUSY searches at ATLAS, the squarks and gluinos will decay into the LSP, and because of colour charge conservation the decay products will include quarks that hadronise into jets, and can include W and Z bosons. The simplest decay conceivable for the squark is thus into a quark and the LSP, and for a gluino into a quark-antiquark pair and the LSP. The LSP escapes the experiment undetected and its presence is inferred by missing transverse energy. These are the "golden" signatures for SUSY and have been exploited from the very first data.

Additional inclusive signatures addressing more complex decay chains for squarks and gluinos include one, two or more leptons accompanied by jets and missing transverse energy. At first, no attempt was made to identify the flavour of jets (the type of quark or gluon that initiated the jet). In the absence of the observation of excess with respect to the SM, the results of these searches have been interpreted as limits on the parameters of SUSY breaking in different models. Initially the model with the most focus was the cMSSM, where all of the SUSY mass hierarchy and couplings can be predicted in terms of four parameters plus one sign. This model had the benefit that it was very simple but has now been essentially excluded by LHC data. The MSSM, which has more than a hundred unknown parameters, is much less constrained. Interpretation of the search results in such a large parameter space is challenging. Therefore, a so-called simplified model approach is used, where a single production and decay chain is chosen, e.g., gluino-gluino with both gluinos decaying into a quark-antiquark pair and the LSP. The kinematics of this chain can be represented solely in terms of two parameters, the masses of the gluino and the LSP, and the results of the search can be expressed in terms of the cross section excluded by a given analysis in the m(gluino) versus m(LSP) parameter space. In the cMSSM the reach of the very inclusive signatures described above is mostly determined by two parameters, m_0 and $m_{1/2}$, the common scalar mass and fermion masses as the energy scale of SUSY breaking. The achieved exclusion limits for the different inclusive analyses using these two different model approaches are shown in Fig. 9.22. The different coloured lines highlight different analyses. All models below the solid lines are excluded at 95% confidence level.

9.4.3. *The quest for third-generation squarks*

The crucial role of the top squark (stop) gained more and more attention after the discovery of the Higgs boson in 2012. Since each particle that couples to the Higgs field has a coupling proportional to its mass, the most significant corrections to the Higgs mass will originate from the heaviest SM particle, the top quark. As a consequence, the superpartners of the third-generation quarks, bottom and top squarks, play a fundamental role in regularising the radiative corrections to the Higgs boson mass and are not expected to exceed the TeV scale. Searches for third-generation squarks had already started in 2010 and have been extended significantly over the next few years. No excess over the SM predictions was found. An extensive and systematic effort was put in place to target pair production of the top squark. Depending on its mass and on its relation to SUSY and SM particles such as chargino and W boson, very different signatures are expected. The sensitivity of those searches was pushed to an unexpected level thanks to the great performance of the detector, its capability to identify b-jets with high efficiency, and novel ideas used in those searches. Figure 9.23 shows an example of the efficiency to select a b-jet versus the number of light-flavour jets that were rejected for algorithms developed during Run-1. An ideal

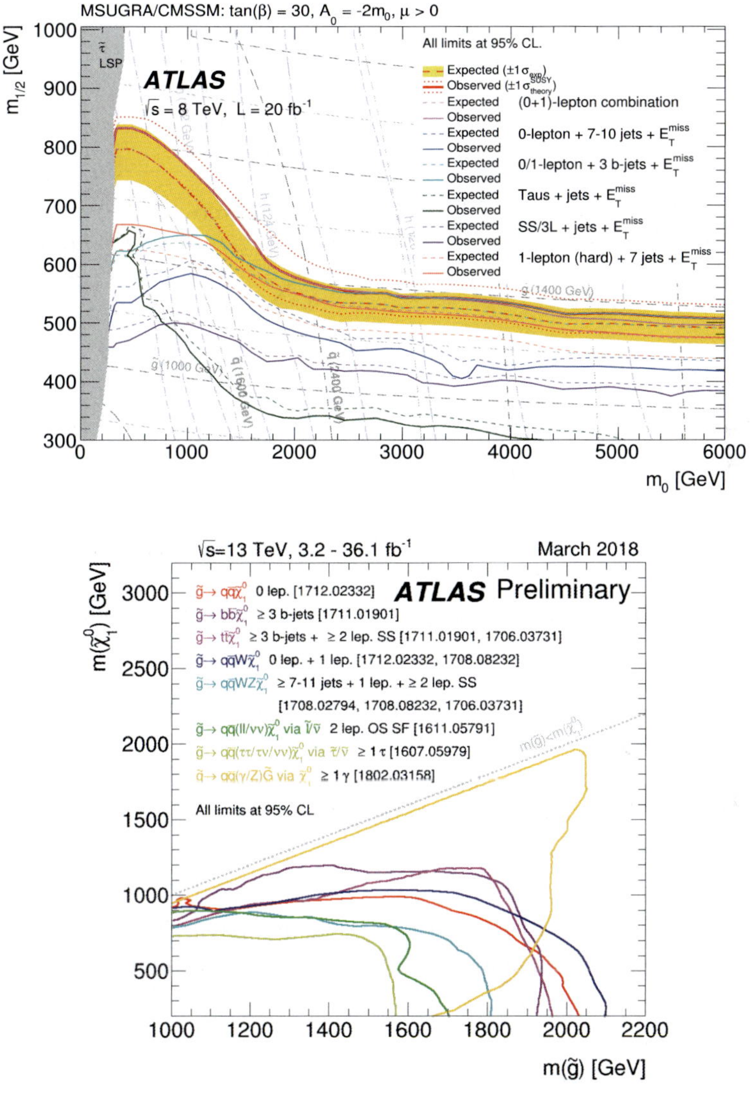

Fig. 9.22. (Top) Exclusion limits in an example SUSY model defined as a function of m0 and m1/2. (Bottom) Exclusion limits for simplified models as a function of m(gluino)-m(LSP).

algorithm would have a b-jet efficiency of one and a light-flavour jet rejection that is very large. The different coloured lines indicate different algorithms. The IP3D (pink) was one of the earliest algorithms to be developed while MV1 (black) was later used for Run-1 analyses. The improvement in the rejection factor for a given efficiency is at least one order of magnitude.

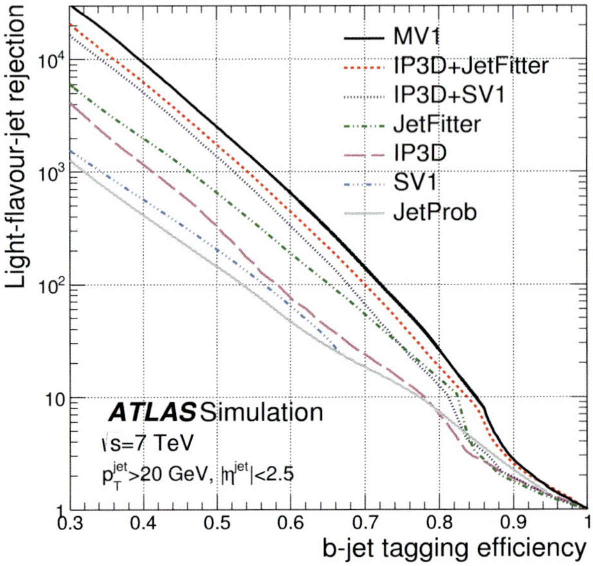

Fig. 9.23. The b-jet efficiency versus light-flavour jet rejection for different algorithms.

Despite the absence of signal, the first comprehensive summaries of top squark mass limits, shown in Fig. 9.24, marked a milestone in the history of SUSY searches at the LHC, excluding top squarks with masses above and below the SM top quark mass. In the following years several searches have been performed, targeting high mass top squarks as well as difficult corners of the parameter space, such as where the top quark and its superpartners have very similar mass. More data will be needed to exclude the existence of a top squark, but it is clear that naturalness arguments are being pushed to "less natural" corners by ATLAS.

9.4.4. *The new challenge: Electroweak SUSY partners*

All the searches described above focus on partners of quarks and gluons that are produced with high cross section at the LHC. There is however another sector in the theory, the partners of the gauge and Higgs bosons of the SM. These partners mix with each other to generate a set of four neutral particles, the neutralinos, and two charged particles, the charginos. The production and decays of these particles are determined by the electroweak and Higgs couplings, and by the four parameters driving the mixing. Naturalness considerations require that some of these particles be light, with a mass scale of a few hundred GeV. The lightest neutralino is often taken as the

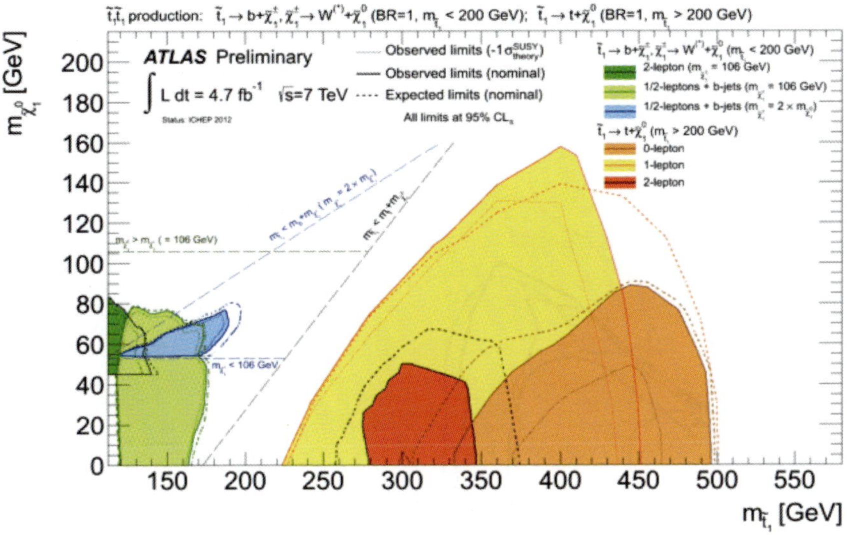

Fig. 9.24. The first summary plot on searches for top squarks, showing the top squark–LSP mass plane.

LSP in LHC studies, hence searches focus on the production of the second-lightest neutralino or the lightest chargino, and their decays to the LSP, which can happen through partners of the leptons, or through the Higgs and the W and Z gauge bosons. The role of the sleptons is very important for the phenomenology and slepton pair production is by itself a target of ATLAS searches. In the assumption that the sleptons are heavier than the second-lightest neutralino, the search for chargino and neutralino production relies on the leptonic decays of the intermediate W and Z bosons, or on the identification of the decays of the Higgs. The results of the searches for the production of a chargino and a neutralino approximately degenerate in mass, or of two charginos, are shown in Fig. 9.25. Hierarchies, by which the higher mass charginos and neutralinos are approximately mass degenerate with the LSP, arise naturally in SUSY. The experimental consequences depend on the amount of degeneracy. For a mass difference down to a few GeV the final state leptons are very soft and the searches require the development of dedicated experimental techniques, based on the detection of soft leptons and/or on the selection of topologies boosted by QCD radiation from the initial state partons. This is a thriving field for ATLAS analyses. For mass differences of the order of a few hundred MeV, particles might decay slowly and become long-lived, such as those described in Sec. 9.4.5.

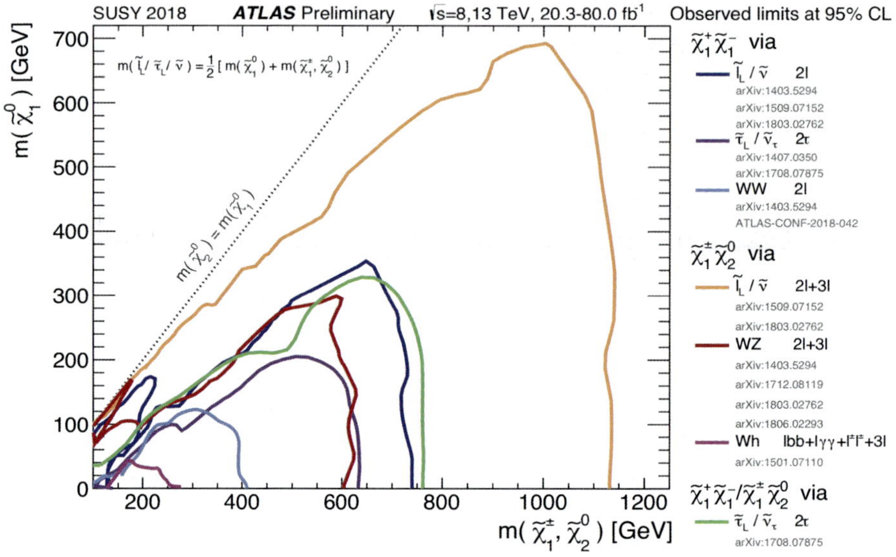

Fig. 9.25. The most recent summary plot on searches for charginos and next-to-lightest neutralinos.

9.4.5. *R-parity violation and long-lived particles*

R-parity conserving SUSY models are characterised by the presence of the LSP that constitutes a dark matter candidate. Nonetheless, R-parity violating (RPV) scenarios are as valid, with appealing features such as the possibility of explaining neutrino mixing. Because R-parity violating couplings might lead to lepton or baryon numbers violation (or both), the LSP can decay into SM particles in many ways. Final state events characterised by high lepton or jet multiplicities as well as little or no missing transverse energy are searched for signs of new physics since the beginning of Run-1. Searches for RPV particles are often challenging because of the lack of the distinctive large missing transverse energy signature. For example, it is very difficult to extract a tiny 2-jet or 3-jet resonance from the overwhelming multi-jet SM background. However, the results obtained exceeded expectations thanks to the ingenuity of experimentalists and theorists who developed dedicated techniques and algorithms in the past years. Stringent constraints have been placed on RPV SUSY models including gluinos, squarks, charginos and neutralinos. If RPV couplings are very small, decays of SUSY particles can be non-prompt, slow, delayed or might happen outside the detector, possibly after many years! Such unconventional signatures are searched for with dedicated analyses exploiting

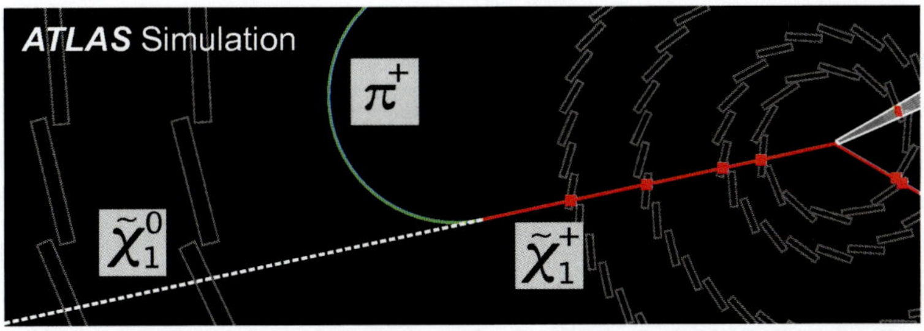

Fig. 9.26. A drawing of the decay of a long-lived chargino, decaying just outside the Pixel Detector.

at best the capabilities of the ATLAS detector. The phenomenology of the expected signal depends on the lifetime of the targeted SUSY particles, which is set by various parameters of the models. It is important to mention that signatures of this kind might also arise in R-parity conserving SUSY scenarios. Long-lived particles arise if the mass hierarchy is very compressed or the decay chains include heavy intermediate particles, or in some of the gauge-mediated SUSY models. For short or medium lifetimes (few nanoseconds), SUSY particles can be identified by the presence of displaced, high mass vertices within the inner ATLAS volume, reconstructed using charged tracks or charged leptons. SUSY particles with medium or long lifetimes could decay in the outer ATLAS volume or outside the detector, behaving as very massive muons if charged. Their imprint is identified from anomalous energy deposits or other unconventional features. Figure 9.26 shows an example of the simulated decay of a long-lived chargino inside the tracker. The green and red lines indicate particles that are detected by the tracker, whereas the white dashed line is a non-detected particle.

9.5. Heavy-ion physics

Measurements of heavy-ion collisions pose completely different experimental challenges compared to proton-proton collisions. These collisions are used to study the quark-gluon plasma, a postulated state of ultra-hot, ultra-dense hadronic matter with unbound quarks and gluons. A collision of two lead ions, with 208 nucleons each, can produce several hundred times more particles than a typical *pp* collision, pushing the capabilities of the detector. However, ATLAS is a powerful device to study these collisions, as it couples hermetic calorimetry with a large-acceptance charged-particle tracker and a precise muon spectrometer. For example, despite the unprecedented particle densities produced in these collisions, the ATLAS Inner Detector is more than capable of measuring thousands of tracks in each event and associating them with a single event vertex. Still, it was realised that this enormous number of produced particles would create backgrounds that differ widely from one event to the next, and that would distort the measurements of the energies of jets, photons and electrons. To overcome this challenge, new techniques were developed to estimate and subtract this "underlying event" from the calorimeter cells used to reconstruct jets, photons and electrons. In addition, to provide a robust experimental trigger and to suppress high-energy electromagnetic (EM) processes, ATLAS installed "zero-degree calorimeters" (ZDCs) about 140 metres downstream in both directions, which provide a coincidence signal from spectator neutrons for inelastic events. They can also be used to tag EM processes by requiring no activity in one or both ZDCs. These photon-photon and photonuclear processes are of intrinsic interest in their own right, as described below.

Before the start of data taking, the heavy-ion programme at ATLAS was often seen as a mere by-product of the main *pp* collision programme. However, several exciting results and observations in the past years have highlighted its importance. One such result is the observation of jet quenching in single events, obtained using the very first ATLAS heavy ion data. This phenomenon is studied by using the asymmetry (A_J) between the transverse energies of pairs of jets. The left panel of Fig. 9.27 shows what turned out to be a typical event — a jet clearly-observed in the calorimeter and the opposite recoiling jet energy spread out over large angles. Proton-proton events typically produce jets that are back-to-back with balanced transverse energies (right, open circles), but when two heavy-ions collide

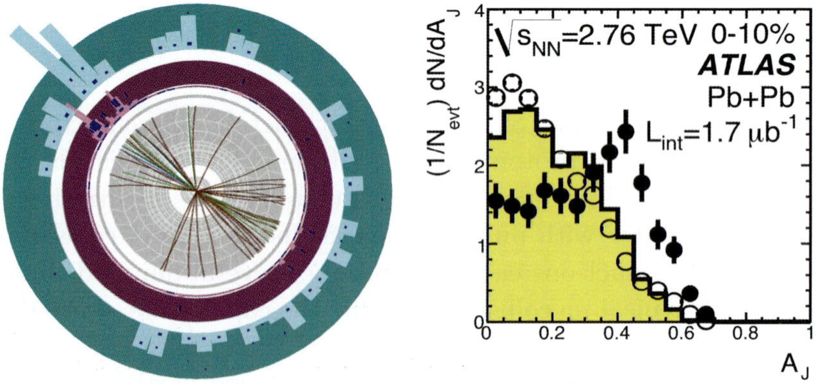

Fig. 9.27. (Left) Event display of a jet-quenching event from a single lead-lead collision. (Right) The dijet asymmetry.

almost head-on, the events show a dramatic widening of the A_J distribution (right, closed circles).

Since then, ATLAS has systematically explored the physics of jet quenching through the study of a wide range of SM processes in lead-lead collisions. The variable typically used to characterise nuclear effects is the ratio $R_{AA} = N_{AA}/\sigma_{pp}T_{AA}$, the measured yield in lead-lead collisions divided by the pp cross section times the average nuclear thickness, which scales linearly with the number of binary nucleon-nucleon collisions. The yields of electroweak bosons are observed to be linear with the nuclear thickness, as shown in Fig. 9.28 by the R_{AA} near unity for Z bosons (black), indicating that these bosons are not much influenced by the surrounding medium. However, a complicated behaviour for high-p_T hadrons (blue), J/Ψ particles (green) and fully reconstructed jets (red) is observed. All of the hadronic observables are found to be suppressed, due to energy loss in the medium, with the hadrons slightly less suppressed for $p_T > 100$ GeV, consistent with direct measurements of jet fragmentation properties.

Typical heavy-ion collisions are not associated with high momentum physics objects like jets, photons or muons, and so are selected by requiring activity in the calorimeters, as well as neutrons emitted in both directions. However, even these typical collisions show a rich set of signatures of collective expansion, observed as short-range azimuthal angular correlations between particles emitted with a large separation in pseudorapidity. This phenomenon is typically referred to as the "ridge". The ridge in large

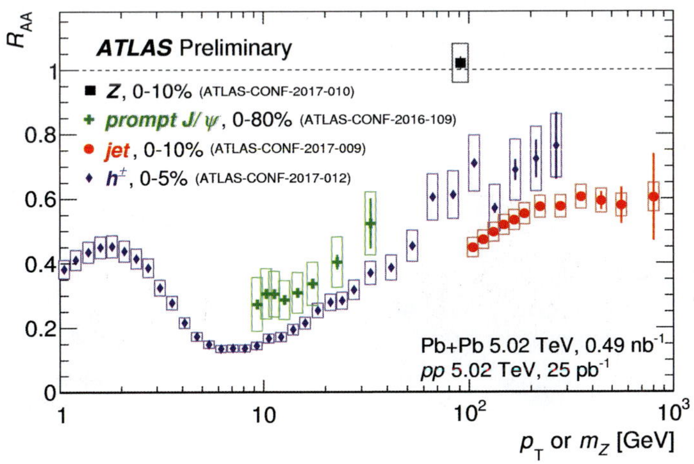

Fig. 9.28. Ratio of the measured yield in lead-lead collisions to the pp cross section, scaled by the nuclear thickness for different processes.

systems is commonly interpreted as a result of the collective hydrodynamic expansion of the hot, dense medium created in the initial moments of the collision. The origin of the ridge in smaller systems, with examples shown for proton-proton and proton-lead data in Fig. 9.29, can be modelled as a smaller volume of the same medium. However, this interpretation continues to generate considerable debate and has led to deeper theoretical questions over the mechanisms for achieving the apparent local equilibrium needed for hydrodynamics to apply. To address this, ATLAS has further probed the collective nature of the ridge phenomena by using multi-particle correlation techniques, particularly using cumulants to remove lower-order correlations involving fewer particles. In particular, an improved method based on correlating particles from separated regions in the detector is shown to reduce contributions from resonance decays and jets, which mask truly collective effects. Results for proton-proton and proton-lead collisions provide robust evidence that multi-particle collective behaviour persists in those systems, even for events with small track multiplicities.

One of the legacies of the LHC heavy-ion programme will be the large sample of minimum bias data collected by all of the detectors from all available energies and collision systems, using special triggers to enhance the more rare events with large particle multiplicities. These samples have already been used to study event-by-event fluctuations in both the initial collision geometry and local energy deposition in heavy-ion collisions, which are thought to seed hydrodynamic evolution in the strongly coupled medium. This evolution imprints strong three-dimensional momentum-space correlations on the final state, which can be observed in both the transverse and longitudinal directions with respect to the collision axis. This has

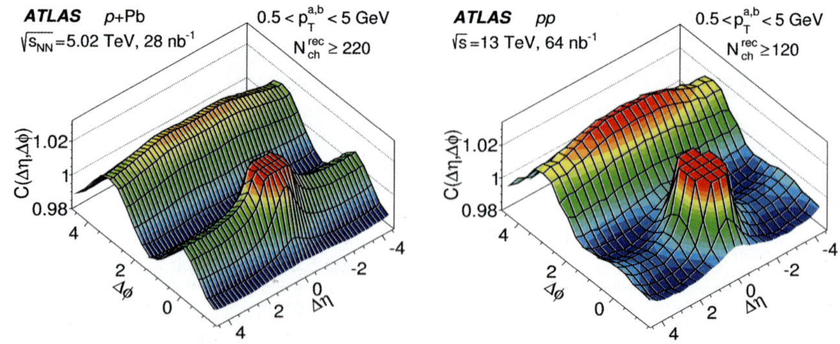

Fig. 9.29. Two-particle correlation functions for inclusive charged particles in high-multiplicity events from proton-lead (left) and *pp* (right) collisions, both showing a "ridge" structure at $\Delta\Phi \approx 0$.

been addressed using an encyclopaedic set of results on flow (of light and heavy flavour hadrons), flow fluctuations, event-plane correlations, event-shape engineering, pseudorapidity correlations and flow decorrelations. More sophisticated techniques involving multi-particle cumulants continue to be developed and provide new uses even for older data.

One of the most striking results obtained using heavy-ion beams is the first direct evidence of light-by-light scattering ($\gamma\gamma \to \gamma\gamma$), one of the oldest predictions of quantum-electrodynamics (QED). This process was observed in "ultra-peripheral collisions", which are events where the two nuclei interact at large impact parameters of at least twice the nuclear radius, such that no strong interactions are possible. In these cases, the primary interactions are electromagnetic, which are enhanced relative to proton-proton collisions, as the electric field strength scales with the square of nuclei charge (Z^2), which for lead-lead collisions is 82 squared. For light-by-light scattering, the cross section is proportional to $\alpha_{\mathrm{EM}}^4 \approx 10^{-9}$, and is thus minuscule, which makes it very challenging to observe experimentally. However, with the large data samples available from the LHC and a carefully designed experimental trigger, ATLAS collected the first significant sample of exclusive two-photon events. Out of 4 billion sampled minimum-bias events, 13 di-photon candidates were observed with no additional activity in the detector (an example is shown in Fig. 9.30). From the expectation of

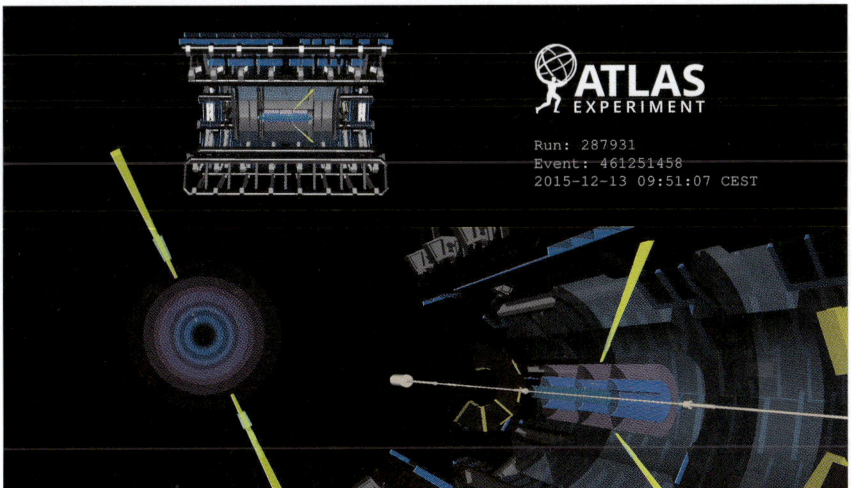

Fig. 9.30. Event display from light-by-light scattering in ultra-peripheral lead-lead collisions.

7.3 signal events and 2.6 background events, a significance of 4.4 standard deviations was obtained; this was the first evidence of this long-awaited process.

The heavy-ion programme will benefit strongly from the increased luminosity expected in the last year of Run-2 and future high-luminosity runs. More differential jet measurements will be used to precisely probe the microscopic structure of the hot, dense medium. Collectivity will be explored in even more exquisite detail. Photon-photon and photo-nuclear processes will be used to study SM processes in a very clean environment and the partonic structure of nuclei in advance of the planned electron-ion collider in the United States.

The road to the first ATLAS 13 TeV results

The year 2015 was one of many challenges. Besides the Phase-0 upgrades, the Collaboration had prepared new simulation and reconstruction software, committed to systematic use of state-of-the-art generators, and changed analysis framework, while striving to harmonise and simplify physics object performance recommendations. These great ambitions brought daunting efforts of commissioning that had to be finalised at the start of 2015, in time for new data!

The LHC started Run-2 with a cautious increase of the luminosity while almost doubling the centre-of-mass energy and operating with a 25 ns bunch spacing. Though a short year for pp collision physics, 2015 had to lay the foundations for operations throughout the future life of the LHC.

The stable beam collision data arrived on schedule in the first week of June with the initial special low luminosity runs to measure the increased charged particle track multiplicities in minimum-bias events and the total cross section. With higher multiplicities and a dedicated trigger, long-range azimuthal correlations (the "ridge" effect) were measured. Measurements of the production of photons, jets, J/ψ, W, Z, and top quark pairs were released at the EPS-HEP conference in Stockholm, days after the higher luminosity data were collected in July.

By mid-August, the machine was successfully ramping up in intensity with 25 ns bunch spacing enabling ATLAS to show the first 13 TeV search results in dijet and multijet final states at the LHCP conference in Saint-Petersburg in late August. ATLAS steadily accumulated data to obtain a dataset with an integrated luminosity of $3.2\,\text{fb}^{-1}$ by the end of November. There followed a short period of PeV Pb-Pb collisions for the heavy-ion program. With the very short doubling time of the luminosity, both ATLAS and CMS went public with their first full dataset results at a special CERN seminar in December. The Collaboration was enthused and set the ambitious goal of preparing more than 30 new results for the occasion. It was an exceptional collaborative challenge to prepare such a large number of results in little more than a month after the data was collected.

The new ATLAS results at 13 TeV were beautiful. They included first measurements of single-top and diboson production, a first glimpse of the Higgs boson and results of searches with a vast number of signatures, for SUSY, additional Higgs bosons and new resonances. ATLAS also paid special attention to those topologies where the most significant excesses were seen at Run-1. No significant excess was found but one result drew a large amount of attention: a modest 2.0σ global (3.6σ local) excess in the diphoton search was obtained at a mass of 750 GeV, approximately where CMS also obtained a slight 1.2σ global (2.6σ local) excess. No excess had been seen at this mass at Run-1 and more data were needed to be conclusive. These results however provoked huge speculation in the community. The 2016 data were eagerly awaited, but ultimately did not confirm the excess as anything other than a statistical fluctuation.

The main conclusion of this exciting year, following the immense efforts and remarkable cooperative spirit of the Collaboration: ATLAS was ready for Run-2!

The road to the first ATLAS W-mass measurement

The W-boson mass project lasted, from first ideas to last iteration with the journal, a little less than 10 years. In 2008, an initial note argued that the measurement wouldn't be too difficult, with the main issues being statistics and control measurements. Shortly afterwards, a study entitled "Measurement of the W boson mass with early data" still showed the same optimism.

When data taking started, it was soon realised that there was a detector to understand, a completely new collision environment to explore, and a surprise under every stone. This was a concern for all ATLAS analyses, but the precision nature of this particular measurement required more detailed understanding than elsewhere. Over the years, a small but extremely dedicated team came together, embarking on this open-ended project with enthusiasm, great collaborative spirit, infinite patience, and no guarantee of success or even a partial result in the medium term. Sadly, several of these talented young scientists indeed left our field before having a chance to see the result they so much contributed to.

Among the steps toward the final result, key milestones were the measurements of the W and Z boson transverse momentum distributions and detailed electron, photon and muon calibration studies, all of which led to publications. Other excellent results from the Standard Model working group also provided crucial input, namely precise measurements of the W and Z cross sections at 7 TeV and of spin correlations in Z production at 8 TeV; the expertise of ex-HERA colleagues in QCD and in proton structure also proved to be absolutely fundamental. All of these studies took their time to complete, each with their own complications, while the larger part of the Collaboration moved on with the continuous excitement of new data and preparations for the next conference.

It was sometimes difficult for the team to be so isolated from the rest of the Collaboration, slowly working out tiny effects on old data, and using obsolete software releases while everyone else was flying through Run-1 and preparations for Run-2. The performance and computing groups never failed them, however, and can be credited with a good part of the final result. In 2013–2014, the team gained some visibility, when it was realised that their e/γ calibration studies could help to resolve the discrepancies in the Run-1 measurements of the Higgs boson mass and consolidate the measurement of that parameter as well.

The measurement converged in the course of 2016, and a convincing and competitive result with an uncertainty of 19 MeV, the first measurement of the W-boson mass obtained at the LHC, was released to great fanfare. But the job wasn't done yet, and the discussions with the journal took another year to conclude. Finally, in February 2018, the result was published and took its place in the long history of measurements of this fundamental parameter.

This first result is only a start of course, and much more remains to be done to reach a final objective of an uncertainty below 8 MeV, the precision of current indirect measurements. Since the main pitfalls can now be anticipated, the next round should be a little faster. But how fast? Time is running, already.

Chapter 10

Towards the High-Luminosity LHC

10.1. Setting the stage

The most important LHC parameters for the ATLAS physics programme are the collision energy and the integrated luminosity. Increasing either or both of these increases the rate of rare processes, which ATLAS explores. An upgrade programme of the LHC and its injectors is in place in order to increase both the energy and the luminosity, employing the periodic 2–3-year-long LHC shutdowns. The full capability will come with the so-called high-luminosity LHC, HL-LHC, operation, which will run for a decade and accumulate $3000\,\mathrm{fb}^{-1}$ or more, at least ten times the originally planned integrated luminosity. ATLAS likewise uses the long shutdowns to consolidate and upgrade the detector in order to maintain and improve its performance, to replace radiation-damaged or aged components, and to cope better with luminosity well beyond the design. The capability of ATLAS to search for heavier particles and rarer physics processes, and to make higher precision measurements of the Higgs boson and other Standard Model particles, is substantially improved by these upgrades.

10.1.1. *LHC and ATLAS upgrade plans*

The long-term schedule for LHC operation is outlined in Fig. 10.1. The first long shutdown ("LS-1") took place in 2013–2014. For the LHC, LS-1 was used to reinforce the superconducting interconnects between the magnets and to consolidate the protection systems, allowing the collision energy to be raised safely from 8 TeV in 2012 to 13 TeV in 2015–2018 ("Run-2"). For ATLAS, the major upgrade during LS-1 was the installation of a fourth pixel detector layer closer to the collision point than the original pixel layers, and with better position resolution. Before LS-1 the LHC had reached about

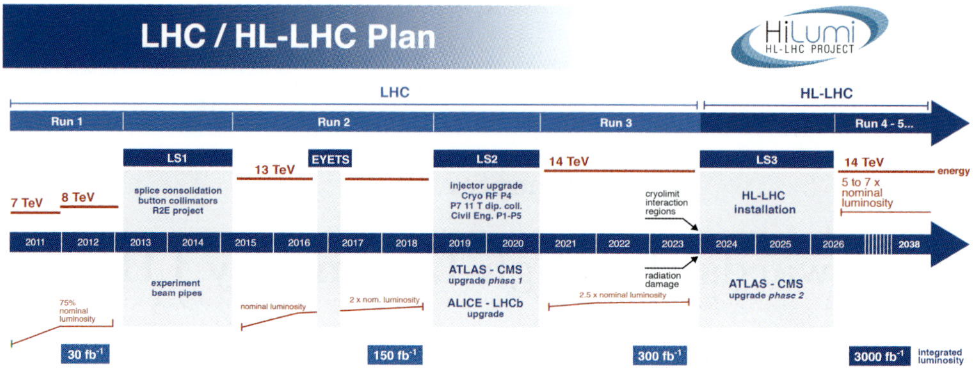

Fig. 10.1. Plan for LHC operation towards the upgrade for the High-Luminosity LHC.

80% of the design luminosity, but was able to double it in Run-2, thanks to ingenious use of both the LHC injectors and the LHC.

The second long shutdown, LS-2, is scheduled for 2019–2020. It will be used to increase the brightness of the beam delivered by the injectors to the LHC by increasing both the number of protons and their density. The maximum luminosity in the LHC will still be limited to about twice the original design due to cooling limitations in beam-focussing magnets located close to the experiment. However, these upgrades will allow the LHC to run at the same high luminosity for long periods (hours) through a process called luminosity levelling, so increasing the data accumulated per day. It is also an important milestone towards preparing for HL-LHC collision rates. In LS-2, ATLAS will make a major upgrade to its trigger system to increase its selectivity. The inner part of the endcap muon detector is also being replaced, to provide increased trigger capabilities at the expected HL-LHC particle fluxes.

The third shutdown, LS-3, is scheduled for 2024–2026 and aims to complete the HL-LHC and detector upgrades. This is a major upgrade for the LHC where the magnets near ATLAS will be replaced to provide a stronger focussing of the beams, giving higher luminosities. In addition, the protection and beam dump systems will be upgraded to handle larger proton intensities. These upgrades will potentially allow luminosities up to twenty times the LHC design, but it is foreseen that the luminosity will be levelled at 5 to 7.5 times the design luminosity to provide better experimental conditions. All of the inner tracker will be replaced during this shutdown, to increase the detector granularity to handle the higher pile-up and to increase the radiation hardness. The read-out electronics and data-acquisition system of almost all systems will also be replaced to cope with the much higher trigger and data rates. The trigger system in turn will be upgraded and expanded to facilitate more advanced algorithms needed for high pile-up. With these upgrades, it is expected that ATLAS will accumulate up to six times more integrated luminosity per year than in the best year so far (2017).

10.1.2. *ATLAS performance at the High-Luminosity LHC*

Reconstructing what happens in a collision at the HL-LHC in order to make physics measurements will be challenging due to the very high pile-up. At the ultimate luminosity, 7.5-times design, an average of 200 proton-proton collisions are expected each time the bunches cross, i.e., every 25 ns. At most, one of the 200 collisions will produce particles of interest for the vast majority of the physics programme: the rest effectively add noise and confusion.

One powerful handle on the pile-up is the separation of collisions along the beam-line. The collisions in any single bunch crossing will be spread out over a distance of more than 10 cm, with an average separation between neighbouring collisions of around 1 mm. The upgraded inner tracker will be able to reconstruct the path of charged particles much more precisely than that over a large part of its angular acceptance, and so can remove charged tracks coming from pile-up. The calorimeters, used to detect uncharged particles and measure the energy of photons, electrons and jets, have little capability on their own to reduce pile-up, but much can be gained by careful use of tracking information together with calorimeter energy deposits. Furthermore a new timing detector is planned for installation in the detector endcaps, which will provide separation of particles from different collisions in a bunch crossing on the basis of their timing.

The expected performance of the upgraded detector to reconstruct, identify and measure different particles and jets in HL-LHC conditions has been studied in some depth. In many cases, performance can be obtained that is almost as good as, or even better than, the original detector at design luminosity. For example, the charged particle momentum will be measured almost a factor of two more precisely, and the efficiency to identify jets which contain B-hadrons (so-called b-jets) improves, in both cases due to the better tracker granularity. The upgraded tracker also allows the identification of b-jets over a wider angular range. For quantities that rely mostly on the calorimeter information such as photon energy or the missing transverse energy, a degradation of up to a factor of two due to pile-up may be seen.

10.1.3. *The ATLAS physics programme at the HL-LHC*

The discovery of the Higgs boson, H, with a mass of 125 GeV, by ATLAS and CMS opened a new chapter in physics at the TeV energy scale. The consistency, so far, of the measured properties of this new object with Standard Model (SM) predictions suggests that this is the last missing piece of the SM. Moreover, no observation of physics beyond the SM has yet been found. This has weakened the case for theories beyond the SM at the TeV scale. However, puzzling observations, some astrophysical, indicate that the SM is not the final theory describing particles and forces, but may be a low-energy "effective" theory contained in some larger, as yet unknown, framework. Physics beyond-the-SM must come in at some higher energy scale that might be directly accessible at the LHC, or that may produce deviations from SM predictions when performing high-precision measurements at the LHC or elsewhere. A balanced strategy of direct searches for new physics and high-precision measurements is key to testing the SM at the energy scale offered by the HL-LHC.

The Higgs sector is a completely new domain containing the only known elementary scalar particle, H. Its study in all possible ways with the highest achievable precision is of paramount importance. The couplings of this particle to the SM fermions and bosons must be investigated. Some new physics models suggest couplings differing from the SM by the order of 10%. First estimates suggest that precisions of the order of 4% are attainable by ATLAS with a 3000 fb^{-1} dataset on the vector boson couplings to the H, while for fermions, an accuracy of around 8% is possible. In practice, with the large data samples, significantly higher precisions may be possible. Very large luminosities are crucial to measure well rare decays of the H, such as $H \rightarrow \mu^+\mu^-$. This decay is of great importance, as it probes the Higgs boson coupling to second-generation leptons. An accuracy of about 10% looks feasible. Other rare decay modes, such as the process $H \rightarrow J/\psi\gamma$, allow constraints on the H coupling to second-generation quarks. Figure 10.2 shows the expected accuracy on so-called "reduced coupling scale factors" as a function of the mass of particle i.

The study of the Higgs potential, which affects how the Higgs boson interacts with other (identical) Higgs bosons (its "self-coupling"), represents a key test of the Standard Model description of the electroweak symmetry breaking mechanism. This can be probed by studying production of pairs of Higgs bosons (HH). The HL-LHC can pioneer this challenging work, as it is expected that the HH rate can be measured with a significance of

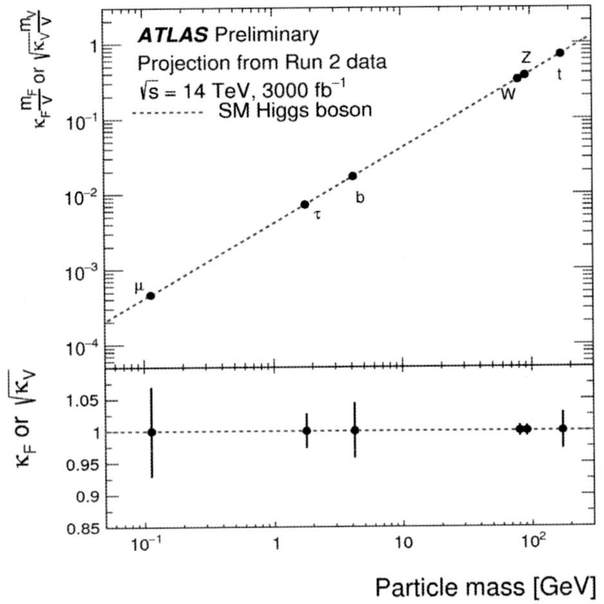

Fig. 10.2. Expected results for reduced coupling scale factors of the Higgs boson to SM fermions, F, or vector bosons, V, as a function of the particle mass, assuming $3000\,\text{fb}^{-1}$ of 14 TeV data and a 125 GeV Standard Model Higgs boson. The diagonal, dashed line indicates the predicted mass dependence.

about three standard deviations (combining results from ATLAS and CMS). Precision measurements of the Higgs boson self-coupling will likely require a new particle accelerator facility after the HL-LHC.

The very large datasets that will be produced at HL-LHC will provide a unique opportunity to probe many Standard Model processes with high accuracy and to test still unexplored sectors in this theory. Observation of deviations in data with respect to the SM would indicate the presence of new physics, at the LHC energy scale or even perhaps beyond. One of the best cases is represented by the so-called vector-boson scattering (VBS) process of the gauge bosons $V = W$, Z, where gauge bosons emitted from the colliding protons scatter off each other. This will provide a key opportunity to study the nature of the electroweak symmetry-breaking mechanism, as well as to probe for physics beyond the SM. It is still to be experimentally established if the Higgs boson preserves the unitarity of longitudinal VV scattering at all energies on its own, or if new physics processes are involved. Events with same-sign $W^{\pm}W^{\pm}$ produced in association with two forward, high-p_T jets, provide one of the best experimental signatures to measure the scattering of two vector bosons. At the time of writing, ATLAS has recently observed $W^{\pm}W^{\pm}jj$ production, using 13 TeV data. At the HL-LHC, the upgraded detector should measure the production cross section at the level of ±4%.

As well as exploring the Standard Model, including the Higgs sector, with its sensitivity to beyond-SM physics, the HL-LHC also increases the opportunities to directly discover new particles and phenomena, ranging from modified Higgs sectors, to searches for dark matter candidates and for new resonances. The discovery potential for new states with low production rates will often strongly benefit from the factor ten increase in luminosity. The electroweak production of neutralino-chargino pairs predicted by SUSY theories provides one such striking example, where the HL-LHC data should increase the mass reach by about 50%. Other scenarios with low rates to conventional and observable final states include SUSY models such as "split supersymmetry", or models with small mass differences between new particle states.

10.2. Improvements during the first long shutdown, LS-1

The first upgrades to ATLAS were made during LS-1 (2013–2015), together with a broad programme of maintenance and consolidation of detector systems. The most significant upgrade was the installation of a new innermost pixel-sensor layer allowing improved tagging of b- and c-flavoured hadrons and tau leptons. A further important change was to the architecture of the High-Level Trigger (HLT): a three-level trigger system was used in Run-1, but the second-level and event filter stages were merged during LS-1 to give a two-level system (known as the first-level, "Level-1", trigger and HLT). Activities aimed at consolidating and improving the reliability of the detector included: improvements to the Inner Detector read-out, refurbishing calorimeter power supplies and electronics, installing new muon chambers in the barrel-endcap transition region and in the feet of the detector, installing an improved LUCID luminometer, and additional improvements to the first- and high-level triggers. Some of these have been touched upon in Chap. 5. The combination of several of these improvements enabled the maximum first-level trigger rate to be increased from 75 to 100 kHz, and to run with larger occupancies coming from the higher pile-up.

10.2.1. *The Pixel Detector upgrade*

The original plan for the Pixel Detector, well before the start of LHC data taking, was to extract and replace the inner layer, located approximately

5 cm from the beam-line, after a few years of running. However, after a detailed study in 2008, this strategy was changed in favour of an ambitious but appealing proposal: namely, to insert a new, additional, innermost layer inside the existing Pixel Detector, thus reducing the beam pipe diameter. This new layer is known as the "IBL" (Insertable B-Layer) and was adopted as a Phase-I upgrade project in 2009. The primary physics driver was to enhance the lifetime tagging of b-flavoured hadrons. The IBL project proceeded rapidly, given the short time before LS-1, and included the new beam pipe and its interface as well as many parallel activities for construction, integration and finally installation.

The IBL upgrade project was possible thanks to the reduction of the inner diameter of the central beam pipe from 58 to 47 mm, with a wall thickness of 0.8 mm. The new beam pipe has a beryllium section 7.1 m long, positioned symmetrically around the interaction point. At the ends, aluminium flanges 10 cm long are welded on, to connect to the next beam-pipe sections. It remains challenging to produce a beryllium vacuum chamber of such a size that will meet the stringent requirements of the LHC beam vacuum. Beryllium is the best choice for the interaction region because of a combination of transparency to particles, high specific stiffness and compatibility with ultra-high vacuum. The beam pipe material was carefully measured, with a few percent precision, by X-ray absorption using a Cd^{109} source. As a result, the total thickness of the beam pipe was just 0.32% radiation lengths (X_0) thick at the centre of the experiment.

The IBL provides a new, high precision, track measurement close to the interaction point, improving the track extrapolation back to the interaction vertex. This is particularly important for low-momentum particles, where multiple scattering effects are more important. In addition, a smaller pixel pitch of the IBL (250 μm) parallel to the beam axis improves the resolution of the vertex position in the longitudinal coordinate, over the full transverse-momentum spectrum.

The IBL is a relatively small detector with a little less than 0.2 m^2 of silicon sensors supported on 14 staves surrounding the beam pipe at an average radius of 3.3 cm from the beam axis. The instrumented length is 64 cm. The IBL was designed so that extraction and insertion is possible without removing the other layers of the Pixel Detector. The radial envelope is less than 10 mm between the beam pipe and the inner support tube of the Pixel Detector, within a 7 m long enclosure. The staves and services are supported along an inner positioning tube, providing an independent mechanical support and a sealed volume with respect to the beam pipe.

The services were laid with sufficient flexibility to cater for a stress-free thermal movement when lowering the temperature to $-40°$C. The staves are supported at the end by the inner positioning tube, while they are bonded together in the centre via a ring made of polyamide-imide, one of the thermo-mechanically most stable plastic materials.

In order to preserve the advantage of the extra layer for tracking, the material budget was targeted to be below $2\% X_0$ for tracks perpendicular to the beam. A value of $1.88\% X_0$ was achieved thanks to design features such as: thinned silicon read-out chip and sensors ($150\,\mu$m and $200\,\mu$m, respectively), a stave service bus made with its power line in aluminium instead of copper, the stave local support structure material choice, a thermo-mechanically optimised cooling pipe, and a thinner inner positioning tube carbon fibre structure in the centre of the detector.

The design, prototyping, production and integration phases took place on a relatively short timescale of approximately four years. This was made possible thanks to good connections between R&D groups and industrial partners, while iterating on the fabrication techniques to comply with the targeted quality assurance. Some novel technologies and ideas were adopted to build the detector in a different way than the original Pixel Detector. For instance, the new front-end read-out electronics chip, the FEI4, was implemented using industry-standard 130 nm technology, and is able to cope with higher occupancy while having lower noise and better radiation hardness. Its active area is almost 6 times that of the original pixel read-out chip. This large area, in combination with the thin silicon layers for the front-end and the sensor, was a real challenge for the fine-pitch bump bonding.

After extensive qualification tests, excellent performance was obtained for two sensor technologies: planar, and 3D, silicon. For the latter, the production yield was not reliably established, and in order to gain experience also for future upgrades, it was decided to build a stave with mixed sensor flavours. IBL staves (Fig. 10.3) were built with 12 planar double-chip modules located in the central region covering the range $|\eta| < 2.5$, while four 3D single-chip modules are mounted at the ends, extending the coverage at up to $|\eta| = 3$.

The complete IBL detector was installed into ATLAS in spring 2014 (see Fig. 10.4). Many technical challenges were dealt with during the construction and installation, providing valuable experience toward the full tracker upgrade for the HL-LHC.

Fig. 10.3. (Top) The final IBL stave being added onto the inner positioning tube. The lower pictures show a planar-sensor module (left) and a 3D-sensor module (right).

Fig. 10.4. Insertion of IBL, together with the beam pipe, inside the pixel inner support tube.

Fig. 10.5. (Left) A new service quarter panel. (Right) Re-installation of the original Pixel Detector.

10.2.2. *New read-out services for the Pixel Detector*

While operating the Pixel Detector during Run-1, a number of defects developed, giving more than 5% dead modules, due to electrical or optical communication failures. In order to mitigate the weaknesses revealed, it was decided to remove the Pixel Detector from ATLAS in LS-1, to replace its read-out services, and to reinstall it. This required the design and construction of so-called "new service quarter panels" (nSQP, see Fig. 10.5) to displace the optical transceiver boards, called opto-boards, outside the inner tracker main volume, to more accessible locations. In addition, anticipating that a bandwidth limitation could arise in the inner layer of the original Pixel Detector at high luminosity, it was decided to install additional cables and fibres to allow the data transmission rate to be doubled. This proved essential for Run-2 operation, as the LHC well exceeded its design luminosity.

The Pixel Detector was extracted and moved to the ATLAS surface building clean rooms (SR1) for about a year. The nSQP were installed and the Pixel Detector fully re-commissioned before re-insertion inside ATLAS at the start of 2014, before the IBL insertion in spring (see Fig. 10.5). With this major consolidation of the Pixel Detector, the system attained a coverage of 98% for Run-2.

10.3. Detector upgrades for increasing luminosities, Phase-I

The outstanding luminosity delivery of the LHC means that peak collision rates above LHC design were already achieved in 2017 and have become routine. Correspondingly, high levels of pile-up, twice the design level or more, are also being delivered. The discovery of the 125 GeV Higgs boson, and at the time of writing, no higher-mass new states, pushes strongly to maintain relatively low transverse momentum inclusive triggers. A series of upgrades is underway, mainly for installation in LS-2, to provide additional selectivity in the trigger. This is being done through improvements at the first-level trigger via new electronics for the calorimeter trigger chain, the introduction of a hardware track-finding processor FTK at the input to the High-Level Trigger, a new detector system — the New Small Wheels — replacing the original muon small wheels for forward muon triggering, and the use of geometric properties in whole-event information ("topological triggering"). In addition, some barrel muon chambers located in the barrel-endcap transition region will be replaced, to incorporate new trigger chambers. The calorimeter and topological triggering, the FTK and the largest Phase-I upgrade project (the New Small Wheels) are each discussed in the following.

10.3.1. *First-level calorimeter and topological trigger upgrades*

A ten-fold increase in granularity of the first-level trigger signals from the LAr Calorimeters is being implemented in Phase-I, with the signals being digitised on the detector, rather than summed as analogue signals and sent off-detector as in the original chain. The improved granularity is schematically illustrated in Fig. 10.6. The new system will also have improved energy resolution at the trigger level. Together, these will allow more

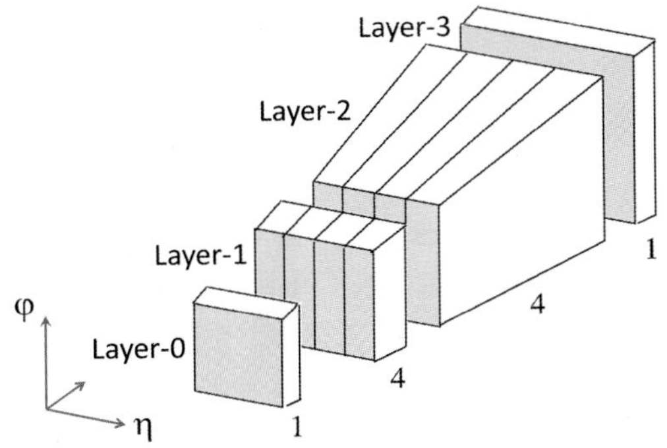

Fig. 10.6. Granularity of new trigger outputs from the LAr Calorimeter: all of these signals were summed into a single 0.1 × 0.1 (η,φ) tower in the original system.

sophisticated selections to be made at the first trigger level, for example, on the activity around electron and photon candidates ("isolation"). The provision of the new trigger input signals is implemented via new electronics cards (installed in the front-end crates mounted on the detector, with a new crate backplane requiring delicate replacement work) and new off-detector digital processing electronics. A first "demonstrator" system of the LAr Calorimeter on-detector part was already installed in LS-1 and the new full system is being prepared for installation in LS-2.

The new fine granularity digital signals from the LAr Calorimeter are sent on optical fibres to new trigger processor boards, sited in the off-detector counting room. Digitised energy sums from the Tile Calorimeter are also combined there, and a large optical plant regroups, duplicates and routes the optical signals to the right instances of the trigger processing cards, each dealing with a specific geometrical region of the calorimetry. There are three types of these processing units, known as "feature extractors" or "FEXes", searching for different trigger object types — for electrons, photons and taus, or hadronic jets, or global event quantities, dubbed, respectively, the "eFEX", "jFEX" and "gFEX". The three types of FEX each takes input signals with different granularity: the most fine-grained algorithms being run on the eFEX. Each of the FEXes is equipped with large field-programmable gate arrays (FPGAs), utilising the major increase in capacity of these devices since the time of the original trigger design. Prototype eFEX and gFEX electronics cards are shown in Fig. 10.7. At the time of writing the FEXes are in the late prototype or pre-production stages.

Fig. 10.7. Prototype eFEX (left) and gFEX (right) cards for the Phase-I Level-1 calorimeter trigger upgrade.

The original first-level calorimeter trigger electronics, the first-level muon trigger electronics, and the eFEX and gFEX cards, are all designed primarily to count the number of triggering objects (electrons, jets, etc.) passing each of several specific energy thresholds: i.e., finding whole-event multiplicities for specific-threshold triggers. However significant further discrimination power can be obtained at this first hardware trigger level for certain physics processes by combining the multiplicity information with geometrical information about the locations of the triggering objects. The topological trigger processor is the Level-1 system that forms such composite triggers, using calorimeter trigger and muon trigger information. An initial topological trigger processor has been commissioned during Run-2, using information from the original, coarser grained, calorimeter trigger. An updated version using the FEX-based trigger objects is being designed for Run-3 and will also use the full first-level muon trigger information.

10.3.2. *A fast-tracking system, FTK, for the trigger*

The fast tracker (FTK) at the trigger level improves the capability of ATLAS to deal with higher LHC luminosity and hence larger pile-up. At higher luminosity it becomes essential to be able to reconstruct precisely, at the High-Level Trigger, the trajectories of charged particles (hereafter referred as tracks), to efficiently select potentially interesting events from all those accepted by the first-level trigger.

For each accepted event, a series of particle tracks must be reconstructed from the individual digitised signals (hits) caused by the charged particles as they pass through the Inner Detector. As the number of hits increases, so too does the complexity of selecting and combining them, and it becomes prohibitively expensive, both in terms of processing times and resources needed, to perform extensive particle track reconstruction on standard processors.

The FTK is a custom electronics system that implements massively parallel processing to reconstruct particle tracks and provide accurate information to the high-level trigger computers, to be used in the event selection algorithms. The key requirements on the system are the capability to produce results in a limited time budget (latency) as well as with good precision, to enable efficient event selection at the trigger level. The system is designed to use the data from twelve layers of the Pixel and SCT Detectors, accessed in two stages. In an initial stage the hits from eight layers are analysed to identify track candidates and to perform a first fit to them. In the second stage the derived tracks are extrapolated to the additional four layers to check whether corresponding hits are found, and a more precise fit is performed on the complete series of hits to derive the track parameters with ultimate precision.

In a standard processor-based approach, the most computationally intensive task would be the identification of track candidates (i.e., series of hits belonging to the same particle trajectory) because of the very large number of hit combinations that should be considered. The complexity of this process would also increase dramatically with pile-up. To cope with this challenge the FTK carries out this selection using an associative memory system (AM) that compares almost simultaneously the hits from an event with hundreds of millions of preloaded patterns (also referred to as roads), corresponding to the various combinations of hits associated to possible tracks in the detector. The more precise this selection should be, the larger the number of patterns to be compared needs to be, and ultimately the larger the size of the AM system will be. The design is thus optimised considering the requirements of good precision and affordable cost.

The FTK is organised as a set of 64 independent engines, each working on a different region of the Inner Detector. The tracker volume is divided in towers, 16 azimuthally by four longitudinally. To avoid inefficiency in detecting tracks at the boundaries of the volumes, an overlap is introduced in the definition of the regions. The complete system implementation comprises several hundreds of custom boards consisting of five types, some of which also house custom mezzanine cards. One type of board receives the hit data from the read-out system on clustering mezzanines and distributes the clustered data by tower, the second passes the hits through the Associative Memory chips, the third performs the first eight-layer fit, the fourth extrapolates tracks to the remaining four layers and performs the final fitting stage, and the last board formats the results and transmits them to the HLT. Figure 10.8 shows the board carrying the AM chips. At the time of writing, the FTK system is well into its commissioning phase.

10.3.3. *New inner endcap muon chambers — the New Small Wheels*

Running at higher luminosities brings the challenge of an increasing rate of "fake" muon triggers in the forward regions. As material in the endcap toroids becomes activated, the big wheels of the Muon Spectrometer (MS) are bombarded with neutrons, creating many tracks apparently originating

Fig. 10.8. Picture of an AM board. The 4 mezzanine cards house the individual AM chips.

from the interaction region. Most of these "fakes" could easily be vetoed in the first-level trigger by checking for a matching track segment in the small wheels, in front of the endcap toroids, but the original forward muon trigger used only the thin gap chambers (TGCs) of the big wheels. Unfortunately, the single layer of TGCs in the existing small wheels, designed to provide an azimuthal coordinate for precision tracking, does not have sufficient resolution in the radial bending direction to match big wheel track segments with the required milliradian resolution. Moreover, the precision detectors (MDT and CSC) were designed to operate at the original design luminosity and start to lose efficiency for luminosities above twice the design.

The solution is a complete replacement of the small muon wheels. The New Small Wheels (NSW) are designed with the same eight-fold symmetry as the precision MDT chambers found in the rest of the MS, allowing their relatively straightforward integration into the existing endcap optical alignment system. Among the many technologies proposed and considered for the NSW, two were selected. Unlike the rest of the MS, the NSW relies primarily on MicroMegas (MM) for precision tracking, and on precision "small-strips" thin gap chambers (sTGC) for triggering. Since it will not be possible to replace individual chambers once they are installed, redundancy

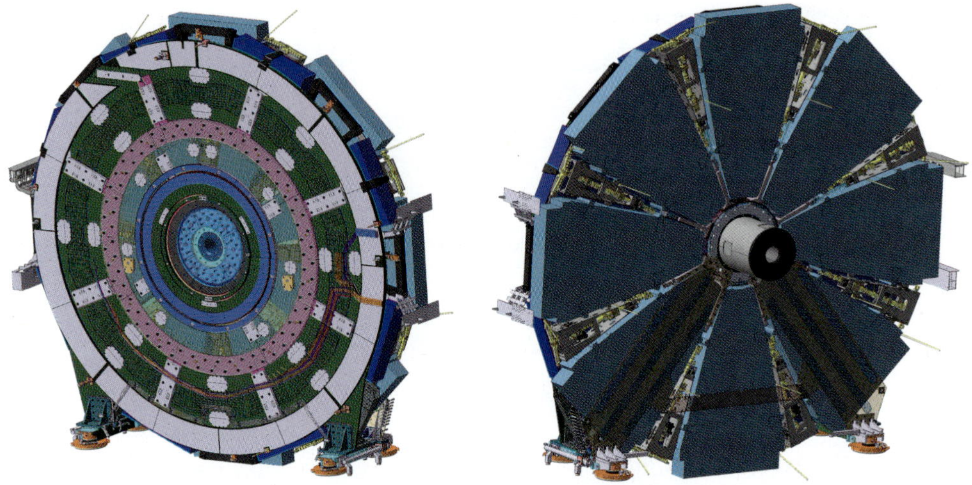

Fig. 10.9. The New Small Wheel design, showing sectors and arrangement of MM and sTGC.

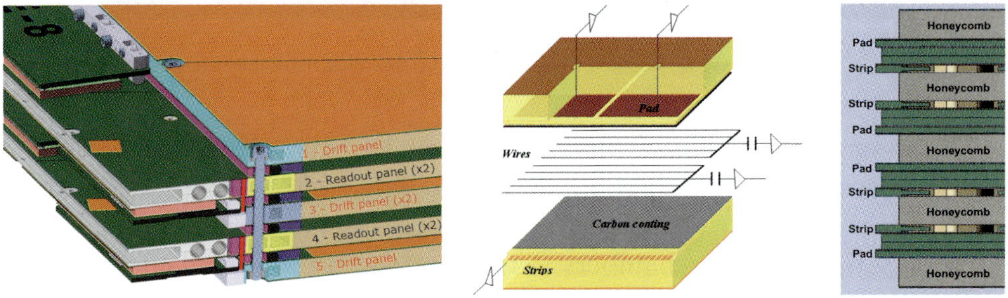

Fig. 10.10. Micromegas (left) and sTGC (right) quadruplet structure.

is an essential design element: both technologies are very fast, and are thus suitable for use in high-rate environments. Both can also give very high-resolution position measurements. Consequently, either detector could do a reasonable job filling in for the other, in case of need. Each NSW (Fig. 10.9) will consist of eight large and eight small wedge-shaped sectors, suspended from a framework mounted on the "JD" shielding disc. Both sTGC and MM are built as quadruplets of gas gaps (Fig. 10.10), providing rigidity and redundancy. Each sector comprises a central spacer frame with a wedge of two quadruplets of MM chambers mounted directly on each side, and wedges of three sTGC quadruplets mounted outside the MM, like sTGC bread on an MM sandwich. Overlaps between layers and sectors eliminate dead regions. A muon traversing the NSW always passes through at least eight sTGC gas gaps and eight MM gas gaps, allowing milliradian pointing to both the interaction point and big wheel hits at the first-level trigger, giving the required fake-muon rejection rate, and reconstructed track resolution better than $100\,\mu$m, comparable to the MDTs, but covering $|\eta| < 2.7$ in a high-rate environment.

MicroMegas are a new technology for ATLAS and have never been produced on this scale anywhere. However, the ATLAS team has been researching and developing MM technology since 2008. The MM for the NSW incorporates resistive strips for spark suppression, and a novel floating mesh attached to the drift panel and offset from the read-out panels by tiny pillars that facilitates construction of very large-area panels. The MM production chain separates the assembly of read-out and drift panels, and the stretching of the mesh on the latter, as well as their final assembly into full gaps and quadruplets. Quadruplets consist of two layers of double-sided read-out panels, with a double-sided drift panel between them, and sandwiched between two single-sided drift panels. They come in four sizes: two for large sectors and two for small. Each is made at different sites, in different countries. One of the greatest challenges in bringing both technologies to production was finding printed circuit board manufacturers to supply the read-out boards and implement non-standard PCB manufacturing. Because MM quadruplets are so large, the read-out and drift panels are assembled from several precisely aligned 60 cm wide printed circuit boards (PCBs), the widest size readily available in industry. The requirements are extraordinary: the read-out panels are covered by a Kapton layer with screen-printed resistive strips, which is in turn covered in tiny pillars, $128\,\mu$m high, to offset the mesh where high voltage is applied, and these must be produced by etching away a coverlay laminated to the PCB.

Two of the five countries producing the six sizes of sTGC quadruplets also made TGCs for the original small and big wheels, while the others started from scratch. However, even for the veterans, the switch to building high-precision chambers entailed many changes in equipment and procedures. For every read-out strip to be locatable within tens of μm, not only did the strips have to be precisely machined or etched, but they also have to be very precisely positioned with respect to the alignment features at the outside edge of the chambers that connects the alignment of the inside of the chamber to the outside world and every other layer in the quadruplet. All assembly steps are performed on flat granite tables, equipped with vacuum systems for gluing, and positioning pins for alignment of gas gaps as they are assembled into quadruplets. Moreover, every board is sprayed with a resistive graphite-resin mixture under carefully controlled temperature and humidity conditions, and hand-polished to obtain a uniform resistivity. The cathode boards are trapezoidal multilayer printed circuit boards measuring up to 2 m in length and over a metre wide, with read-out strips that must be laid out to a precision of 75 μm. Manufacturing such boards requires specialist expertise, and extensive exchanges of personnel and expertise were needed to produce boards with the required precision and accuracy. The manufacturing is being partly done by a specialist company and partly by a consortium of others, including the original-detector TGC PCB supplier.

To increase the interoperability of the two technologies, it was decided to use common electronics, developed by a third set of institutes, who are developing and testing the common ASICs and dedicated front-end boards for the sTGC and MM.

Once chamber quadruplets of either type arrive at CERN, they will be assembled into wedges and integrated into sectors of the NSW. This must be done with the same precision as the quadruplet assembly. Preparation of tooling and facilities at CERN to perform the sector assembly began in 2015. In parallel, a new set of alignment bars is being made — these are necessary to establish precise alignment of the chambers. Whereas the original small wheels have eight bars each, the NSW have sixteen, as each sTGC wedge is a separately alignable object and can move independently with respect to the central spacer frame holding the MM. Electronics integration and testing are also done at CERN: quadruplets are tested with X-rays and cosmic rays in the countries where they are built, using test electronics, and then shipped to CERN, where final front-end boards can be installed and tested, along with other wedge services and cabling. As the sectors are ready, they will be mounted on the wheel structures (see Fig. 10.11).

At the time of writing, the NSW project has entered its production phase, with the goal being to install the wheels into ATLAS in LS-2.

10.4. Major detector upgrades for HL-LHC, Phase-II

The largest changes to the ATLAS detector come with the Phase-II upgrade, in readying the detectors for the challenging environment during the HL-LHC period. The HL-LHC will operate at an instantaneous peak luminosity of up to 7.5 times the design luminosity, for a total integrated luminosity of 3000 to 4000 fb^{-1}, ten times more than expected before this upgrade. As a consequence, the detectors will have to handle higher particle densities, leading to increased occupancies in the detector components and

Fig. 10.11. NSW support structure and shielding disc assembly in a surface building at CERN.

higher radiation levels. The number of pile-up events per collision will increase up to 200. The primary detector challenges in such an environment are to maintain the excellent performance, in particular the vertex and track reconstruction, lepton identification and heavy flavour tagging. These will be addressed via three major detector upgrades: a complete replacement of the inner tracking system, new radiation-tolerant read-out electronics using state of the art technologies for the tracking, calorimeter and muon detector systems, and an upgraded Trigger and Data Acquisition (TDAQ) architecture that will cope with the increasing rates. These upgrades are necessary to offset the effects of high rates and radiation damage imposed by the HL-LHC environment and are designed to fully exploit the physics potential. All upgrades have to fit existing geometrical envelopes and re-use (as far as possible) the existing ATLAS infrastructure to reduce cost.

10.4.1. *The inner tracker upgrade (ITk)*

The ATLAS inner tracking system plays an essential role in the reconstruction and identification of leptons, photons, hadronic decays and in the tagging of b-jets. The key role of the tracking detector in the physics programme becomes even more pronounced during the high-luminosity phase. The current tracking detector has been designed for ten years of operation at a peak luminosity of $10^{34}\,\mathrm{cm^{-2}s^{-1}}$ (the LHC design value) with 25 pile-up events per 25 ns bunch crossing, a first-level trigger rate of 100 kHz, and a radiation tolerance equivalent to an integrated luminosity of $700\,\mathrm{fb^{-1}}$ for the SCT and $850\,\mathrm{fb^{-1}}$ for the newly inserted IBL. For an optimal tracking performance up to $4000\,\mathrm{fb^{-1}}$, new and much more radiation-tolerant, sensing elements and read-out electronics are required. Due to a factor four higher track density, the granularity of the tracking detector has to be increased to keep the occupancies no more than 1%. The design of the layout of the new tracking detector has been a large effort over several years.

The baseline design for the new tracking detector ("ITk"), shown in Fig. 10.12, consists of an all-silicon active-element detector with pixel sensors at the inner radii around the beam pipe, surrounded by strip sensors. The Strip Detector has four barrel layers and six endcap petal-design disks, both equipped with modules on both surfaces of the layers, each with a small stereo angle to add $z(R)$ resolution, respectively. The Strip Detector, covering $|\eta| < 2.7$, surrounds a five-layer Pixel Detector extending the coverage to $|\eta| < 4$ to also provide measurements for particles at a shallow angle to the beam. In addition, and due to the harsh radiation environment

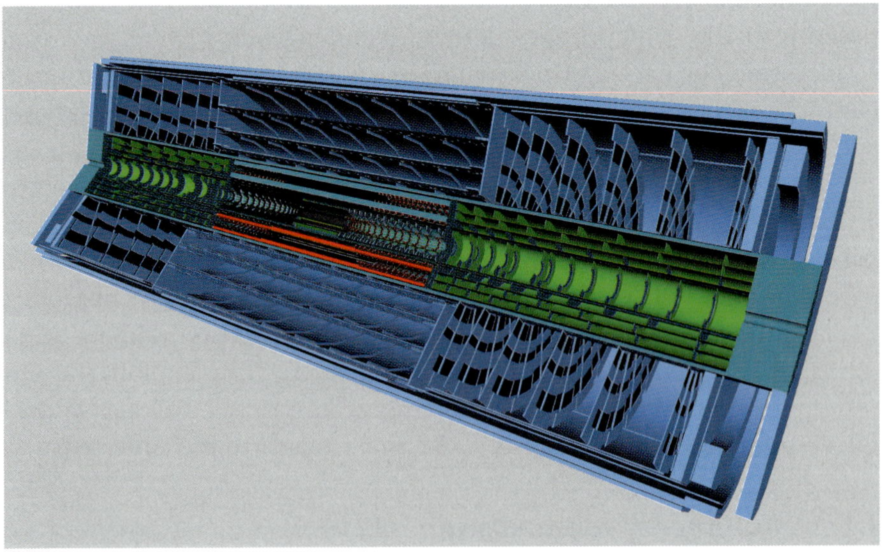

Fig. 10.12. The layout of the new all-silicon tracking detector, ITk, for operation at HL-LHC.

expected, the inner two layers of the Pixel Detector, which will be exposed to the largest radiation fluences, equivalent to 10^{16} 1 MeV-neutron equivalent per cm^2, are replaceable. The layout has been arranged to maximise the length of the trajectory of the particles inside the solenoid, providing a total of 13 hits for $|\eta| < 2.6$. The ITk is planned to have about half as much material as the present tracker, thereby minimising the effects of losses due to hadronic interactions and Bremsstrahlung.

Based on experience during the construction of the current SemiConductor Tracker (SCT), simplicity in the design of the Strip Detector was pursued to ease the production of about 18,000 modules. Modules are glued directly onto CO_2 cooled carbon fibre planks that are designed to reduce material in the detector. These mechanical support structures are called "staves" for the barrel section and "petals" for the endcap section. The advantage of this concept is that all components are independently testable prior to assembly, thus allowing for early systems-like testing.

The four barrel layers consist of 392 staves with modules on the top and bottom side of the staves, each populated with 28 silicon strip modules. In the endcaps each disk is populated with 32 identical petals. Each petal has nine modules on each side with six different sensor geometries to cover the wedge-shaped petal surface. The choice of the petal geometry for the endcaps was driven by the decision to implement strips that point towards the centre of the beam pipe, thus allowing a constant φ coordinate along the strip length.

A strip module consists of a silicon sensor and 1–2 low-mass PCBs, called hybrids, hosting the read-out ASICs designed in 130 nm chip technology. The module design is equally focused on material reduction with the elimination of a substrate between the hybrid and the sensor compared to the SCT module of the current ATLAS tracker. The modules are constructed by directly gluing Kapton flex hybrids to silicon sensors. Various different strip lengths and geometries are foreseen, depending on the planned location of the module within the detector. A photo of a strip module prototype is shown in Fig. 10.13.

The layout of the ITk Pixel Detector is based on a short barrel part followed by inclined modules to cover the intermediate $|\eta|$-region. The high-$|\eta|$ region is instead covered by rings perpendicular to the beam direction. This arrangement covers the pseudo-rapidity region of $|\eta| < 4$. The baseline concept for all layers is to use hybrid pixel modules comprising two parts: a high resistivity silicon sensor and a front-end read-out chip fabricated in CMOS technology. The two parts are joined using the high-density

Fig. 10.13. Prototype strip module with two hybrids and a power board.

flip-chip connection technique with solder bumps called "bump-bonding". This assembly is then glued onto a flex circuit providing the connections to the read-out and power distribution systems.

The foreseen pixel sensor technologies are developed to withstand the harsh HL-LHC radiation environment and to meet expected industrial production capabilities. Different types of pixel sensors, with different levels of radiation tolerance, will be used in the different parts of the detector: 3D sensors and thin (100 μm) planar sensors in the inner section, 150 μm planar sensors in the outer three layers and pixel endcaps, all to be bump-bonded to radiation-tolerant front-end electronics designed in 65 nm chip technology. The outer barrel layer may employ CMOS monolithic sensors if this technology is proven early enough.

In 3D pixel sensors, the charge collecting electrodes are columns etched into the sensor volume and oriented perpendicularly to the sensor surface. 3D sensors are intrinsically more radiation tolerant than planar sensors due to shorter drift distances. The feasibility to successfully use this technology in an LHC experiment was shown with the IBL. Planar pixel sensors were already used for the current Pixel Detector and further developed for use in the HL-LHC environment. While n-in-n technology was used for the current detector, the planar sensors will use n-in-p technology featuring an increased radiation tolerance and the simplification of the production process. An advantage of the n-in-p technology is single-sided processing and reduced handling complexity.

The same front-end chip is used throughout the entire Pixel Detector. It will be implemented in 65 nm CMOS technology, which has the gate density large enough to define and implement all the building blocks needed for an HL-LHC pixel detector with 40,000 pixel cells per cm^2. The chip size will be $20 \times 19.2\,\text{mm}^2$, and the assumed pixel cell size $50 \times 50\,\mu\text{m}^2$. A common effort between ATLAS and CMS was begun in 2013 to build a radiation-hardened front-end chip for the new Phase-II pixel detectors.

The mechanical system is made of local supports, organised in cylinders or rings and connected to a global support. The local supports provide mechanical support, alignment, routing of services and thermal management of the pixel modules. They are made of low mass, high rigidity and high thermal conductivity materials, including carbon-fibre laminates, pyrolytic graphite and other carbon-based composites. A small diameter titanium pipe is embedded in the structure to allow heat extraction via a CO_2 evaporative cooling system. The outer barrel local supports are "staves" that are positioned along the beam direction at various radii. The outer endcaps local supports are "rings" positioned perpendicular to the beam direction at different z locations, chosen to guarantee hermeticity. Each ring is composed of two carbon fibre skins with thermal conductive carbon foam in-between. The cooling pipe and the bus tape for power and data transmission are embedded in the carbon foam.

The production of the ITk Detector is scheduled to take place over a few years from 2020, starting with the strip modules due to the large number of parts. Installation and commissioning is planned in 2025 and the ITK is targeted to be ready for the HL-LHC start-up in 2026.

10.4.2. *The upgrade of the calorimeter read-out electronics*

The front-end electronics of both the LAr and the Tile Calorimeters will be upgraded for HL-LHC operation. The motivation is two-fold: firstly, the on-detector front-ends of both calorimeters will be exposed to significantly higher radiation levels, but several of the ASIC and commercial components, using now-obsolete technologies, were designed to sustain radiation levels only corresponding to a maximum integrated luminosity of 500–1000 fb^{-1}. Secondly, the existing read-out architecture is not compatible with the HL-LHC trigger requirements, specifically a read-out rate of 1 MHz and a latency (time from collision to trigger decision) of up to a maximum of $10\,\mu\text{s}$.

After a stage of amplification and pulse shaping, the detector signals for every calorimeter cell will be digitised at 40 MHz, synchronously to the bunch crossings, and the resulting digitised samples transmitted through fast optical links to processing units in the off-detector counting room, where the raw data will be converted to fully calibrated energies applying digital filtering in highly performant FPGAs. The upgraded read-out architecture will offer the advantage of providing full granularity and full resolution information to the hardware-based Level-0/Level-1 trigger system, significantly improving its performance.

The Phase-II upgrade of the LAr Calorimeter foresees the replacement of 1524 Front-End Boards (FEBs), installed on the detector in custom front-end crates, as indicated in Fig. 10.14. Each FEB processes 128 signals from the calorimeter cells, splitting the amplified signals into two overlapping linear gain scales and applying a bipolar (RC)-(CR)^2 shaping function. Off-detector, 372 LAr Signal Processor Boards will be installed in 8 racks in the underground electronics cavern, receiving the front-end data through approximately 31,900 fibres providing the signals to the trigger and DAQ

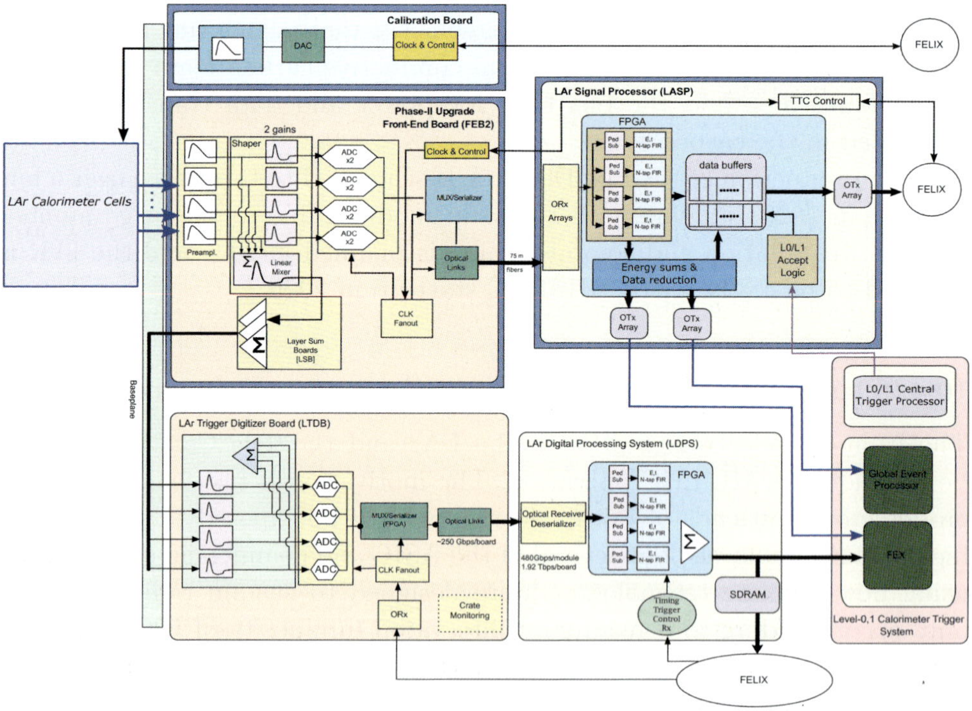

Fig. 10.14. Schematic of the Phase-II LAr Calorimeter read-out architecture. The LAr Trigger Digitiser Board (LTDB) and Digital Processing System (LDPS) are part of the Phase-I upgrade.

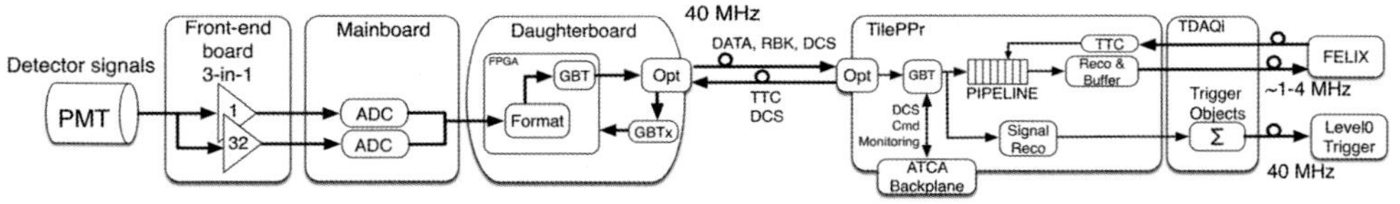

Fig. 10.15. Block diagram of the upgraded read-out of the Tile Calorimeter.

systems. In addition, calibration boards will allow the injection of accurate calibrated signals — with a precision better than 0.1% over the full 16-bit dynamic range — directly into the calorimeter cells.

For the Tile Calorimeter, the pulse of each photomultiplier (PMT) will be amplified and shaped in the front-end boards, digitised at the LHC bunch crossing rate on the main-boards, and then transferred to the off-detector Pre-Processor (PPr) modules using fully redundant optical links installed on dedicated daughter-boards, see Fig. 10.15. The off-detector PPr modules process and stream the data to the DAQ system and to the trigger processors through the TDAQi interface modules. A total of 4096 fibres, each running at 9.6 Gb/s bandwidth, i.e., for a total of 1.3 Tb/s throughput, will interface the on-detector front-end to 32 PPr modules.

An important aspect of the HL-LHC Tile Calorimeter upgrades is the reorganisation of the mechanical units that support the PMTs, the read out electronics and the distribution of the power, cooling and monitoring/control services. Each module will be serviced by four independent units, the "mini-drawers", easing the installation and the maintenance operations of the 12 PMTs and of the corresponding read-out electronics in any detector opening scenario, and minimising the impact of component failures on the overall ATLAS data quality.

10.4.3. *A new High Granularity Timing Detector in the forward region*

With the very high pile-up levels at the HL-LHC, it is challenging to associate tracks and composite objects to the correct primary vertex, important in extracting physics results. Tracking information is crucial for this, but is less powerful for forward-going tracks than in the bulk of the detector acceptance. However, the spread of collision times within a single bunch-crossing can also be utilised with high-precision timing information if a resolution of tens of ps can be obtained. To make use of this, ATLAS will install a precision timing system based on Low Gain Avalanche Detectors (LGADs) in the region $2.4 < |\eta| < 4.0$, where tracks will have poorer longitudinal impact parameter resolution. The baseline configuration of the High Granularity Timing Detector (HGTD) comprises two layers of LGAD pixelated sensors ($1.3 \times 1.3\,\text{mm}^2$) installed in the gap between the ITk and the endcap cryostat, for a total active area of $6.3\,\text{m}^2$ and approximately 3.5 million read-out channels. Beam tests of small-scale LGAD prototypes have demonstrated an intrinsic time resolution better than 30 ps for minimum-ionising particles.

10.4.4. *The Muon Spectrometer upgrade*

The main challenge for the Muon Spectrometer (MS) at the HL-LHC is to cope with the much harsher experimental conditions in terms of particle rates, pile-up and integrated radiation doses, while preserving its excellent identification, selection and tracking capability. The MS Phase-II upgrades include the installation of new Resistive Plate Chambers (RPC) with increased rate capability in the inner barrel layer (BI) to improve the acceptance and robustness of the barrel muon trigger. This will help address limitations under HL-LHC conditions of the existing RPC detectors, which will require running at lower voltage, reducing the gas gain and losing up to 35% efficiency in the worst regions. Adding new BI RPC chambers is challenging in terms of space, in particular in the small sectors of the MS, where the installation of the new BI RPCs is possible only by replacing the existing Monitored Drift Tubes (MDT) with new smaller diameter MDTs. Selected RPC chambers in the middle and outer barrel layers, in the areas of highest rate, i.e., at $|\eta| > 0.8$, will also be refurbished after LS-3, replacing electronics and read-out planes, so that the chambers can be operated at reduced high voltage without efficiency loss.

In the barrel-endcap transition region, new triplet layers of TGCs will replace the current TGC doublets. The triplets, with finer read-out granularity, will allow the implementation of a more robust majority logic in the trigger, with a smaller coincidence window, suppressing the rate of random coincidences generated by low-p_T charged particles produced inside the endcap toroid cryostats.

A large fraction of the front-end and on- and off-detector read-out and trigger electronics will be replaced to make them compatible with the higher trigger rates and longer latencies necessary for the new Level-0 trigger, discussed below. The trigger and read-out chain of the RPC and TGC chambers will be completely redesigned for improved performance. All data from each bunch crossing from the TGC and RPC detectors will be transmitted off-detector, and will be available for trigger processing. More powerful and adaptable algorithms will be implemented compared to the present ones based on simple coincidences. The muon trigger will also be enhanced by new MDT electronics, which will send the raw MDT data to dedicated processors that will make a precise calculation of the hit coordinates for use in the hardware trigger, resulting in sharper trigger efficiency turn-on curves.

10.4.5. *The trigger and data acquisition upgrade*

The Phase-II upgrade of the TDAQ architecture is motivated by the need to support the broad physics programme of the HL-LHC, at the much higher luminosities expected. The architecture design specifies, in its baseline configuration, a single Level-0 hardware trigger with a read-out rate of 1 MHz and a maximum latency of $10\,\mu$s, and is shown in Fig. 10.16. The Level-0 trigger decision is formed using information from the calorimeter and muon detector systems.

The Level-0 muon trigger comprises a range of new trigger processors: for muon candidates from the barrel RPCs and endcap TGCs, for muons in the NSW detectors, and for the newly available MDT information that can provide a precise measurement of the muon transverse momentum.

The calorimeter trigger processors built for the Phase-I upgrades will be maintained during HL-LHC operation, with an updated firmware matching the expected pile-up conditions and increased latency. The processors will be complemented by additional feature extractors to process information from the forward calorimeter, and by a Global Trigger system that can access full-granularity calorimeter information and muon information from the whole detector, allowing it to refine the trigger objects calculated earlier, calculate event-level quantities and execute whole-event algorithms that are more sophisticated than in today's trigger.

Following the Level-0 trigger decision, detector data are transmitted over point-to-point serial links to the Front-End Link eXchange (FELIX) nodes, the first element of the read-out system. FELIX provides a common interface between the detector-specific read-out links and a commodity multi-gigabit data network downstream. Along the network, data are received by the data handlers, implemented as commodity servers, where detector-specific processing, e.g., formatting and/or monitoring, can be implemented before being transferred to the dataflow system, whose main function is to transport, aggregate, buffer and compress event data for use in the Event

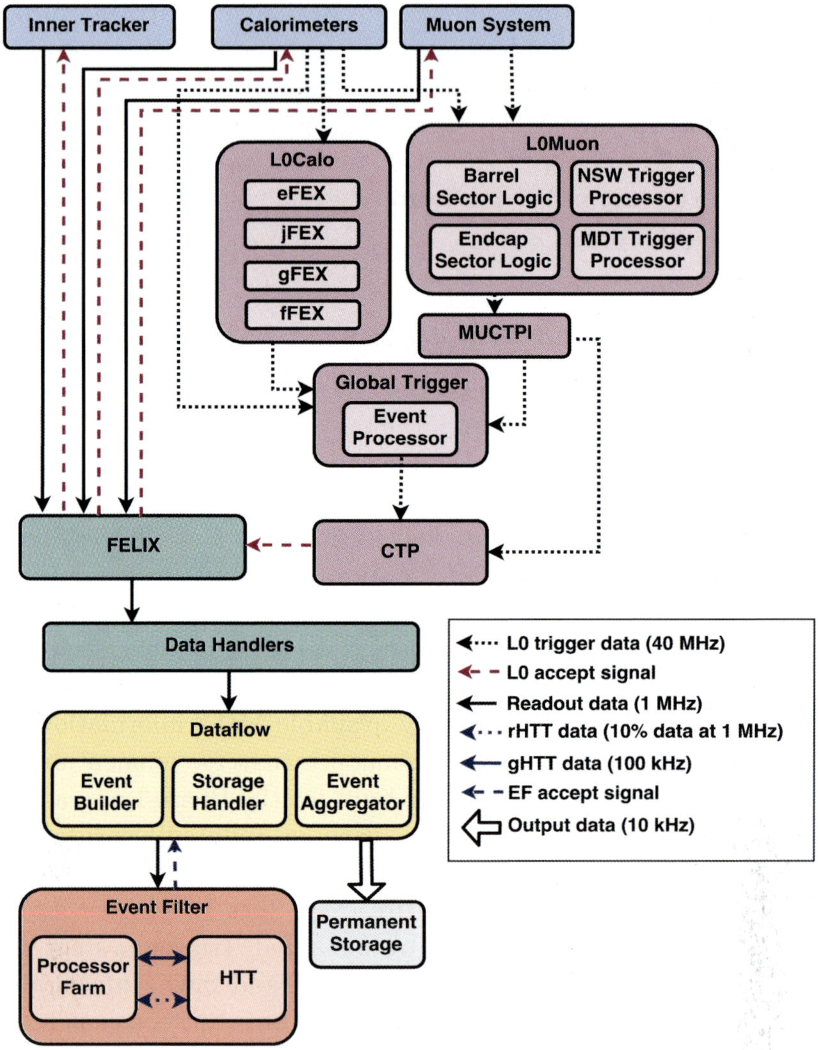

Fig. 10.16. Design of the Phase-II TDAQ architecture.

Filter (EF) system where the final trigger event selection is made. The read-out and dataflow system are designed to handle a total bandwidth of 5.2 TB/s.

The EF system will continue to employ commodity CPUs, and will be upgraded to match the detector upgrades and maintain the rejection by employing offline-type particle reconstruction. New selection software will exploit the increased computing power available. The EF farm will be aided by a custom-electronics hardware trigger-tracking (HTT) system. The HTT will, in the baseline configuration, implement two levels of tracking — an initial fast reconstruction of relatively high transverse momentum tracks ($p_T > 2\,\text{GeV}$) in regions around Level-0 triggered objects, and a global scan across the whole tracker volume reconstructing tracks with $p_T > 1\,\text{GeV}$. The former will run at the full 1 MHz Level-0 trigger rate, the latter at around 100 kHz, after further selection, and should produce tracks close to offline quality, suitable for b-jet tagging, soft-term calculation for the missing transverse energy algorithms, reconstruction of relatively soft jets, and pile-up suppression. Events selected by the EF are transferred to permanent storage. The final physics output rate can be up to 10 kHz, giving a potential aggregate data rate to storage of 50 GB/s.

Each element of the upgraded architecture is designed to be able to evolve to a dual-level hardware-based trigger architecture, in case pile-up conditions or physics requirements so dictate. In the two-level configuration, part of the HTT will operate within the Level-1 trigger and the reconstructed tracks will be combined with calorimeter- and muon-based trigger objects in the Global Trigger to further improve selectivity, and eventually form the Level-1 decision.

ATLAS: Readying for the future

While ATLAS is enjoying the excellent LHC Run-2 data-taking conditions, it is also getting ready for an even better version itself... For the high-luminosity phase of the LHC, ATLAS will improve or even replace many of its sub-detectors, to be ready for the challenge of collecting up to $4000\,\text{fb}^{-1}$ of data.

Some important milestones in this endeavour are the detailed documentation of the upgrade projects as Technical Design Reports (TDRs) and the approval of these documents both internally, as well as externally with the CERN LHC Experiments Committee. In total, six TDRs had to be approved within just one year — a huge task even for such a large collaboration as ATLAS.

The stages of producing any given TDR are very similar: a dedicated team starts with great enthusiasm collecting material, then setting deadlines for the various chapters, sending constant reminders that the first (then second, then third...) deadline is coming ever closer, postponing the deadline to give people just a bit more time to deliver, discussing — again — the structure of the document to make it better, but eventually lowering the expectation of what will "make it in", then breathing new life into old efforts to get the results, moderating lively exchanges about American English versus British English, investing a lot of time making the tables and figures beautiful — and in the end, collecting thousands of comments from the Collaboration making the document perfect. Many late-night hours went into the successful completion of these epic documents.

Then, there is the added complication that everyone has, naturally, a different opinion of what should be covered in such a technical document. In the case of the very first ATLAS TDR for the high-luminosity phase of ATLAS (ITk Strip Detector) — which also happened to be the very first TDR of all LHC experiments and thus the guinea pig — the reviewing committee was not too sure of what to expect and thus provided limited guidance. Hence, the ITk team wrote the document as it thought it should and ended up with an encyclopedia of more than 500 pages; a 1.8 kg book was delivered, much to the dismay of the reviewing body. The later TDR teams were then given more explicit guidance, to make the documents still portable. However, scientists like to document their work and so the final tally of all TDRs still was 2428 pages.

In total, 15,963 litres of coffee were needed to document the upgrade projects, resulting in a set of nicely coloured books vaguely reminiscent of Easter eggs.

Six colourful TDRs for ATLAS upgrades.

Chapter 11

ATLAS Collaboration: Life and its Place in Society

11.1. ATLAS Collaboration today

The ATLAS Collaboration in 2018 consists of ∼5500 physicists, students, engineers and technical staff from 183 institutions (including 2 international laboratories) and 38 countries (see Table 11.1). However, when considering the origin of the ATLAS members, the number of nationalities represented is closer to 100, as shown in Fig. 11.1!

The number of members who author ATLAS Collaboration papers currently total ∼2900, of whom 2400 are physicists. As can be seen in Fig. 11.2, the people that make up the ATLAS Collaboration are in large part (44%) PhD-holding physicists; however, 22% of the ATLAS members, i.e., 1200 people, are PhD students. In fact, the number of PhD students has more than doubled, as compared to the start of the data taking era in 2009,

Table 11.1. The composition of the ATLAS Collaboration in 2018.

The ATLAS Collaboration today				
Argentina	Armenia	Australia	Austria	Azerbaijan
Belarus	Brazil	Canada	Chile	China
Colombia	Czech Republic	Denmark	France	Georgia
Germany	Greece	Israel	Italy	Japan
Morocco	Norway	Poland	Portugal	Romania
Russia	Serbia	Slovakia	Slovenia	South Africa
Spain	Sweden	Switzerland	Taiwan	The Netherlands
Turkey	United Kingdom	United States of America	CERN	JINR

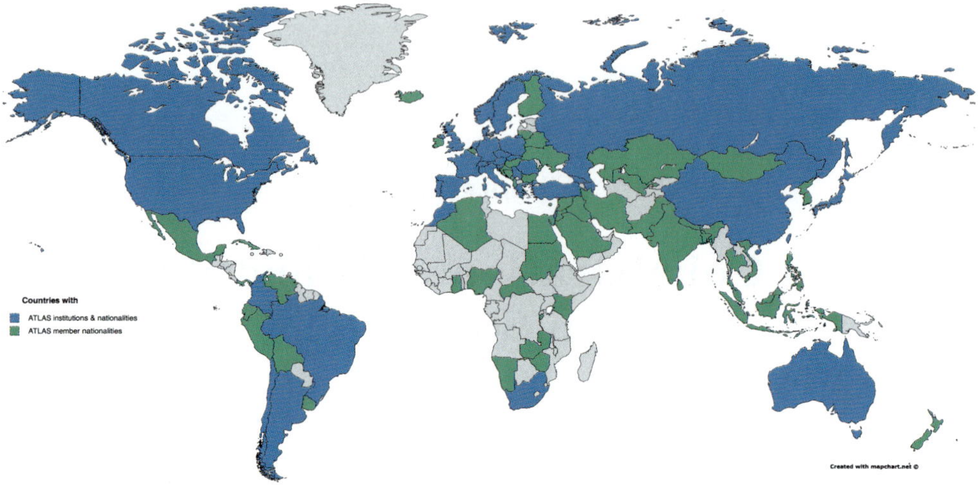

Fig. 11.1. Countries with member institutions of the ATLAS Collaboration are shown in blue, while countries where at least one ATLAS Collaboration member holds that nationality are shown in green. In total, 94 nationalities are represented (from 2016).

as is shown in Fig. 11.3, whereas the fraction of physicists has increased by 11% over this same period of time.

The members in the ATLAS Collaboration are relatively young. In May 2016, it was determined that 50% of the Collaboration was under 35 years of age. With this in mind, the ATLAS Early Career Scientist Board (ECSB) was formed in 2017 to help provide information on any matters concerning the work and experience of early-career scientists. The ECSB advises the ATLAS Management and Collaboration Board with the intent to integrate and support early career scientists to prepare them for future roles in experimental particle physics or beyond.

It is a strong goal of the ATLAS Collaboration to be as diverse and inclusive as possible. Yet, the composition of ATLAS is, e.g., not gender balanced, though the situation seems to be slowly improving. In 2016, a total of 19% of the members were women. Figure 11.4 shows the fraction of women as a function of the age of the person, with a high of 25% between 20–25 years of age and dropping to 11% for 65 years and above. In February 2017, the Collaboration Board formally endorsed the first contacts for Diversity and Inclusion, with whom ATLAS members can discuss concerns related to diversity and inclusion. These contacts also communicate with diversity groups or contacts in the other high-energy physics experiments and at CERN.

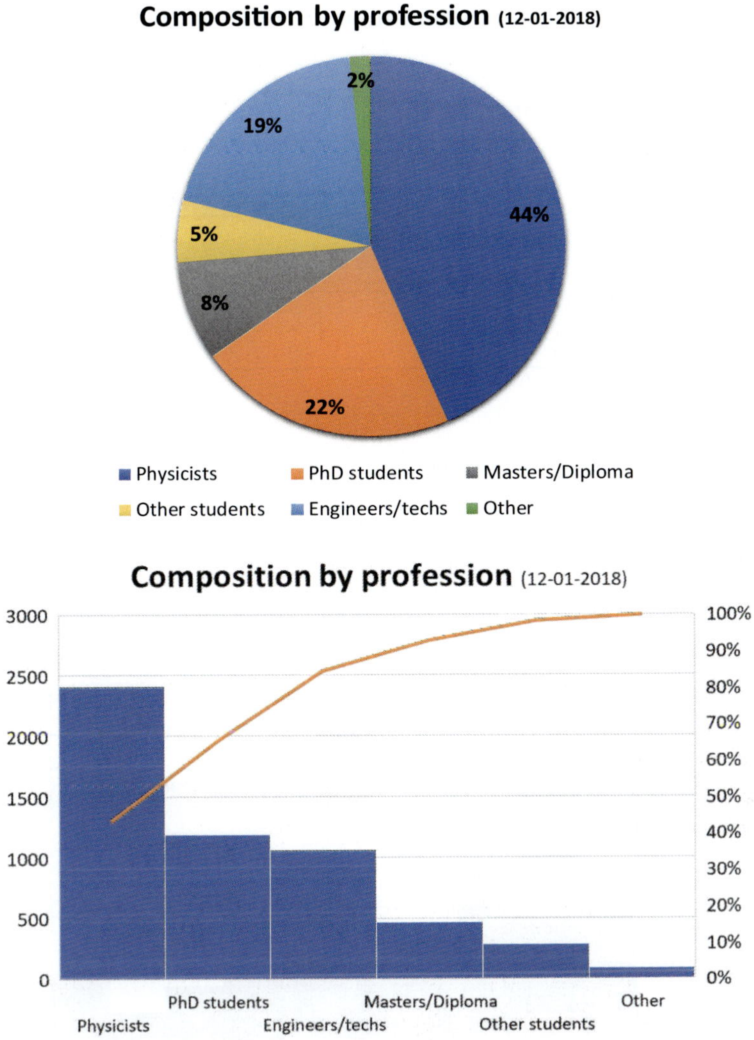

Fig. 11.2. Current composition of the ATLAS Collaboration by profession (12 January 2018). The orange line represents the running sum, in percent.

11.2. ATLAS Weeks in the world and times for celebration

ATLAS is a large collaboration with a broad spectrum of activities related to the design and construction of detector elements, their installation and commissioning, the operation of the detector and data acquisition, and data processing and production of physics results. Every area involves large communities of ATLAS members that need to meet regularly to keep their project focused. To get an overview of the complete project, general

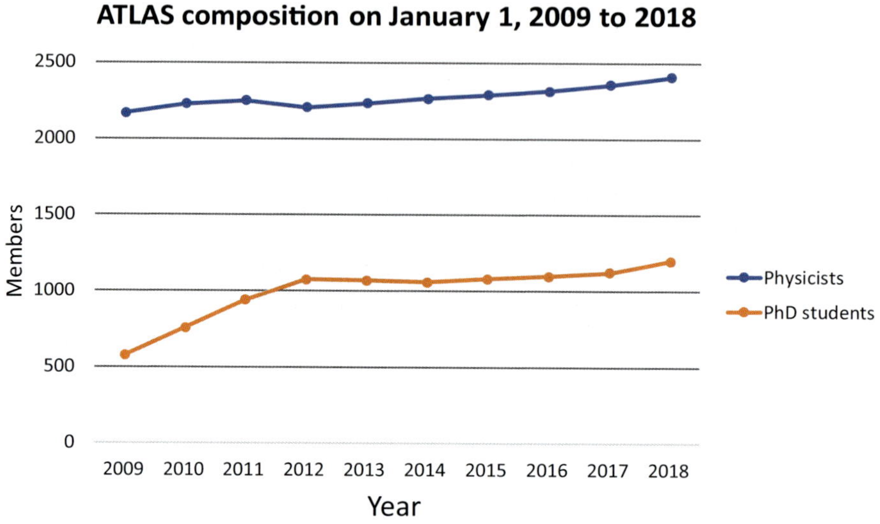

Fig. 11.3. Number of physicists and PhD students on January 1 of each year from 2009 to 2018.

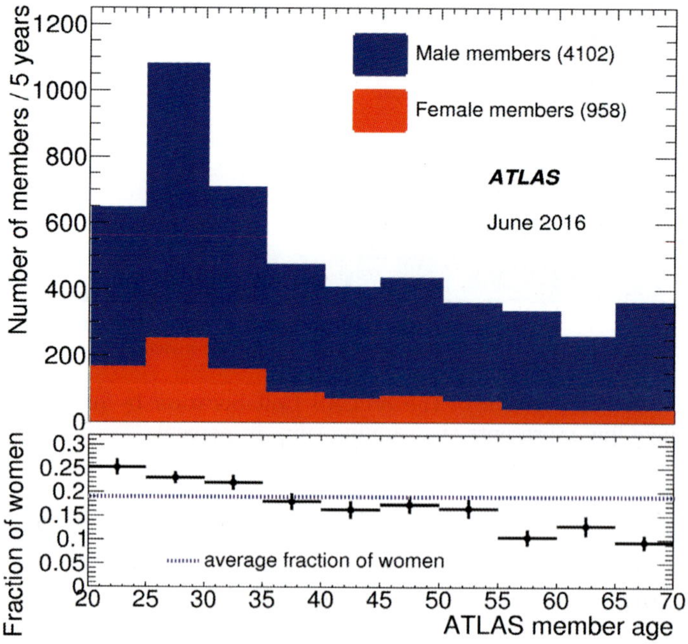

Fig. 11.4. Age distribution of ATLAS members. The bottom panel shows the fraction of women as a function of age. The dotted line shows the average fraction of female ATLAS members of 19% (from 2016).

collaboration meetings take place thrice per year. There is a special one, the ATLAS Overview Week, that takes place outside CERN on a yearly basis and is organised by one (or a few related) member institution(s). Since 2000, there have been 19 Overview Weeks — 15 in Europe, two in America and one in Africa; the most recent being in Asia (for the first time) in 2018. Fig. 11.5 shows a collection of some of the posters announcing the ATLAS Overview Weeks.

The main purpose of these ATLAS Overview Weeks is to draw everyone's attention to the experiment as a whole, beyond the area of one's direct activity. It is also meant to provide its members with a stimulating and motivating time in the beautiful and original settings provided by the hosts.

During its 25 years of existence, the experiment went through many phases and the Collaboration attended many aspects in parallel. Here, we recall the various locations where the Overview Weeks took place and the evolving focus of interest through these different periods. The first Overview Week took place at the Joint Institute of Nuclear Research (JINR) in **Dubna** (2000), and the second one in **Brookhaven National Laboratory** (2001). At that time most detector systems were already well into the construction phase, while some were getting close to it. During these weeks, the Collaboration learned about the progress of the systems and they visited construction sites. In addition, it was important to stimulate attention towards the experiment as a whole and to consolidate the strong commitment of ATLAS to readiness for the first collisions, expected at that time for summer 2006. This raised the Collaboration's awareness to the crucial importance of Technical Coordination and the integration work.

The next two editions took place in **Clermont-Ferrand** (2002) and **Prague** (2003). Highlights included the evolving LHC schedule, the issue of missing resources addressed by the Cost-to-Completion process and detector staging. By then, the first infrastructure elements were installed in the cavern. There was increasing awareness of the big challenge that the Computing needs represented. Physics preparation, a widespread effort in ATLAS, was intensively discussed. The Collaboration met in **Freiburg** (2004) and **Paris** (2005) to review the culmination of the detector installation and discuss the planning of the next huge task ahead in depth, the commissioning of the detector, as well as establishing the Operation Model for the data taking era.

In **Stockholm** (2006), **Glasgow** (2007) and **Bern** (2008), detector integration and commissioning issues were subjects of particular attention. The last LHC magnet was installed in 2007 and the LHC beam pipe ring was

Fig. 11.5. Posters of the ATLAS Overview Weeks.

closed in the ATLAS detector in June 2008. The whole data movement from the pit to the surface HLT/DAQ farm and to the Tier-0, Tier-1s and Tier-2s started to take shape and operate. The preparation for the data collection, the distributed analysis and the physics were discussed. They were crucial elements for the readiness of ATLAS at large for the start-up of the LHC. It was also already time to start planning for the longer-term future and thus the ATLAS Upgrade organisation was set up.

In September 2008, the first proton beam circulated in the LHC, but the LHC incident occurred shortly afterwards. The first proton-proton collisions at injection energy (0.9 TeV) took place on 23 November 2009, just after the Overview Week in **Barcelona** (2009). Detector consolidation work and cosmic rays data useful for commissioning were reviewed. Brainstorming sessions took place to discuss ATLAS procedures for data preparation and reconstruction, physics analysis, and the release of results. By the **Copenhagen** Week (2010), the experiment had observed the first W and Z boson candidates produced in 7 TeV collisions, and the first ATLAS paper with proton-proton collision data was published.

Geneva CICG (2011), **Montreux** (2012) and **Marrakech** (2013) were the scenarios of the presentation of the wealth of LHC Run-1 physics results. There was a huge amount of detector activity and physics results to review. The rapidly increasing luminosity was a challenge. There were further discussions on Operation Task Planning, Authorship, Analysis and the Publication process. At the CERN End of the Year event in 2011, ATLAS was able to restrict the most likely region for the Higgs boson to 116–130 GeV, and an intriguing excess was seen at 125 GeV. By July 2012, the discovery of a new particle consistent with the Higgs boson predicted by the Standard Model was confirmed by ATLAS and CMS, just a couple of months before the Montreux Week began. In Marrakech, we reviewed the many results presented at the 2013 summer conferences, but also discussed the planning for the work during the Long Shutdown 1, with the Pixel Detector refurbishment, the preparation for IBL installation, the software upgrade, milestone weeks for ATLAS restart and the preparation for Phase-I upgrade TDRs. The announcement of the 2013 Nobel Prize in Physics, awarded jointly to François Englert and Peter W. Higgs *"for the theoretical discovery of a mechanism that contributes to our understanding of the origin of mass of subatomic particles, and which recently was confirmed through the discovery of the predicted fundamental particle, by the ATLAS and CMS experiments at CERN's Large Hadron Collider"* was followed in real-time by the Collaboration during the Marrakech meeting.

The 13 TeV Run-2 was the focus of the Sibiu, Lecce, New York, Bratislava, and Tokyo meetings. By the time of the **Sibiu** Week (2014), the IBL was installed and the ATLAS detector was closing. The 13 TeV run preparation was discussed. Many Run-1 results were also presented. The **Lecce** Week (2015) took place just after the LHC started collisions at the new 13 TeV energy frontier, with the **New York** Week (2016) happening after the CERN Council approved the High Luminosity LHC project. During the **Bratislava** (2017) and **Tokyo** (2018) weeks, Run-2 was well underway. Many physics results were presented, of which dealing with the high average pile-up was one of the subjects of interest. Phase-I and Phase-II upgrades were intensively discussed during these weeks. These were crucial times for Phase-II decisions and to collect the commitment of the institutions for the Phase-II upgrade.

Since a few years ago, a new tradition has emerged. Poster sessions are organised where students are given the opportunity to present their work. It is a good opportunity for the Collaboration to get to know their younger scientists better.

The organisation of ATLAS Weeks is an excellent opportunity for the institutions to establish useful contacts with the university, funding agencies and political authorities. Welcome addresses and receptions were often organised in historic buildings with the participation of local authorities (to name a few, the historic Town Halls of Stockholm, Bern or Copenhagen). The local press usually publicised the event and outreach talks were organised in parallel to the week. The organisers always provide a rich social program with visits to architectural sites of great interest (for example, Prague, Freiburg, Lecce and Marrakech), or special visits to art museums (The Cluny museum in Paris, the Fundació Juan Miró in Barcelona and the Paul Klee museum in Bern), and Science museums (the Hunterian Museum — Lord Kelvin — in Glasgow, the Einstein Museum in Bern). Further social activities include classical music concerts (Brookhaven, Paris, Prague) or jazz music (Montreux). There is also room for sport; a football match between two ATLAS teams was organised and held at the Stade de Suisse. Last but not least, a great collaboration dinner was organised during every ATLAS Week, in wonderful places as spectacular as the Vasa museum in Stockholm, the Rothmayer hall of the Prague castle, a Louis XIII castle near Clermont-Ferrand, or a pavilion with a panoramic view of the 5th Avenue and the New York skyline.

In the past 25 years, there have been various important milestones that have been celebrated, for example, the start up of the ATLAS experiment was celebrated on 4 October 2008 in the SM18 building at CERN with various speeches recording the path covered so far and entertainment activities like classical, traditional and rock music and dancing. The discovery of the Higgs boson was celebrated with a drink on 30 August 2012 in the same building. Everybody received a commemorative T-shirt. Shortly after, the ATLAS' 20th anniversary was celebrated during the Montreux ATLAS week. Participants to the meeting could enjoy their share of the cake. During the Bratislava ATLAS Week in 2017, a big event was organised to celebrate ATLAS' 25-year-long journey. Former members of the Collaboration were invited. We could all recall the origins of ATLAS from its first ideas to the Higgs discovery, as seen and told by the people who were present, and the many stories and anecdotes giving different points of view — generational, geographic and responsibilities. It was to be a very enjoyable event, in which a celebration cake could not be missing. Figure 11.6 shows a few pictures taken during these various events.

11.3. Reaching out and sharing: ATLAS education and outreach

With physicists from nearly 100 nations, ATLAS has an unparalleled opportunity to share the excitement of new physics with a worldwide audience. With its extensive experience, the Collaboration can have an impact on the public, students and their teachers, as well as policy makers and the news media. ATLAS wishes to inform and intrigue these audiences about the experiment and particle physics, in general.

The great exploration of particle physics into the unknown realms of the universe are funded by people from many nations and their governments, and ATLAS wants to share the wonderful excitement and fascination that this research generates. There is a great desire to participate in educating the next generation of scientific leaders by involving students at many levels of research.

The early years

During its early years (1999–2005), ATLAS produced brochures, an extensive website, posters, CD-ROMs, a movie, long animations (see Fig. 11.7), a 3D greeting card showing the detector and an exhibition in the building above the ATLAS cavern. It helped set up tours of ATLAS sites and created an ATLAS segment for the CERN Open House. Four web cameras were

Fig. 11.6. Various celebrations of ATLAS: the top four pictures are from the celebration of ATLAS' start up in 2008 in the SM18 hall at CERN, the pictures in the middle row are from the celebration of the Higgs discovery in 2012 in the same place, followed by the 20th ATLAS anniversary party in Montreux. The bottom two photos were taken at the celebration of its 25th anniversary in 2017 in the hall and the Aula of the Law Faculty of the Comenius University in Bratislava.

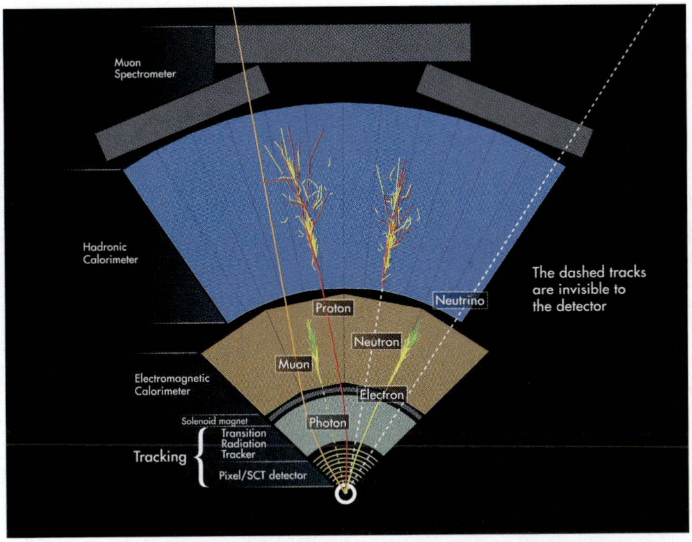

Fig. 11.7. View of an early ATLAS animation.

installed in the cavern that showed the construction of the detector. One of the brochures mentioned above focused on the extensive technology transfer that has occurred as a result of work on ATLAS.

The website, since its development, has so far received more than forty million hits and its site on YouTube has very high viewership numbers. The ATLAS movie won four gold medals at international film festivals (in the science category). These projects were made possible because scientists from ATLAS collaborated with professional film producers, graphic designers and web designers.

Members of the Collaboration have translated many of these brochures, films and posters into different languages; the movie alone was translated into 10 languages! Many talks and presentations were given to both students and the general public.

The start-up era

The ATLAS turn-on and the first collisions provided great opportunities to strengthen the image of ATLAS worldwide, leading it to collaborate closely with media experts and science writers.

Its outreach group promoted ATLAS' science with programmes connected with the release of the Hollywood film Angels & Demons in 2009. There were also two trips to the Sony Pictures headquarters in Los Angeles

Fig. 11.8. Mural depicting the ATLAS Detector.

to negotiate access to film, photos, etc., for their use on its websites and print materials, and in a nationwide lecture series. ATLAS also collaborated with Sony Pictures for a Sony-produced extra on the DVD release of Angels & Demons about CERN and ATLAS. It helped organise events at CERN for 200–300 entertainment journalists and brought them on a tour through ATLAS. A booklet about antimatter was produced, and the movie's portrayal of it was produced.

In 2010, ATLAS commissioned mural artist Josef Kristofoletti to produce a three-story mural of the detector near the control room (see Fig. 11.8). It garnered much news coverage, and there also exists a time-lapse movie of its production. An ATLAS member designed a model of the experiment made entirely of LEGO bricks (see Fig. 11.9) in 2011. It had almost 10,000 pieces!

In July 2009, a multimedia contest open to the general public was announced. Entries were to be in the form of short multimedia pieces about ATLAS, and the grand prize was to be an internship at CERN during the time of the LHC re-start. The contest was a great success! After careful consideration by a panel of physicists, engineers, animators,

Fig. 11.9. ATLAS representation from a LEGO set.

photographers and other ATLAS personnel, the overall winner and recipient of the multimedia internship was declared and five other entries also won prizes.

The data taking era

With results coming in thick and fast, ATLAS communications were up to the task! News articles, blogs and official statements were published on the public website to inform the public and the press about breaking news, Collaboration updates and milestone results, such as the Higgs boson discovery. In order to highlight specific physics results, it began publishing physics briefings in 2014, aimed at non-expert enthusiasts, which describe individual physics results (see e.g., Fig. 11.10). Later that year a modern new-look public website was released, providing easy access to updates from the experiment, resources and evergreen material about the physics, detector and Collaboration.

Over the past decade, social media has become a dominant force in public communications. ATLAS has thrived on such platforms, developing content and multimedia on Facebook, Twitter, Google+, YouTube and Instagram, thus accessing a large and varied audience.

The ATLAS Virtual Visit service was pioneered by the outreach team in 2012, bringing students from all over the world to interact directly with physicists in the control room and occasionally the experimental

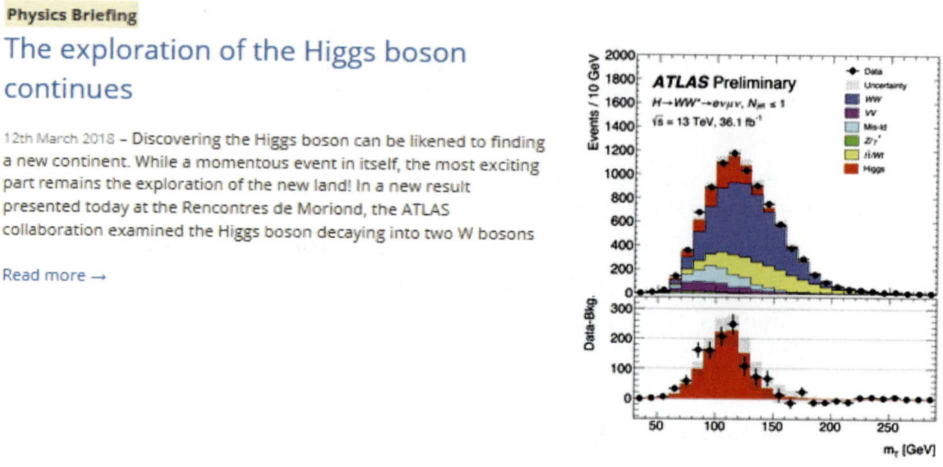

Fig. 11.10. Publicising a physics briefing on the on-going studies on the Higgs boson.

cavern. Over 400 of such visits have been hosted from classrooms, public venues and partnering experiments across all seven continents. ATLAS also pioneered other virtual experiences to interact with the experiment, including "ATLASRift", "ATLASCraft", "ATLAS in your Pocket" and the new "LHCPanorama", an ATLAS/CMS joint project.

Resources for education grew in 2016 with the release of the ATLAS Colouring Book aimed at children aged 5–7 years old, and kids from 8 to adults as old as 80 have also enjoyed constructing the ATLAS-made LEGO models of the LHC and its experiments.

During 2014–16, three ATLAS physicists collaborated to produce a state of the art planetarium show on dark matter. It featured spectacular scenes of the LHC and of ATLAS, as well as of the LUX Experiment, and astrophysics. The show was translated into 17 languages and presented in more than 56 countries to great reviews. It even won a prize at an international full dome festival.

ATLAS Masterclasses in collaboration with the International Particle Physics Outreach Group have grown year-on-year to an unprecedented level. In 2018 alone, over 144 of such masterclasses took place in around 39 countries worldwide, with many more scheduled to take place in the upcoming years.

The outreach group went further in bringing ATLAS' physics to undergraduate and advanced high school students, with the release of the ATLAS Open Data project in 2016, which made 1 fb^{-1} of 8 TeV ATLAS data, simulated samples and developed analysis tools, freely available. Webpages were developed to host this full package alongside videos and full documentation. This allowed, in particular, ATLAS members to develop courses and projects for their students. In 2017, the ATLAS Open Key project allowed the full data, tools and documentation to be hosted on a USB key, for distribution to academic institutions with little internet access.

ATLAS turned 25 on 1 October 2017, a quarter-century after the submission of its letter of intent to the LHC Experiments Committee in 1992. To celebrate this occasion, a number of specially developed resources were created including a new logo and banner, a website, videos (one celebrating the history of the Collaboration and another with birthday greetings) and interviews with members of the Collaboration that were shared on social media.

Concluding Remarks

25 years already... I vividly remember the Evian workshop in 1992, and the incredibly enthusiastic and enriching atmosphere. I was a post-doc at the time, and was fascinated by the effervescent environment, the many ideas presented on the physics and the detectors, and the optimism with which people were ready to embrace what appeared to be insurmountable technical challenges. From an outside perspective, we must have looked as if we were crazy, but we were doing nothing more than pursuing our dreams with great determination.

From the detector R&D work in the early 1990s, through the detector design and construction, the preparation of the online and offline analysis tools, to the first minimum-bias collisions at 7 TeV, the first W, Z and top-quark pair events, the discovery of the Higgs boson and the extraordinary results from Run-2, ATLAS has been a fantastic scientific and human adventure for many of us. Generations of physicists, engineers and technicians from all over the world have grown up in a unique, highly-stimulating open and inclusive international environment, offering great opportunities for learning and professional development. And it's not yet over. We can anticipate that, whether or not Run-3 and the High-Luminosity LHC phase will bring new major discoveries, the LHC, over its almost 30 years of operation, will have had a profound impact on our understanding of fundamental physics.

The accomplishments and successes of the LHC project (including the accelerator, the experiments and the computing infrastructure) are the demonstration of the ingenuity, competence, passion and perseverance of the worldwide particle physics community and a brilliant example of what humanity can do when we set aside our differences and we focus on a common goal. Very often, in meetings with people from other fields, policy makers or politicians, I witness the great respect and admiration they have for what our community has been able to achieve. I think we can all be proud of what we have done together, and look forward to future projects and challenges in our field with confidence in success.

Fabiola Gianotti
Director-General CERN, October 2018

Further Readings

A comprehensive description of the overall LHC project, with an emphasis on its technological aspects, is available in:

L. Evans (ed.), The Large Hadron Collider, a Marvel of Technology, EPFL Press, 2009, and updated edition 2018.

The reference article for the ATLAS detector as constructed with ample technical details and initial performance characteristics as installed can be found in:

The ATLAS Collaboration, The ATLAS Experiment at the LHC, JINST 3 (2008) S08003.

The original Higgs boson discovery publications from the ATLAS and CMS Collaborations, given below, contain references to the pioneering articles laying the foundations of the Standard Model and the Brout–Englert–Higgs mechanism:

The ATLAS Collaboration, Observation of a new particle in the search for the Standard Model Higgs boson with the ATLAS detector at the LHC, Phys. Lett. B 716 (2012) 1;

The CMS Collaboration, Observation of a new boson at a mass of 125 GeV with the CMS experiment at the LHC, Phys. Lett. B 716 (2012) 30.

The ATLAS Collaboration published an account of the Higgs boson discovery for non-expert science readers, available in open access on the web:

The ATLAS Collaboration, A particle consistent with the Higgs Boson observed with the ATLAS detector at the Large Hadron Collider, Science 338 (2012) 1576, http://science.sciencemag.org/content/338/6114/1576.

Two web-based articles on Scholarpedia provide an easy access to comprehensive summaries of the ATLAS detector and of the Higgs boson discovery:

M. L. Dunford and P. Jenni, The ATLAS experiment, Scholarpedia, 9(10):32147, http://www.scholarpedia.org/article/The_ATLAS_experiment;

K. Jakobs and C. Seez, The Higgs Boson discovery, Scholarpedia, 10(9):32413, http://www.scholarpedia.org/article/The_Higgs_Boson_discovery.

The reader is invited to visit the ATLAS public webpages with ample up-to-date information about physics from, and events in the Collaboration: http://atlas.cern/ and http://atlas.cern/updates.

Finally, a complete list of ATLAS publications, kept up-to-date, is available at: https://twiki.cern.ch/twiki/bin/view/AtlasPublic.

Institutions of the ATLAS Collaboration and ATLAS Computing Centres

Institutions of the ATLAS Collaboration

(Status May 2018)

Argentina

Departamento de Física, Universidad de Buenos Aires, Buenos Aires
Instituto de Física La Plata, Universidad Nacional de La Plata and CONICET, La Plata

Armenia

Yerevan Physics Institute, Yerevan

Australia

Department of Physics, University of Adelaide, Adelaide
School of Physics, University of Melbourne, Victoria
School of Physics, University of Sydney, Sydney

Austria

Institut für Astro- und Teilchenphysik, Leopold-Franzens-Universität, Innsbruck
Fachhochschule Wiener Neustadt, Wiener Neustadt

Azerbaijan

Institute of Physics, Azerbaijan Academy of Sciences, Baku

Belarus

B.I. Stepanov Institute of Physics, National Academy of Sciences of Belarus, Minsk
Research Institute for Nuclear Problems of Byelorussian State University, Minsk

Brazil

Brazil cluster (Universidade Federal do Rio de Janeiro; Universidade Federal de Juiz de Fora; Universidade Federal de São João del Rei; Universidade de São Paulo)

Canada

Department of Physics, University of Alberta, Edmonton AB
Department of Physics, Carleton University, Ottawa ON
Department of Physics, McGill University, Montreal QC
Group of Particle Physics, University of Montreal, Montreal QC
Department of Physics, Simon Fraser University, Burnaby BC
Department of Physics, University of Toronto, Toronto ON
TRIUMF, Vancouver BC and York University, Toronto ON
Department of Physics, University of British Columbia, Vancouver BC
Department of Physics and Astronomy, University of Victoria, Victoria BC

CERN

European Organization for Nuclear Research, Geneva, Switzerland

Chile

Chile cluster (Pontificia Universidad Católica de Chile, Santiago; Universidad Técnica Federico Santa María, Valparaíso)

China

China IHEP-NJU-THU cluster (IHEP, Beijing; Nanjing University, Jiangsu; Tsinghua University, Beijing)
China USTC-SDU-SJTU cluster (University of Science and Technology of China, Hefei; Shandong University, Shandong; Jiao Tong University and T.D. Lee Institute, Shanghai)
Hong Kong cluster (Chinese University of Hong Kong; University of Hong Kong; Hong Kong University of Science and Technology)

Colombia

Centro de Investigaciónes, Universidad Antonio Nariño, Bogota

Czech Republic

Palacký University, RCPTM, Olomouc
Institute of Physics, Academy of Sciences of the Czech Republic, Prague
Charles University, Faculty of Mathematics and Physics, Prague
Czech Technical University in Prague, Prague

Denmark

Niels Bohr Institute, University of Copenhagen, Copenhagen

France

LAPP, CNRS/IN2P3 et Université Savoie Mont Blanc, Annecy-le-Vieux
Université Clermont Auvergne, CNRS/IN2P3, LPC, Clermont-Ferrand
LPSC, Université Grenoble-Alpes, CNRS/IN2P3, Grenoble
CPPM, Aix-Marseille Université et CNRS/IN2P3, Marseille
LAL, Université Paris-Sud, CNRS/IN2P3, Université Paris-Saclay, Orsay
LPNHE, UPMC et Université Paris-Diderot et CNRS/IN2P3, Paris
DSM/IRFU (Institut de Recherches sur les Lois Fondamentales de l'Univers), CEA, Saclay

Georgia

Georgia cluster (Georgian Academy of Science; Tbilisi State University, Tbilisi)

Germany

Institut für Physik, Humboldt-Universität zu Berlin, Berlin
Physikalisches Institut, Universität Bonn, Bonn
Deutsches Elektronen-Synchrotron DESY, Hamburg und Zeuthen
Lehrstuhl für Experimentelle Physik IV, Technische Universität Dortmund, Dortmund
Institut für Kern- und Teilchenphysik, Technische Universität Dresden, Dresden
Physikalisches Institut, Albert-Ludwigs-Universität, Freiburg
II Physikalisches Institut, Justus-Liebig-Universität Giessen, Giessen
II Physikalisches Institut, Georg-August-Universität, Göttingen
Kirchhoff-Institut und Physikalisches Institut, Ruprecht-Karls-Universität Heidelberg
Institut für Physik, Universität Mainz, Mainz
Fakultät für Physik, Ludwig-Maximilians-Universität München, München
Max-Planck-Institut für Physik (Werner-Heisenberg-Institut), München
Department Physik, Universität Siegen, Siegen
Fakultät für Mathematik und Naturwissenschaften, Bergische Universität Wuppertal
Fakultät für Physik und Astronomie, Julius-Maximilians-Universität, Würzburg

Greece

Physics Department, National and Kapodistrian University of Athens, Athens
Physics Department, National Technical University of Athens, Zografou
National Center for Scientific Research "Democritos", Agia Paraskevi
Department of Physics, Aristotle University of Thessaloniki, Thessaloniki

Israel

Department of Physics, Technion, Israel Institute of Technology, Haifa
School of Physics and Astronomy, Tel Aviv University, Tel Aviv
Department of Particle Physics, The Weizmann Institute of Science, Rehovot

Italy

INFN Sezione di Bologna; Dip. di Fisica e Astronomia, Università di Bologna, Bologna
INFN Gruppo Collegato di Cosenza, Università della Calabria, Rende
INFN e Laboratori Nazionali di Frascati, Frascati
INFN Sezione di Genova; Dipartimento di Fisica, Università di Genova, Genova
INFN Sezione di Lecce; Dip. di Matematica e Fisica, Università del Salento, Lecce
INFN Sezione di Milano; Dipartimento di Fisica, Università di Milano, Milano
INFN Sezione di Napoli; Dipartimento di Fisica, Università di Napoli, Napoli
INFN Sezione di Pavia; Dipartimento di Fisica, Università di Pavia, Pavia
INFN Sezione di Pisa; Dipartimento di Fisica E. Fermi, Università di Pisa, Pisa
INFN Sezione di Roma; Dipartimento di Fisica, Sapienza Università di Roma, Roma
INFN Sezione di Roma Tor Vergata; Dip. di Fisica, Università di Roma Tor Vergata, Roma
INFN Sezione di Roma Tre; Dip. di Matematica e Fisica, Università Roma Tre, Roma
INFN-TIFPA; Università di Trento, Trento
INFN Gruppo Collegato di Udine, Sezione di Trieste; ICTP; Università di Udine, Udine

Japan

Faculty of Applied Information Science, Hiroshima Institute of Technology, Hiroshima
KEK, High Energy Accelerator Research Organization, Tsukuba
Graduate School of Science, Kobe University, Kobe
Faculty of Science, Kyoto University, Kyoto
Kyoto University of Education, Kyoto
Research Center for Advanced Particle Physics, Kyushu University, Fukuoka
Nagasaki Institute of Applied Science, Nagasaki
Graduate School of Science and KMI, Nagoya University, Nagoya
Faculty of Science, Okayama University, Okayama
Graduate School of Science, Osaka University, Osaka
Department of Physics, Shinshu University, Nagano
ICEPP and Department of Physics, University of Tokyo, Tokyo
Graduate School of Science and Technology, Tokyo Metropolitan University, Tokyo
Department of Physics, Tokyo Institute of Technology, Tokyo
Faculty of Pure and Applied Sciences, University of Tsukuba, Tsukuba
Waseda University, Tokyo

Morocco

Morocco cluster (Université Hassan II, Casablanca; Université Mohammed V, Rabat; Université Cadi Ayyad, Marrakech; Université Mohamed I, Oujda; CNESTEN, Rabat)

Netherlands

Nikhef National Institute for Subatomic Physics and University of Amsterdam, Amsterdam
Radboud University Nijmegen and Nikhef, Nijmegen

Norway

Department for Physics and Technology, University of Bergen, Bergen
Department of Physics, University of Oslo, Oslo

Poland

AGH University of Science and Technology, Krakow; Jagiellonian University, Krakow
Institute of Nuclear Physics Polish Academy of Sciences, Krakow

Portugal

Portugal cluster (LIP, Lisboa; Universidade de Lisboa; University Coimbra; Universidade do Minho, Braga; Universidade Nova de Lisboa; Universidad de Granada)

Romania

Romania cluster (Horia Hulubei National Institute, Bucharest; Politehnica Bucharest; University of Brasov; INCDTIM, Cluj; University Iasi; West University Timisoara)

Russia

P.N. Lebedev Physical Institute of the Russian Academy of Sciences, Moscow
Institute for Theoretical and Experimental Physics (ITEP), Moscow
National Research Nuclear University MEPhI, Moscow
Institute of Nuclear Physics, M.V. Lomonosov Moscow State University, Moscow
Budker Institute Nuclear Physics, SB RAS, and Novosibirsk State University, Novosibirsk

NRC "Kurchatov Institute" Petersburg Nuclear Physics Institute, St. Petersburg
State Research Center Institute for High Energy Physics, NRC KI, Protvino
Tomsk State University, Tomsk

JINR

Joint Institute for Nuclear Research, Dubna, Moscow region, Russia

Serbia

Institute of Physics, University of Belgrade, Belgrade

Slovak Republic

Slovak cluster (Comenius University, Bratislava; Institute of Experimental Physics of the Slovak Academy of Sciences, Kosice)

Slovenia

Jozef Stefan Institute and University of Ljubljana, Ljubljana

South Africa

South Africa cluster (University of Cape Town, Cape Town; University of Johannesburg, Johannesburg; University of the Witwatersrand, Johannesburg)

Spain

Institut de Física d'Altes Energies, BIST, Barcelona
Departamento de Física Teórica C-15, Universidad Autónoma de Madrid, Madrid
Instituto de Física Corpuscular, Centro Mixto Universidad de Valencia — CSIC, Valencia

Sweden

Fysiska institutionen, Lunds universitet, Lund
Department of Physics, Stockholm University; Oskar Klein Centre, Stockholm
Physics Department, Royal Institute of Technology, Stockholm
Department of Physics and Astronomy, University of Uppsala, Uppsala

Switzerland

AEC und Laboratorium für Hochenergie Physik, Universität Bern, Bern
Département de Physique Nucléaire et Corpusculaire, Université de Genève, Genève

Taiwan

Department of Physics, National Tsing Hua University, Hsinchu
Institute of Physics, Academia Sinica, Taipei

Turkey

Ankara cluster (Ankara University, Ankara; Istanbul Aydin University, Istanbul; TOBB University of Economics and Technology, Ankara)
Bogazici cluster (Bogazici University, Istanbul; Gaziantep University, Gaziantep; Istanbul Bilgi University, Istanbul; Bahcesehir University, Istanbul)

United Kingdom

School of Physics and Astronomy, University of Birmingham, Birmingham
Cavendish Laboratory, University of Cambridge, Cambridge
SUPA — School of Physics and Astronomy, University of Edinburgh, Edinburgh
SUPA — School of Physics and Astronomy, University of Glasgow, Glasgow
Physics Department, Lancaster University, Lancaster
Department of Physics, University of Liverpool, Liverpool
School of Physics and Astronomy, Queen Mary University of London, London
Department of Physics, Royal Holloway University of London, Egham
Department of Physics and Astronomy, University College London, London
School of Physics and Astronomy, University of Manchester, Manchester
Department of Physics, Oxford University, Oxford
Particle Physics Department, Rutherford Appleton Laboratory, Didcot
Department of Physics and Astronomy, University of Sheffield, Sheffield
Department of Physics and Astronomy, University of Sussex, Brighton
Department of Physics, University of Warwick, Coventry

United States of America

Physics Department, SUNY Albany, Albany NY
High Energy Physics Division, Argonne National Laboratory, Argonne IL
Department of Physics, University of Arizona, Tucson AZ
Department of Physics, University of Texas at Arlington, Arlington TX
Department of Physics, University of Texas at Austin, Austin TX
Physics Division, Lawrence Berkeley Nat. Lab. and University of California, Berkeley CA
Department of Physics, Boston University, Boston MA
Department of Physics, Brandeis University, Waltham MA
Physics Department, Brookhaven National Laboratory, Upton NY
Enrico Fermi Institute, University of Chicago, Chicago IL
Nevis Laboratory, Columbia University, Irvington NY
Physics Department, Southern Methodist University, Dallas TX
Physics Department, University of Texas at Dallas, Richardson TX
Department of Physics, Duke University, Durham NC
Laboratory for Particle Physics and Cosmology, Harvard University, Cambridge MA
Department of Physics, Indiana University, Bloomington IN
University of Iowa, Iowa City IA
Department of Physics and Astronomy, Iowa State University, Ames IA
Louisiana Tech University, Ruston LA
Department of Physics, University of Massachusetts, Amherst MA
Department of Physics, University of Michigan, Ann Arbor MI
Department of Physics and Astronomy, Michigan State University, East Lansing MI
Department of Physics and Astronomy, University of New Mexico, Albuquerque NM
Department of Physics, New York University, New York NY
Department of Physics, Northern Illinois University, DeKalb IL
Ohio State University, Columbus OH
Department of Physics and Astronomy, University of Oklahoma, Norman OK
Department of Physics, Oklahoma State University, Stillwater OK
Center for High Energy Physics, University of Oregon, Eugene OR
Department of Physics, University of Pennsylvania, Philadelphia PA
Department of Physics and Astronomy, University of Pittsburgh, Pittsburgh PA

Institute for Particle Physics, University of California Santa Cruz, Santa Cruz CA
Department of Physics, University of Washington, Seattle WA
SLAC National Accelerator Laboratory, Stanford CA
Departments of Physics & Astronomy and Chemistry, Stony Brook University, NY
Department of Physics and Astronomy, Tufts University, Medford MA
Department of Physics and Astronomy, University of California Irvine, Irvine CA
Department of Physics, University of Illinois, Urbana IL
Department of Physics, University of Wisconsin, Madison WI
Department of Physics, Yale University, New Haven CT

Tier-0 and Tier-1 computing centres of the ATLAS Collaboration

(Status May 2018)

CERN Tier-0, Switzerland
SARA Reken- en Netwerkdiensten, Amsterdam, Netherlands
PIC, Universidad Autónoma de Barcelona, Bellaterra, Spain
INFN-CNAF, Bologna, Italy
UK-T1-RAL Tier-1, Rutherford Appleton Laboratory, Didcot, United Kingdom
GridKA Tier-1 FZK, FZ Karlsruhe GmbH, SCC, Eggenstein-Leopoldshafen, Germany
Nordic Data Grid Facility, NORDUnet A/S, Kastrup, Denmark
RHIC and ATLAS Computing Facility, BNL, Upton, New York, U.S.A.
Academia Sinica Grid Computing, Institute of Physics, Academia Sinica, Taipei, Taiwan
ATLAS-Canada Tier-1 Data Centre, TRIUMF, Vancouver, Canada
Centre de Calcul de l'IN2P3, Villeurbanne, France

Contributors

P.P. Allport, O. Arnaez, D. Barberis, R.M. Barnett, C.P. Bee, A. Blue, M. Bomben, M. Boonekamp, M. Bosman, J. Boyd, M. Bruschi, S. Campana, A. Catinaccio, A. Cerri, F. Cerutti, D.G. Charlton, M.A. Chelstowska, A. Ciocio, A. Clark, P.J. Clark, C. Clement, S.H. Connell, M. Delmastro, A. Di Ciaccio, F. Dittus, M. D'Onofrio, M.T. Dova, J. Dowell, M. Dunford, K. Einsweiler, N. Ellis, M. Elsing, E. Etzion, A. Fenyuk, D. Ferrere, D. Fournier, D. Froidevaux, K.K. Gan, C.N.P. Gee, H.A. Gordon, B. Gorini, I. Grabowska-Bold, P. Grafstrom, H.M. Gray, I.M. Gregor, C. Guyot, B. Heinemann, A.M. Henriques Correia, G. Herten, L. Hervas, A. Hoecker, H.F. Hoffmann, K. Jakobs, S. Jakobsen, P. Jenni, S. Jin, M. Kado, T. Kawamoto, T. Kobayashi, T. Kondo, W. Kozanecki, J. Kretzschmar, P. Krieger, E.S. Kuwertz, F. Lanni, M. Lefebvre, G. Lehmann Miotto, R. Leitner, A. Lister, P. Loch, P. Maettig, B. Mansoulie, S. Malyukov, G. Mikenberg, E. Monnier, A.K. Morley, M. Nessi, A. Nisati, M. Nordberg, A. Oh, M.A. Parker, B.A. Petersen, J.E. Pilcher, G. Polesello, L. Pontecorvo, K. Pretzl, J. Proudfoot, M. Raymond, I. Riu, A. Romaniouk, L.P. Rossi, N.A. Rusakovich, J.P. Rutherfoord, P. Savard, P. Schacht, S. Schramm, K. Shaw, G. Spigo, P. Steinberg, H. Stenzel, S. Strandberg, D.M. Strom, K. Tackmann, G.N. Taylor, H. ten Kate, K. Terashi, E. Torrence, D.R. Tovey, I.M. Trigger, C. Troncon, M.G. Vincter, P.S. Wells, T. Wenaus, T. Wengler, I. Wingerter-Seez, C.J.S. Young, A.M. Zaitsev, B. Zhou.